Abyss

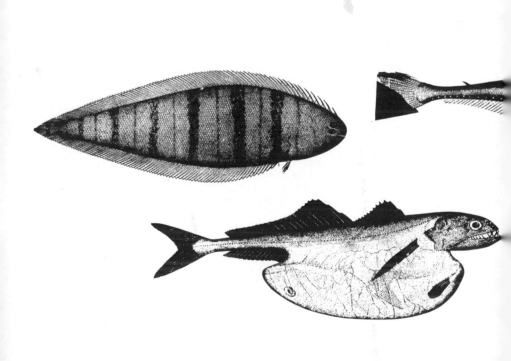

Abyss

The Deep Sea

and the Creatures

That Live in It

NEW, UPDATED EDITION

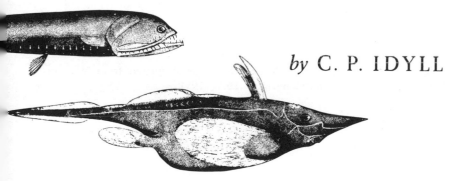

by C. P. IDYLL

THOMAS Y. CROWELL COMPANY

New York, Established 1834

Library of Congress Cataloging in Publication Data

Idyll, Clarence P
   Abyss: the deep sea and the creatures that live in it.

   Bibliography: p.
   Includes index.
   1. Marine biology.   2. Ocean.   3. Abyssal zone.
   4. Fishes, Deep-sea.   I. Title.
   QH91.I3   1976   574.92   75-46081
   ISBN 0-690-01175-X
   ISBN 0-8152-0400-0 (Apollo ed.)

1 2 3 4 5 6 7 8 9 10

Acknowledgment is made to the following:

Duell, Sloan & Pearce, New York, for material quoted from William Beebe, *Half Mile Down* (Copyright, 1934, 1951).

Houghton Mifflin Company, New York, for material quoted and Figures 4-2, 4-6, 5-3, 5-5, 8-4, 8-6, 11-3, from Sir Alister C. Hardy, *The Open Sea: Its Natural History. Part 1. The World of Plankton; Part 2. Fish and Fisheries* (1956, 1959).

Hutchinson and Company, London, for Figures 5-4, 5-6, 6-2, 6-4, 7-3, 7-5, 7-6, 10-8, 10-10, 10-12, 10-13, 13-8, 13-10, 14-5, 16-1, from N. B. Marshall, *Aspects of Deep Sea Biology* (1954).

Macmillan and Company, Ltd., London, for Figures 7-1, 8-3, 9-4, 12-1, 13-1, 13-3, from John Murray and Johan Hjort, *The Depths of the Ocean* (1912).

Masson et Cie, Éditeurs, Paris, for Figure 12-4, from A. V. Ivanov, "Pogonophora," *Traité de Zoologie* (Vol. 5, 1960). National Geographic Society, Washington, D.C., for Figure 10-14, from Paul A. Zahl, "Hatchet Fishes, Torchbearers of the Deep," *National Geographic Magazine* (May, 1958); Figures 10-7, 10-9, 13-4, from Paul A. Zahl, "Fishing in the Whirlpool of Charybdis," *National Geographic Magazine* (November, 1953); and Figure 17-5, by Harold E. Edgerton, from Jacques Cousteau, "Calypso Explores an Undersea Canyon," *National Geographic Magazine* (March, 1958).

Scientific American, New York, for Figures 2-8 and 16-5, from John L. Mero, "Minerals on the Ocean Floor," *Scientific American* (December, 1960).

Sea Frontiers, for Figure 11-2, from F. G. Walton Smith, "Marine Monsters," *Sea Frontiers* (Vol. 2, No. 1).

Report on the Scientific Results of the Exploring Voyage of H.M.S. Challenger During the Years 1873–1876 (London, H.M. Stationery Office, 1880–95), for Figures 6-3, 6-6, 6-7.

The Danish Dana-Expeditions 1920–1922 in the North Atlantic and Gulf of Panama (Copenhagen, Carlsberg Foundation, 1925–29), for Figures 10-3, 10-4, 10-5, 10-6, 13-11, 14-3.

The Carlsberg Foundation's Oceanographical Expedition Round the World, 1928–1930 (Copenhagen, Carlsberg Foundation, 1934–63), for Figures 10-1 and 10-2.

Universitetets Zoologiske Museum, Denmark, for Figures 2-10, 6-9, 7-4, 9-1, 12-3, 14-4, from *The Galathea Deep Sea Expedition, 1950–1952* (New York, The Macmillan Company, 1956).

Vanderbilt Marine Museum, Long Island, New York, for Figure 13-5, from Lee Boone, "Scientific Results of the Cruises of the Yachts Eagle and Ara, 1921–1928," *Bull. Vanderbilt Marine Museum* (Vol. 3, 1930).

Verlag von Gustav Fischer, Jena, Germany, for Figures 2-9, 6-1, 6-8, 8-2, 8-5, 9-7, 12-5, 12-7, 13-2, from *Wissenschaftliche Ergebnisse der Deutschen Tiefsee Expedition auf dem Dampfer Valdivia 1898–1899* (published from 1902–32).

To Marion with love

Foreword

This is a good book to read about the sea because it relates the oceans to land, air, and space, and emphasizes that we cannot look at part of our environment without paying regard to the whole. The formation of the ocean basins is a part of the story of the birth of sun, moon, and earth. Ocean deeps have counterparts in mountain peaks; continental crests reflect the ocean troughs.

It is a good book about the life in the sea because it describes the physical, chemical, and dynamic characteristics of the ocean water and its boundaries which mold and influence all life in them.

It's a good book from which to learn what is known because it emphasizes also what is *not* known. The author tells us the fascinating theories of the ocean's origin, behavior, and its life, but emphasizes the *different* ideas still to be weighed. And the book concerns mainly the eighty percent of the ocean about which we know *least* rather than the three percent of the shallow waters on the shelf (which is submerged land) about which we know most.

As he communicates the excitement of working at sea and thinking about the sea, the author takes us on numerous excursions from the main story. He connects past pioneers in oceanography with present discoverers, and prehistory and history with the future of the oceans.

ATHELSTAN SPILHAUS

Preface

I have attempted in this book to give a general account of the deep sea for the nonscientist. This has made necessary excursions into cosmology, geology, physics, chemistry, botany, zoology, engineering, and several other areas of human knowledge. Then, to describe the deep sea and the living animals and plants that inhabit it requires first some account of the shallow sea and its living creatures, and this means that I have undertaken the presumptuous and impossible task of encompassing nearly all science. Clearly I have omitted an enormous amount of interesting and important material.

I have been greatly assisted in my task by many people. I am deeply grateful for the loving assistance given by my wife. Her patience and forbearance during the long months while the book was being prepared were remarkable; her skillful help in editing and correcting the manuscript have been of great help, and I have leaned heavily on her advice.

The help of scientists in many fields has been generously given, and I have gained enormously from their expert knowledge. I am especially grateful to my colleagues at the School of Marine and Atmospheric Science, of the University of Miami, who have courteously given me the benefit of their special knowledge. Many of them have read parts of the manuscript, and have enabled me to avoid a large number of errors. Despite their best efforts I am sure I have made many mistakes, and I will be glad to have them drawn to my attention.

Scientists at the Institute who were kind enough to read various parts of the manuscript are Dr. F. G. Walton Smith, Dr. F. F. Koczy, Dr. Charles E. Lane, Dr. Gilbert L. Voss, Dr. Cesare Emiliani, Dr. Gene A. Rusnak, Dr. Hilary B. Moore, Dr. Carl H. Oppenheimer,

Dr. C. Richard Robins, Dr. Donald P. de Sylva, Dr. Robert J. Hurley, Dr. Frederick M. Bayer, and Dr. Eugene Corcoran. I have also had the valuable advice of Dr. Oscar Owre of the Department of Zoology of the University of Miami, Mr. K. M. Rae of the University of Alaska, Dr. Paul S. Galtsoff, Bureau of Commercial Fisheries, Woods Hole, Massachusetts, Commander J. R. Lumby, formerly on the staff of our Institute, and Dr. John Randall of the University of Puerto Rico. I extend my deep gratitude to them all for their willing and friendly help.

I am indebted to many individuals and organizations for permission to reproduce material and illustrations. I must mention two of these to whom I am particularly grateful for ready and courteous compliance to my requests. These are Dr. N. B. Marshall of the British Museum and Sir Alister Hardy of the University of Oxford. I have drawn heavily on their excellent books (Dr. Marshall's *Aspects of Deep Sea Biology* and Sir Alister's two-volume *The Open Sea*) in the preparation of the present volume. Acknowledgment of illustrations from these and other sources is made in the captions and on pages iv-v.

For the third edition I have had the benefit of advice from a number of scientists in the National Oceanic and Atmospheric Administration. Of these, I am particularily indebted to Dr. Robert S. Dietz of the Atlantic Oceanographic and Meteorological Laboratory in Miami, Florida.

Contents

Illustrations

Abyss

Introduction

There is more excitement and turmoil in ocean affairs at the present moment in history than at any time in the past. The excitement results from new vistas of knowledge about our earth and its inhabitants opened up in the last decade by ocean science, especially in the field of marine geology. The turmoil comes largely from the rapid crumbling of a traditional "freedom of the seas"—license to use ocean resources without constraint. This change has come about because we have been shocked into the realization that fish and other marine resources are not inexhaustible. Earnest attempts to replace the often wasteful "freedom" with a system of world order have been made at several sessions of the United Nations Law of the Sea Conference starting in 1974—with only partial success.

Much of the new interest in the ocean relates to the deep sea—the abyss that is the subject of this book.

Nearly three quarters of the surface of the earth is covered by the sea. Of the 197 million square miles of the earth's area, 139 million are the world ocean—the great interconnected complex of the Atlantic, the Pacific, the Indian and the Arctic oceans, the Mediterranean and the Baltic, the Red, Black, and the White seas, and all the rest. Seventy-one percent of the world is ocean. Of that, 97 percent is more than 200 meters or 656 feet, in depth. This is the abyss, the "deep sea," which covers almost two-thirds of the earth's surface.

Even these figures do not reflect the true immensity of the world of water. Remember that the oceans are three-dimensional, averaging 12,566 feet (2.4 miles) in depth, with a volume of 331 million cubic miles. The greatest part of this huge mass is the deep sea. The inhabitable land is a thin film of living space for creatures who run on its surface—or less often fly above it or burrow underground. But sea creatures swim and crawl and drift through the whole enormous mass of the ocean water, making the deep sea by far the most extensive environment in our planet.

[1]

All of us—man and all the animals and plants with which he shares the land—are creatures of the sea, once removed. Life originated in the seas, and without water no life could exist. Every living cell of every plant and animal contains water, and all life processes depend on it. The possession of an ocean makes the earth unique among the planets of our solar system, and among the members of the sun's family it is only on our relatively small and insignificant globe that life exists as we know it. But life could probably not start its existence on our seas today as it did aeons ago. The presence of bacteria, the high concentration of oxygen, and other alterations in primeval conditions have made the environment unsuitable.

There is vast practical and intellectual gain to be obtained from a better understanding of the ocean. We can probably learn to double the harvest of the kinds of fish we now take from the sea, and at least double that amount again if we learn to use less familiar seafoods such as krill and squids—and deep-sea lantern fishes. Other material resources of the deep sea include enormous amounts of copper, manganese, and other minerals. And it is of great practical importance to control the damaging pollution of the sea by pesticides, oil, and other human garbage.

Among the intellectual gains is the further development of the geological theory of plate tectonics. This describes the crust of the earth as consisting of a series of plates, like those on a turtle's back, but moving ponderously in relation to each other, and producing the world's most spectacular natural events—volcanoes, earthquakes, mountain-building, and mountain destruction. This theory has produced a revolution in human thought, and it has been compared to that of Darwin on evolution in its intellectual importance.

One of the few areas of activity where men over the whole world have a genuine mutual interest, and therefore the best reasons for cooperation, is in the sea. It is a common resource, to be used and preserved through collaborative effort and study. Under the persistent probing of scientists, the sea is gradually yielding its wealth. But as we push aside a small piece of the unknown we reveal a larger mystery, so that ocean science repeatedly scales to a new level more stirring and more productive than the last. Perhaps this self-interest, and this continuously fulfilled promise of satisfaction, can serve to bind men of the nations together, where so many other activities tear them apart.

1 The Birth of the Sea

If the bumpy surface of the earth were rounded and smoothed into a perfect sphere, the oceans would cover the entire globe to a uniform depth of 8,000 feet. Fortunately for the earth's terrestrial inhabitants, however, the land masses are unevenly distributed, and great continents rise above the surface of the water. What shaped the continents and scooped the vast ocean basins? Where did the water come from to fill those basins? How were the mountains built and the valleys carved? What lies beneath the water of the oceans? For answers we must go back to the origin of the earth itself.

WHEN AND HOW

In the seventeenth century an eminent Irish churchman, Archbishop James Ussher, declared with enviable assurance that the earth was created in 4004 B.C. He even named the exact day and hour: Sunday, October 23, at 9 A.M. Modern scientists have pushed the time back considerably. They have not ventured to be nearly so precise, but several lines of evidence agree as to the approximate age of the planet we live on. The most trustworthy dating method is based on the rate at which radioactive elements transform themselves— "decay"—into other elements. Uranium, for example, decays slowly, and at a steady rate that has been measured precisely, to form lead and helium gas. By analyzing rock samples to determine the proportion of uranium to the breakdown products, geologists have measured the age of the oldest rocks known. On this basis the age of the earth accepted today is 4.6 to 5 billion years.

A number of theories of the origin of the solar system have sug-

[3]

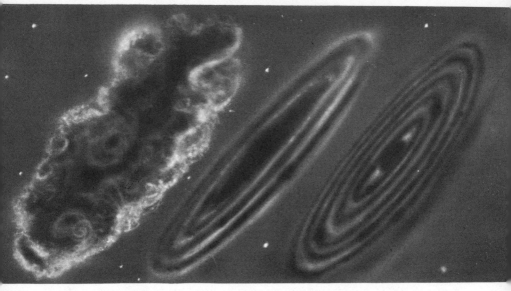

FIGURE 1−1. *Birth of the solar system. Modern scientists think that the solar system began as an immense disc of gas and dust.*

gested the near or actual collision of two stars. By the middle of the nineteenth century the public had been presented with various descriptions of the origin of the earth that involved titanic forces of unimaginable violence together with incandescent temperatures. A French naturalist, the Comte de Buffon, in 1749 suggested that the earth was torn from the fiery breast of the sun. As the sun and a runaway star collided, some of the material from the surface of each was supposed to have been scattered through the space between them; this white-hot material gradually cooled and formed the planets. Thus two systems were created, one our solar system and the other associated with the second star. These theories are now suspect, since the velocities required of the colliding stars are greater than seem possible and the chance of collision is now considered unbelievably small.

In the eighteenth century the German philosopher Immanuel Kant suggested that the solar system was formed from a great envelope of gas and dust. A large mass of matter concentrated at the center became the sun, and lesser masses at varying distances from the center became the planets. Kant's theory was crude but remarkably similar to an explanation widely accepted today. In 1796 the noted

[4]

This contracted, and the sun formed at its center, with the planets spinning at varying distances farther out.

French mathematician, the Marquis de Laplace, developed such a "condensation" theory. He supposed that during its formative stage the sun was contracting while its speed of rotation was increasing. Successive rings of matter, he proposed, would have been thrown out as a result of the centrifugal force thus created. Each ring would have coalesced into a sphere to become one of the planets or smaller bodies in the solar system. More detailed analysis of the motion involved, however, made this theory untenable: it appeared that the sun could not have lost any of its matter with the mass and the low speed of rotation that Laplace assumed.

An explanation that has much support now among scientists has been presented by the American chemist Harold Urey and also by the German physicist C. F. von Weizsaecker and the American astronomer G. P. Kuiper. This theory starts with a bow toward Kant. It supposes that the sun and the planets, including the earth, were formed together. The whole solar system began as a vast, cold, globular cloud of gas and dust. This contracted, at first slowly, then at an accelerating rate. The sun formed at the center, with great sheets of matter wheeling around it in the form of a flattened disc. At this

[5]

early stage the temperature of the system was very low, probably as low as that of interstellar space at the present time. The disc was unstable, and it broke into pieces that would later form planets. Water and liquid ammonia must have been present at this time to act as cement for the particles of dust, which now began to coalesce and form small masses of material. These were added to the embryonic planets through gravitational attraction and collision. As the planets increased in size, the mass of the material, thrusting heavily downward, compressed their centers. This raised the temperature of the interiors of the planets.

The theories of both Buffon and Kant supposed that the sun and the earth were originally enormously hot, and that some of this heat has slowly been lost from the surface by radiation, leaving the centers still at high temperatures. But the modern theories hold that the temperature of the primordial sun and earth was very low: perhaps 370° below zero Fahrenheit (F.). The present high temperature of the sun is thought to result in part from the gain in energy due to shrinking and the consequent rise in pressure in the center, and in part from radioactivity, which was once fifteen times more intense than it is now.

When a radioactive substance decays, its central nucleus breaks apart, throwing out fragments. Some of these fragments are heavy, fast-moving alpha particles, which tear through surrounding matter, reacting with the electrons of other atoms and pulling them out of their orbits. Several thousand atoms may be disrupted by the flight of a single alpha particle, and the energy of their "excited" electrons is transformed into heat.

Presumably, a gaseous earth, at first very cold and a thousand times its present size, gradually coalesced and became nearly spherical in shape. Free hydrogen and helium were at first present in the atmosphere, but they gradually escaped into space. The sun, with its great mass and consequent gravitational attraction, kept these light gases. At present hydrogen and helium make up only a small fraction of the mass of the earth.

As the temperature of the planets increased, other elements of low mass escaped. These included compounds of silicon and oxygen, among the lightest components of rock. The heavier rocks were left, increasing the average density of the planets. In the planets with the greatest mass, like the earth, the heavy rocks settled toward the

[6]

center, so that a core of iron and a shell of silicates were formed. Planets like Mars, with smaller mass and therefore less gravity, remained of almost uniform density throughout.

THE STRUCTURE OF THE EARTH

The heaviest substances, which sank to form the core of the earth, are very dense indeed. The core is believed to consist of two parts, the inner having the properties of a solid and the outer being liquid. The temperature may be as hot as 6,552° F. and the pressure some 4 million times as great as atmospheric pressure. This core is about 4,000 miles in diameter, or nearly twice that of the moon. It is believed to be about 60 per cent iron, 30 per cent silicon, and 10 per cent nickel. The next heaviest layer, surrounding the dense core, is the mantle. There is a sharp boundary between the core of the earth and the mantle, the two having quite different physical properties. The existence of this boundary, like most other information on the structure of the earth, has been inferred by measuring earthquake, or seismic, waves passing through the earth. The mantle is composed of a heavy greenish mineral consisting mostly of oxygen, magnesium, silicon, and iron. It is about 1,800 miles in thickness. Enclosing the blazing core as it does, and being under enormous pressure, the mantle itself is white hot at its deepest part.

Around the mantle is the crust—the thin outermost skin of the earth. The division between these two layers is a sharp one. This abrupt change was first noted by a Yugoslav seismologist, Professor Andrija Mohorovicic, when he was studying seismographic records of Balkan earthquakes. It is called the Mohorovicic Discontinuity, or Moho. Under the continents the Moho is about 18 to 50 miles from the surface, while under the sea it comes much closer to the surface, lying about 6 miles beneath the ocean floor. Compared to the 8,000-mile diameter of the earth, the crust is a very thin covering, like the skin of an apple.

The blazing iron core and the green rock mantle of our big apple are hidden from us; little is known about them, and that indirectly. From the surface the apple-skin crust seems very irregular. Certainly in relation to our own puny size the mountain ranges, thrusting 5½ miles into the air, and the ocean basins, gouging troughs as deep as 7 miles from the sea surface, are of enormous dimensions.

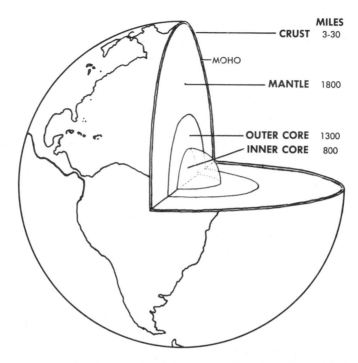

MILES

CRUST 3-30

MOHO

MANTLE 1800

OUTER CORE 1300
INNER CORE 800

FIGURE 1−2. *Layers of the earth. The core, about 4,000 miles in diameter, consists of exceedingly hot, dense material. It is thought to have two parts: a solid inner center and a liquid outer layer. Surrounding the core is the mantle, about 1,800 miles of dense, greenish rock. Next is the crust, and between it and the mantle is a sharp boundary called the Moho, for Andrija Mohorovicic, the Yugoslav seismologist who first described it. The crust, which is the outermost layer of the earth, is in comparison very thin—as little as 3, and at most 50, miles thick.*

As a matter of fact, the earth is very nearly smooth, relatively speaking. It shows fewer bumps and hollows in relation to its diameter than a billiard ball, which is often put forward as the standard of smoothness. The wrinkles on the face of the land are the consequence of mountain-building by vulcanism and the thrusting together of the huge plates that make up the earth's crust. Over the eons of time these mountains are then smoothed out again by the erosion of wind and water.

[8]

SHAPING THE SURFACE

The mechanism by which the wrinkling of the earth's skin took place has been explained in quite as many ways as the origin of the planet itself. Perhaps the most intriguing idea about the genesis of the great ocean basins is the theory that they were formed when a titanic piece of the earth's surface was torn away and flung into space, leaving a gaping hole that became the Pacific Ocean basin. Then the surface of the globe on the opposite side cracked apart, and the pieces gradually slid into new positions, re-establishing some sort of balance over the face of the earth. Meanwhile the part torn away was wrenched into orbit around the earth, became spherical, and was later named the "moon." Sir George Darwin, Cambridge don and son of the great naturalist Charles Darwin, was the first to propose this idea, in 1878. He assumed that the substance of the earth at the time was still hot enough to be fluid. The molten rocks were raised in great tides as the sun—assumed to have been much nearer the earth than it is now—exerted a powerful pull. As these tides rolled over the face of the globe, they grew higher and higher and then gradually began to flow in time with the period of free oscillation of the molten earth. This forced them to still more towering heights, just as repeated pushes send a child on a swing ever higher. The tides increased steadily over hundreds of years until their height was so great that the gravity of the earth could not hold the molten rock, and it was flung into space. By this time the liquid mass of rock on the surface of the earth had cooled sufficiently so that the hole was not entirely filled. At the same time, the piece forming the moon presumably remained plastic enough to become a sphere as it swung into orbit around the earth.

Several facts support this theory. The Pacific Ocean basin has only a very thin layer of light rock covering its "basement" of heavier material, and in many places no such covering at all. By contrast, the continents are covered by light rocks such as granite and granodiorite, and some rocks beneath the sediments in the other oceans may also be composed of these. It is as though the primeval rock of the Pacific had been stripped away, leaving exposed the heavier basalt and peridotite basement substance of the earth beneath. Then, too, there is the coincidence that the mean density of the moon, about 3.3, is roughly that of the outer layers of rock on our planet, but considerably less than the average density of the earth, which is about 5.2.

But there are a number of difficulties ranged against this Darwin theory. One may already have occurred to you from the present discussion: if the material wrested from the earth's surface was fluid enough to be shaped into the nearly perfect sphere of the moon, why was the material surrounding the Pacific Ocean basin so viscid that the hole maintained its shape? You would expect, rather, that the greater bulk of the earth would be more likely to retain heat than would the moon, and that the face of the earth would smooth over. In addition some scientists, including the eminent geologist Sir Harold Jeffries, have concluded that the solar tides on the surface of the primeval earth could not have exceeded 200 feet in height, far smaller than the tides that would have allowed the enormous wave to break free (in the astronauts' jargon, to reach orbital velocity). Instead, the modern theories of Urey and other scientists hold that the earth and the moon are partners in one of the twin-planet systems that are common in the universe; one partner is considerably smaller than the other, it is true, but they were probably both formed at the same time, and not one from the "rib" of the other, like a cosmic Adam and Eve.

So it seems that the moon had no part in the formation of the ocean basins. Instead, the deeps and the ridges of the oceans, together with the mountains and the plains of the continents, were probably formed on top of the original surface of the earth. Through fissures in this primitive surface the molten rock of the interior welled up, spread out, was broken away again, eroded, re-formed, and rearranged over the billions of years of the earth's history. The continents assumed their present shapes, according to a recent idea, because the first hot material erupting from the molten earth appeared in doughnut-shaped masses. Where two doughnuts touched, the light rock fused to form embryonic continents. In the holes of the doughnuts the ocean basins were left, lined with heavy basalt rock or with only thin overlying layers of granite.

THE ORIGIN OF THE OCEAN

Just as the present crust of the earth is derived from its hot interior, it is likely that the gases of the atmosphere and the waters of the ocean were once likewise locked up there. Millions of years of earthquake and volcanic activity not only released molten rock but also flung out clouds of carbon dioxide, nitrogen, and water vapor.

The first water vapor that hissed out of the boiling surface of the earth joined with other gases to form a thick atmosphere, which wrapped a dark pall around the planet. When temperatures above the earth became cool enough to allow the vapor to condense into rain, it fell toward the earth but was quickly boiled back into the gaseous state as it approached the still scorching earth. Dark clouds, making the new earth pitch-black except for the shifting scarlet glow of the molten rocks, probably persisted for millions of years. Ever so slowly the earth cooled; when the rocks were cooler than the boiling point of water, some rain fell to the surface and stayed there, collecting in depressions. No longer was the water from the thick clouds immediately vaporized when it fell to the earth. The infant seas were born. Then such rains fell as to make the biblical Deluge of "forty days and forty nights" seem as a light and transitory mist. The water that had welled up from the new planet and into the atmosphere as vapor finally condensed and poured out of the skies. At last the clouds thinned slowly and then broke, and the sun glinted on the fresh new sea.

The ocean levels at that time were much lower than they are today. Since the end of that first vast rainfall a very considerable additional volume of water has welled out of fissures in the crust of the earth, raising the water to a high enough level so that it overflows the ocean basins and laps the edges of the continents. Some scientists calculate that about a quarter of the water on earth may have entered the ocean in about the last fortieth of the geological history of the planet. If this rate continues, the rising surface of the sea will leave no land dry in another 100 million years. There is ample evidence of lower ocean levels in past geological times, as well as evidence of the sinking and the uplifting of land, including the remains of marine animals found in rocks on lofty mountaintops. But the sea has never covered all the land, or even most of it.

The sea level in recent eras appears to have been determined chiefly by the amount of water trapped in glaciers and by the slow rise of land masses as the crust of the earth adjusts itself to stresses. During the four great glacial periods of recent geological history the level of the sea has gone up and down in response to the ebb and flow of the glaciers. Even now, in an interglacial period, enormous volumes of water are still bound up in the ice—especially on Antarctica, which contains 92 per cent of the world's ice. If all this glacial ice were to melt, the sea level would rise some 200 to 500 feet (according to various estimates), inundating a very considerable portion of our civilization. As a matter of fact the ice has been melting. The gradual warming of our climate since about 1850 has been raising the sea level all over the world at a rate of about 4½ inches per century. The reverse was the case during the glacial periods, when more water was frozen: the sea level at the height of the last ice age was 300 to 450 feet lower than it is now, so that what was dry land about 17,000 years ago is now the submerged edge of the continental shelf, as far as 100 miles from our shores.

As the ice receded, the water released from the thawing glaciers invaded the land, sometimes at a slow rate when temperatures were low, sometimes rapidly as temperatures rose. Between 17,000 and 6,000 years ago the rate was at its height and the margins of the land were drowned. Some such catastrophe may have been recorded in the Bible as the Deluge. The story has often been regarded as legend, but scholars are struck by the recurrence of similar tales in the literature of many cultures. In addition to the Hebrew story, there are ancient legends of overwhelming floods from the Babylonian, Greek, Hittite, Welsh, Icelandic, Burmese, Chinese, Malayan, Polynesian, and North and South American Indian traditions. Perhaps these ancient memories record the same catastrophic rise of the sea over the entire globe, blotting out the land along the water's edge, where much of civilization has always clustered.

LOST WORLDS

If the Deluge may have a basis in fact, what about that most persistent of the myths of the sea, the story of a land that sank beneath the waves? The legendary land most often mentioned is Atlantis, while others such as St. Brenden's Island, Avalon, and Lyonnesse occur repeatedly in ancient legends. There are few ideas more romantic than that of an ancient people swallowed by a cataclysm, leaving nothing but a dim ancestral memory—and who can resist the fascinating notion of ancient temples, palaces, and treasure stores preserved beneath the waves? The grip of such tales on the minds of men is surely a principal reason for the durability of the Atlantis legend in the face of little encouragement from the facts.

Many people have seen confirmation of the story in the discovery of the Mid-Atlantic Ridge, the great submarine plateau that slices north and south through the Atlantic Ocean. The idea is that a portion of the ridge may once have been above the surface and may have been submerged within at least the subconscious memory of man. But geologists believe that with the exception of a few small islands the ridge has not been above water for millions of years—far longer than man has existed.

But if Atlantis never existed, other lands now far beneath the sea once did thrust mountain ranges above the surface; and they have names just as euphonic, and stories, in their way, just as epic. These stories have a solid basis in geological fact—they are not legends, but the truth reconstructed from bits and pieces of rocky evidence collected by geologist-detectives. Appalachia, Cascadia, Scandia, Melanesia—names with a melodious and evocative cadence—denote land areas that may once have existed and are now at least partly submerged.

Consider Appalachia as an example. Many a citizen of Boston or Baltimore would be astonished to realize that a soaring mountain range once existed *east* of their cities, where the Atlantic now swells. During the Triassic and Jurassic periods, some 160 to 200 million years ago, the part of the North American continent now occupied by the Laurentian Mountains was under an inland sea. It is believed by many geologists that to the east of this, and extending about 200 miles from the present coastline, there was the mountainous land of Appalachia. Rivers from these mountains carried sediments westward into the shallow sea. This great piece of the continent filled the indentation in the present outline of North America from Florida on the south to New England and Newfoundland on the north.

Then the face of the earth changed. The great inland sea disappeared as the land beneath it rose; meanwhile Appalachia foundered and warped and sank beneath the Atlantic Ocean. The most easterly edge of the former high land now lies under 10,000 to 13,000 feet of sediments. The coastal areas of Nova Scotia are the only remains of the land of Appalachia.

Cascadia was a similar land in the Pacific Ocean off the present western shore of North America. The Coast Range of British Columbia and Washington and the Sierra Nevadas of California were formed under water from marine sediments, laid down in shallow seas perhaps 150 million years ago. At that time the Rocky Mountains were mud plains beneath an inland sea lying between Cascadia and the continental mass. The mud and sand of the sediments was torn from Cascadia's high mountains. Erosion smoothed them off, the land sank

[13]

seaward, and the muddy plains formed from the inland sea rose slowly 2 miles in the air to form our majestic western mountain ranges. Similarly, Scandia, once high land off Spitsbergen, vanished into the Atlantic in Tertiary times about 50 million years ago, leaving the Hebrides and northwest Scotland as its sole visible remains.

Perhaps the most spectacular instance of a great drowned continent is Melanesia. The name is commonly applied to a vast assemblage of islands in the Central and South Pacific. Geologists now think there was an ancient land mass that encompassed such islands as Yap and Truk in the Carolines, the Fiji and Tonga islands, Guadalcanal, Australia, New Zealand, and New Guinea. All these land masses still rise above the sea. The remainder of Melanesia, a former major continent, has subsided to considerable depths; in fact, the greatest ocean troughs in the world are in this area. Even the yawning Philippines Trench has islands composed of granite and other "continental" rocks on both sides of it. A great belt of volcanoes runs along the southeast border of this former land; it probably marks a fault plane whose activity caused Melanesia to sink into the sea, perhaps 70 million years ago.

What is the geologist's evidence for this rising and subsiding of land? The occurrence of marine animals in the fossils on mountain peaks reveals one of the most astonishing facts about the history of our earth. Mount Everest, for example, soaring 29,028 feet, is composed of limestone—and limestone is formed only in a shallow sea! Of course, the sea was never high enough to cover such mountains in their present positions; the peaks were lifted by majestic and monumental processes whose forces are intimately connected with stresses in the ocean basins. It is not coincidence that mountain ranges are on the borders of the continents, or that the deepest places in the oceans are close to the land rather than in the middle of the basins. The mountains rose and the deeps yawned in rhythm, counterpoints in a vast and awe-inspiring symphony.

2 The Land Beneath the Sea

"The floor of the ocean"—what a misleading phrase that is. Floors are flat and featureless, but the land beneath the sea has more ups and downs than Colorado. Beginning at the edge of dry land, a man walking out into the sea would first traverse the gently sloping continental shelf. Then he would come to the steep slope that tilts suddenly downward into the deep sea, to the abyssal plain onto which sediments have been deposited for thousands of years. But our submarine explorer would find smooth plains the exception. Everywhere he would see individual mountain peaks and the enormous mountain ranges that extend the entire length of oceans. He would peer down into submarine canyon systems looking like the branched river systems on the continents and in some cases connecting with river valleys on the land. And perhaps most astonishing of all, he would come upon deep ocean trenches gaping miles below him.

DOWN TO THE DEEP

The continental slopes are the true "edges" of the continental land masses, and they form the walls of the ocean basins. But these slopes are usually a considerable distance from the edge of the sea where the tides lap at beaches, and in some cases they are hundreds of miles away. The continental shelf, between the coastline and the continental slope, varies greatly in width, depending on the recent geological history of the area. Where young mountain ranges border the sea the land often drops away quickly into the depths of the ocean, and the shelf may be very narrow; this is the case along the western edges of the Americas. In other regions, such as the great shallows

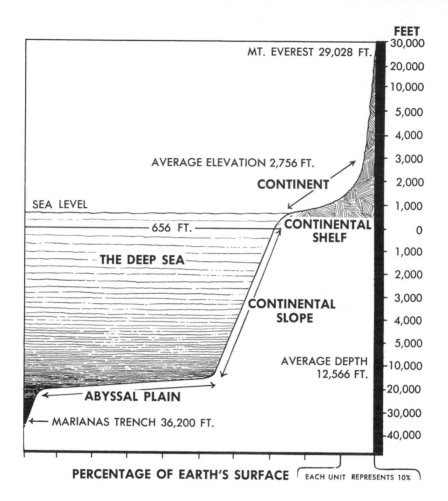

FIGURE 2-1. *The face of the earth. Far more of the earth's surface is beneath the sea than above it, and dips deeper beneath sea level than it climbs above. The highest mountain reaches only to 29,028 feet, while the greatest ocean depth is 36,200. Even figures for the average reflect this disparity: the average height of land is 2,756 feet and the ocean depth 12,566.*

north and east of Asia, the Grand Banks of eastern North America, and the North Sea area of western Europe, the continental shelf extends for miles. The largest of such areas are off the coasts of Asia. The shelf is some 800 miles wide in the Arctic Ocean off Russia, and it has an area of a million square miles off the Pacific coasts of China

[16]

and Russia. These great submerged flats, which drop to 250 and 600 feet deep before they meet the continental slopes, are very important to the economy of the sea. In their shallow, sunlit waters most of the life of the sea is concentrated, so it is here that most of the fish and other useful creatures are caught.

The transition from the continental shelf to the deep sea is abrupt. After sloping gently down from the shore line to an average depth of some 400 feet, the face of the earth plunges into the abyss, down to depths of 10,000, 20,000, and even 30,000 feet. Anyone who has peered over the cliff at Glacier Point in Yosemite National Park, into the heart-stopping depth there, can appreciate these enormous drops: the distance from that mountaintop to the valley floor is only 3,254 feet. Just as no continental valley plunges as deep as the underwater slope, no bluff on land is as long as some in the sea, which may stretch for hundreds of miles in an unbroken and stupendous cliff. One of the greatest of these is the steep, straight precipice in the Gulf of Mexico off the west coast of Florida. It extends for 500 miles and drops a mile to the bottom in less than 2 miles of horizontal distance.

The shallow areas of the sea, because they are close to the continents and easy to observe and because they shelter schools of fish and hide ship-destroying rocks, have claimed the attention of man far beyond their relative size. Actually, the shelf areas of the world constitute only about 8 per cent of the total area covered by the oceans. Over four fifths of the seas is more than 10,000 feet deep; over one third of the oceans is more than 13,000 feet in depth, and the average depth of all the oceans, including the continental shelf, is a tremendous 12,566 feet. (By contrast, the average height of all the land in the world is only 2,756 feet above sea level.) So as we leave the shelf and dip down the continental slope into the deep sea, we are approaching by far the largest environment in the world.

This means that all the rich life familiar to humans, the bustling and buzzing crowds of mammals, birds, and insects and the trees, grass, and flowers of the land, as well as the crabs, snails, fish, and algae of the shallow seas, are crowded into a relatively minor portion of the earth's living space. The far vaster, unseen—and, until recently, unapproachable—area of the deep sea has its own fauna, an assemblage representing nearly all branches of the animal kingdom from the simplest one-celled protozoans to the higher vertebrates. The darkness, cold, and pressure of the abyssal regions have left their

FIGURE 2 – 2. *Undersea mountains. These are grander than those on land because there is much less erosion in the sea. Here is an artist's conception of what the Mid-Pacific mountains might look like if the water were drained away. In the background are flat-topped mountains called guyots.*

This painting by Chesley Bonestell is reproduced through the courtesy of Edwin L. Hamilton.

imprint on the creatures that have chosen it as home, so that these animals include some of the most bizarre and incredible of beings. But before describing these curious animals and the way they live, there is more to be said about their living space and the terrain that underlies it.

VALLEYS AND CANYONS

And rugged terrain it is. The mountains loom menacingly, their crags and escarpments unsoftened by any covering mantle of forests, their canyons steep sided and rough, all their dimensions and qualities grander, rawer, more elemental, and more awe-inspiring than those on land.

The reason is that there is much less erosion than on land. Isaiah maintained that "Every valley shall be exalted, and every mountain and hill shall be made low: and the crooked shall be made straight, and the rough places plain." The erosive forces of flowing water, glaciers, wind, waves, and freezing, which tear down the most adamantine of rocks on land, are greatly inhibited in the sea by the protecting mantle of water. Valleys and canyons cut the slopes of all the oceans in tree-shaped networks, some being over a thousand miles long. Some of the valleys interconnect like the river systems on the land, and look remarkably like them. One that has been mapped in the Atlantic, south of Greenland, is vaster than that of the Mississippi and all its tributaries. Underwater valley systems often empty onto a plain, like a river mouth opening onto a delta.

The origin of submarine valleys and canyons is one of the intriguing mysteries of oceanography. Some of them are associated with existing river systems on the land, as, for example, the great valley that runs seaward from the mouth of the Hudson River. It is 180 miles long; at one point it cuts a groove 4,000 feet deep; at its seaward end, its floor is 14,000 feet below sea level. Another submarine canyon closely associated with a river lies off the coast of Corsica; the best example is the one opposite the Congo River.

Because of this connection between land rivers and many of the sea canyons, it was once suggested that the valleys, although now submerged, were cut by the rivers when the sea level was much lower than at present. If the icecaps were big enough at some stage of glaciation, it was thought they might have borrowed sufficient water to lower the sea level while the rivers did their work; then, in interglacial periods, the old coastline and its valleys were drowned again. This explanation has had to be abandoned. The great depth of the canyons and the age of rocks and sediments on submerged sea mounts and other relatively shallow areas argue against it. In order to expose the deepest of the canyon systems, the water would have had to lower as much as 3 miles. This is more than the most ardent exponent of the theory will accept. Besides, where could all that water have gone? It could not have been frozen into glaciers; the earth could not carry so much ice and still maintain the present character of the land masses.

In the 1930's, another theory was proposed to account for the origin of submarine canyons. It was suggested that they had been cut under water by great currents bearing loads of sediment. When mud or sand is stirred up in water, it produces a heavy mixture that flows down the continental slope. By this mechanism, sand and even gravel can be transported into deep-water areas in the sea. It has now been shown that powerful and rapid currents transporting loads of abrasive sand and gravel do move over the sea floor and that they can carve out undersea canyons.

Dramatic support for this idea was provided by the events which followed a great submarine earthquake of 1929 south of Newfoundland. A number of underwater telephone cables, many of which cross the area, were broken. The sequence in which cables at various locations broke suggested that turbidity currents were responsible, since the cables snapped in regular succession farther and farther from the epicenter of the quake. The progression continued for more than 13 hours after the disturbance, with each break down-slope from the last. The time that each occurred was recorded by monitoring devices of the telephone system, and speeds of 50 miles an hour were calculated for the turbidity currents supposed to be moving down the slope. The speed of the current gradually decreased at increasing distances from the center of the earthquake, but an average speed of 25 miles an hour was calculated over a distance of 300 miles.

Some scientists have been sceptical that turbidity currents attain high speeds, and two other possible explanations have been proposed for the sequence of breaks in the cables: (1) local slides induced by the energy of the earthquake as it moved out from the epicenter, or

(2) the temporary liquification of bottom sediments, which caused the cables to sink and break. But the evidence is sufficient to convince most geologists that turbidity currents produced the breaks in the cables.

According to Dr. Francis Shepard of Scripps, "the final word has not yet been written about the origin of submarine canyons. Probably some canyons have one origin and some another." Many have apparently been cut by turbidity currents, while others, including those around the Hawaiian Islands, were formed from the submergence of the land. Still others, like the valley that comes out of Sagami Bay, Japan, are the result of down-faulting of the earth's crust.

THE GREAT TRENCHES

The submarine canyons, impressive as they are, are mere scratches on the surface of the earth compared with the deep trenches that plunge almost 7 miles below the surface of the sea. One might have expected to find the ocean basins in the shape of bowls sloping down from the continents, with the greatest depth at the center of each basin. Instead, most of the deep trenches are near the edges of the basins, quite close to land and associated with the arc-shaped island chains. Along island arcs like Indonesia, the Aleutians, and the Philippines, and off coastlines like the Pacific coast of Central and South America, the earth is shaken by earthquakes, and volcanoes throw up new land.

The long, narrow, arc-shaped trenches are at one edge of the crustal plates. Here the outer shell of the earth descends steeply into the mantle; its rocks are melted and some of them are returned to the surface as magmas and lavas that build up the island arcs behind the trenches. Trenches and their island arcs are the most unstable parts of the earth, exhibiting violent earthquakes and volcanoes.

Most of the great trenches of the ocean are in the Pacific. The principal trenches there are the Kermadec, Kurile-Kamchatka, Japan, Marianas, Tonga, and Philippines trenches. Until a few years ago it was thought that the Philippines Trench was the deepest, but its depth is now known to be exceeded by three other vast furrows, the Marianas, Kurile-Kamchatka, and Tonga trenches. The depth of the deepest part of the ocean, in the Marianas, or Challenger Deep, is measured at 36,198 feet. The Tonga Trench is 35,700 feet deep and takes second prize for maximum depth. Some of the deep trenches are remarkably long, extending to 2,000 miles along the ocean floor.

In the Atlantic the deepest trench is the Puerto Rico Trench, to the north of the island; its depth has been recorded as 27,500 feet. The only one of the same order of magnitude in the Indian Ocean is the Sunda Trench. A table gives depths assigned in recent years to the principal ocean trenches.

FIGURE 2 – 3 . *Great trenches of the Pacific Ocean. The floor of the sea is gouged by deep cuts that plunge almost 7 miles straight down. Most of them are close to land.*

Edwin L. Hamilton, U.S. Navy Electronics Laboratory, San Diego

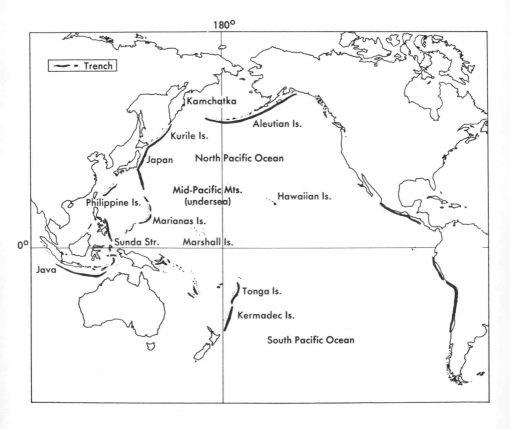

The trenches are similar in their depths. The six great trenches of the Pacific range in maximum depth from 32,800 feet for the Izu Trench to 36,198 feet for the Marianas Trench. The three deep Atlantic trenches range from 25,800 to 27,500 feet. The similarities are not coincidental, but result from isostacy, the equilibrium established in the earth's crust by the slow flow of rock material beneath the surface under the stress of gravity, and by the approximate equality of mass of each column of the earth from the surface to a depth of about 70 miles.

Depths of Deep-Sea Trenches (in feet)

PACIFIC OCEAN

Marianas Trench (Challenger Deep), western Pacific	36,198
Tonga Trench, southwest Pacific (Horizon Deep)	35,700
Kurile-Kamchatka Trench, northwest Pacific	34,600
Philippines Trench (Mindanao Trench), western Pacific	34,400
Kermadec Trench, southwest Pacific	33,000
Izu Trench, western Pacific	32,800
North Solomons Trench (New Britain Trench), western Pacific	30,000
New Hebrides Trench, southwest Pacific	30,100
Japan Trench, western Pacific	28,000
Palau Trench, western Pacific	26,500
Peru-Chile Trench, southeast Pacific	26,400
Yap Trench, western Pacific	26,300
Aleutian Trench, north Pacific	25,200
Ryukyu Trench, western Pacific	21,000

ATLANTIC OCEAN

Puerto Rico Trench, western Atlantic	27,500
South Sandwich Trench, southwest Atlantic	27,113
Romanche Fracture Zone, eastern Atlantic	25,800

INDIAN OCEAN

Sunda Trench (Java Trench)	25,344

Mount Everest would be drowned beneath well over a mile of water if it were dropped into the Challenger Deep. It may further hurt the pride of landsmen to know that several mountains rise from the sea floor higher than Everest. The greatest peak on earth is Mauna Kea in the Hawaiian Island group, which soars 33,476 feet from the ocean bed to its summit, the last 13,796 feet of the mountain protruding above the water.

It was dimly realized more than a century ago that mountain ranges existed in the middle of the ocean basins, and that mid-ocean islands were the peaks of some of these mountains thrusting above the waves. This realization was greatly strengthened in the 1870's as a result of the observations made by the *Challenger* Expedition. Led by a remarkable and enthusiastic scientist, Wyville Thomson, the *Challenger* ranged the oceans of the world, traveling 68,890 nautical miles in 3½ years, bringing back a wealth of biological, chemical, and physical information about the seas. Thomson and his associates on this famous voyage may have done too good a job, because they so impressed scientists with their accomplishments that for a long time it seemed that the last word had been said about the sea, including the bottom topography. Some observations of the deep-sea floor were made in the years following the *Challenger* Expedition by the German *Meteor* and by American vessels like the *Baird* and the *Horizon* and others; but the next comprehensive investigation came in 1947–48, when the Swedish *Albatross* sailed around the world on a scientific cruise under the leadership of Hans Pettersson, making valuable observations of the sea floor and its sediments. Then, during the International Geophysical Year of 1957–58, undersea geology spurted forward, and it has been running hard ever since. Fascinated geologists are now seeing the accumulating information about the mid-ocean ridges support the once discredited theory of drifting continents.

The Mid-Atlantic Ridge was the first to be mapped, on the basis of soundings made in preparation for laying the first transatlantic cable in the second half of the nineteenth century. At first the soundings were scarce, and the mountains appeared to be grouped in isolated ranges. As the number of soundings accumulated, scientists slowly realized that the ridge was continuous down the whole length of the Atlantic Ocean, and that it was an immense natural feature of the globe. The ridge twists down the center of the Atlantic for over 12,000 miles, from above Jan Mayen Island north of Iceland to beyond Tristan da Cunha and Bouvet islands near the Antarctic Continent. This enormous mountain range is known to be broken only once—by the Romanche Trench, near the equator. Most of its peaks are drowned beneath thousands of feet of water, but a few are high enough to break the surface of the lonely Atlantic and become familiar as islands: the Azores, Ascension Island, St. Helena, the Rocks of St. Peter and St. Paul. Early oceanographers knew that

Legend

⊱ Mid-ocean ridges

Ocean depths less than 1500 fathoms

Land

FIGURE 2–4. *The mid-ocean ridge. This is the most im-mense feature on the earth's crust. Occupying more than a quarter of the ocean, the ridge is at least 40,000 miles long and from 300 to 1,200 miles wide. A huge steep-sided, flat-bottomed valley splits most of its length.*

J. T. Wilson, University of Toronto

mid-ocean ridges exist in the Pacific and Indian oceans too. In the north-central Pacific there is a great undersea mountain range extending in an east-west direction. Some of its peaks are 3,000 to 6,600 feet below the surface of the sea. It extends from the Hawaiian Islands nearly to Wake Island. The Marshalls, the Tonga Islands, and many more are emergent peaks of great undersea ridges.

It was not until the increased activity of the International Geophysical Year that it was realized not only that the ridges over the whole ocean are far more immense than had been thought but that many of them are interconnected, forming a long, virtually unbroken ocean mountain system that winds over the entire globe. At its southern end the Mid-Atlantic Ridge turns between Antarctica and Africa to join a range in the Indian Ocean. Branches thrust into the Gulf of Aden and toward Pakistan, while the main ridge passes south of Australia and joins the East Pacific Rise. From Easter Island, the Pacific mountains extend through the Line Islands to Kamchatka. Its final link seems to be the Lomonosov Ridge across the Arctic Ocean, mapped by Soviet oceanographers. As one geophysicist says, "The discovery (only in 1957) of the continuity of this, by far the greatest mountain range of the earth, is an interesting commentary upon the state of our knowledge about some part of the earth."

The great ocean ridge is the most immense feature on the face of the earth. It is at least 40,000 miles long and from 300 to 1,200 miles wide! Its peaks thrust up miles from the floor of the ocean, many of them exceeding in height the highest mountains on land. Its bulk is so great that it occupies between a quarter and a third of the ocean. The mid-sea mountains are the focus of steady volcanic and earthquake activity. Volcanic in origin, they apparently consist almost exclusively of basalt lava.

The most intriguing feature of the mid-ocean mountain ridge is a colossal rift that splits most of its length. This is not a single crack but a creviced zone 3 to 30 miles wide and between ½ and 1½ miles deep. It is a steep-sided valley with a flat floor. It runs nearly the whole length of the ridge in the Atlantic, but it may not be present to as great an extent in the Pacific or in other oceans.

DRIFTING CONTINENTS

This great crack in the bottom of the sea is a major feature of the most stupendous of all earthly pheonomena—the drifting of the continents.

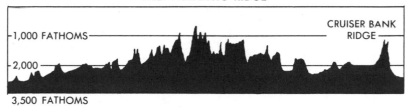

FIGURE 2−5. *Profile of the mid-ocean ridge. These mountains, whose tops extend upward for miles from the ocean floor, are the focus of earthquake activity.*

Ever since maps of the Americas were accurate enough to reveal their outlines, men have been tantalized by the matching shapes of the land masses on opposite shores of the Atlantic Ocean. Sir Francis Bacon remarked on it in 1620; so did Benjamin Franklin, and many others. With their continental shelves added, the continents can be fitted together like the pieces of a jigsaw puzzle. To this coincidence there is now added the fact that the Mid-Atlantic Ridge, twisting and turning as it bisects the ocean, follows the outlines of the continental edges.

In 1912, Alfred Wegener, a German astronomer, suggested that the present continents were once joined as a single immense land mass, which he called Pangaea. About 200 million years ago, he believed, this supercontinent split, with the sections drifting apart, eventually to become the present continents. Between them the Atlantic and Indian oceans gradually were formed, while the original water area of the globe, the Pacific Ocean, slowly became smaller. After a generally favorable initial reaction, Wegener's theory ran into trouble, particularly in Britain and the United States. This was partly because geologists resented an astronomer presuming to encroach upon their province; but more fundamentally, it was hard to conceive how such immense quantities of continental rock could slide over the mantle—"like the rind of a very ripe fruit over the soft flesh," as one account put it.

But in very recent years Wegener has been vindicated, and while the mechanism he proposed has been shown to be wrong, convincing proof has been produced that the continents do indeed drift—in the case of North America as fast as one inch a year.

The new geological theory of plate tectonics proposes that the crust of the earth is composed of nine thin rigid plates, joined together like the sections of a turtle's carapace. Unlike a turtle's skeletal plates, however, those of the earth do not remain passive, but ex-

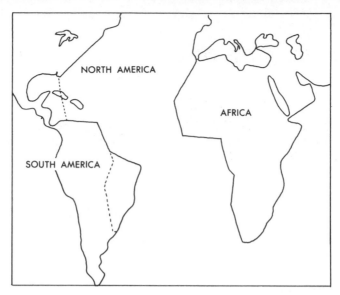

FIGURE 2-6. *The drifting continents. Bulges in the continents on one side of the Atlantic Ocean conform remarkably to indentations in the land on the other side—like matching pieces of a jigsaw puzzle. Recent evidence, chiefly the alignment of magnetized particles in the ancient rocks, suggests strongly that North America was indeed once joined to Europe, and South America to Africa.*

hibit inexorable if ponderously slow motion. The crucial areas of the phenomenon are at the mid-ocean ridges. Here the hot mantle rock of the earth wells up, filling the rift formed as the plates move. As the plates move so must the continents also, since they are perched on top of the plates. Hence, a difficulty encountered with Wegener's theory is overcome—the continental rock does not have to plow through the mantle, but instead rides atop it.

The new material added to the edges of the plates does not increase the size of the earth, since the plates are destroyed by being pushed into the mantle on the far side, tipping down into the great ocean trenches. Here the rock is melted back into the mantle material.

Many of the most spectacular earthly events take place at the edges of the plates. Earthquakes result from the grinding of the plates together—as in the San Andreas Fault of California, which represents part of the edge of the Pacific Plate—and mountains are built on the edges of continents as two plates push against each other, crumpling the land upward.

The theory of moving continents is supported by the remarkable

[28]

similarity of many geological formations of the land on opposite sides of the ocean, as though they had once been parts of the same area. There are also similarities in fossils, suggesting that the same plants and animals flourished long ago in regions now separated by thousands of miles of open sea.

But the strongest evidence for plate tectonics and continental drift comes from precise measurements of earth magnetism. As molten rock pours from the mantle, iron particles line up pointing toward the magnetic pole. When the rock hardens, it freezes millions of tiny compasses which point to the pole. With newly developed highly sensitive instruments it has been discovered that rocks from different continents but from the same geologic age point to different polar positions. Thus, the continents must have moved in relation to each other.

Moreover, the magnetic poles weaken and then actually reverse every few thousands of years, the North Pole becoming the South, and vice versa. These changes can be revealed by measuring the strength of the magnetic fields adjacent to the mid–ocean ridge. A magnetometer towed behind a plane reveals matching zebralike patterns of alternating strong and weak magnetism of the rocks on either side of the ridge. These stripes can only be explained as bands of crustal rock that welled out of the mid-ocean ridge, alternately exhibiting strong and weak magnetism. The result is something like the rings on a section of a tree trunk.

Plate tectonics explains another mystery of the ocean floor—the fact that its rocks are younger than those of the land. If the oceanic plates are renewed at one side by upwelling of rock through the mid-ocean ridges, and shredded at the other side by the stupendous "garbage disposal" of the trenches, they cannot be as old as rocks on land. Moreover, recent measurements reveal that the age of the basalt rocks increases the farther they are from the mid-ocean ridges.

Finally, the perplexing circumstance that the thickness of ocean sediments is considerably less than it should be if they had been accumulating since the earth was formed is explained by the plate tectonics theory, because the burden of sediments is carried slowly into the trenches, to be destroyed along with the plates themselves.

The theory of plate tectonics and of continental drift has been called by Robert Dietz, a geologist of the National Oceanic and Atmospheric Administration in Miami, as important to geology as Darwin's theory of evolution was to biology in the last century.

GUYOTS—PEAKS WITH FLAT TOPS

The theory of plate tectonics also provides an answer to another intriguing mystery, the origin and fate of ranges of sunken volcanoes in the Pacific. Some of these mountains have their tops neatly sliced off (Figure 2-2). These strange flat-topped mountains were discovered by a Princton University oceanographer, Harry Hess, during World War II when the naval vessel on which he was serving as navigating officer recorded one of them on its echo sounder. He was astounded at the flatness of the peak, and took pains to examine echo traces of the Pacific underwater landscape at every opportunity. Before long he had mapped nineteen flat-topped peaks, which were named "guyots" in honor of the distinguished nineteenth–century geologist, Arnold Guyot.

The peaks of these truncated cones could have been planed off only if they had once stood above the surface of the ocean. But many of these mountains now lie 6,000 feet below the surface. Dredging on their tops brings up pebbles and boulders, rounded as by surf on shallow beaches, as well as sandstone that must have cemented on ancient beaches, and coral reefs that grow only in shallow warm waters.

Following Hess's initial discovery, many undersea volcanoes have been found. Soundings now reveal that a great line of sunken mountains, the Hawaiian Archipelago, stretches 1,500 miles across the mid-Pacific from Hawaii to Midway, then after bending to the north, another 2,200 miles to the juncture of the Kuril and Aleutian Trenches to form the Emperor Seamount chain. Geologists now believe that these mountains were all formed in one place, at Hawaii, and that they were carried to their present positions by a titanic conveyer belt, the spreading Pacific floor. The material of the mountains welled up in a succession of periods of volcanism from a "hot spot," or plume of superheated lava lying in a fixed position under the sea floor. One such period of activity is occurring now, producing the island of Hawaii. Oahu, about 200 miles to the northwest, was formed an estimated 3½ million years ago, and Midway Island, 1,650 miles away, about 30 million years ago. These, and the many other seamounts in the chain, were moved from their position of origin by the inexorable drift of the sea floor at the rate of about four inches a year. Eventually the big island of Hawaii will drift north and some new mountain will take its place over the plume of superheated lava.

The guyots on this chain were probably decapitated as the rock

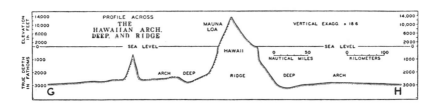

FIGURE 2 − 7. *This profile of Hawaii shows how the big island's weight has pressed down the earth's crust. It is now sunk in a very distinct depression in the ocean floor.*

Edwin L. Hamilton, U.S. Navy Electronics Laboratory, San Diego

was eroded by the waves when the peaks stood above the surface of the sea. Then the mountains moved northward into deeper water and were gradually drowned. In some cases the rate of subsidence was slow enough that coral reefs formed; in other cases the sinking was too rapid or the water too cold for corals to take hold.

One of the most remarkable features of the Pacific seamounts is their youthfulness—none is older than 175 million years. There are none representing the Paleozoic, from 225 million to 600 million years ago, or earlier ages, yet it seems certain that mountains must have risen from the sea floor then as they did later. Erosion is so slow in the sea that these mountains would not have been destroyed by that mechanism, as many ancient peaks have been on land.

The explanation for the disappearance of mountains earlier than those of the Cretaceous age (beginning 115 million years ago) is also provided by the theory of sea floor spreading. We are now offered the astounding concept that after a steady succession of seamounts are formed by the intermittent activity of the Pacific hot spot, to be moved on top of the Pacific Plate to the brink of the trenches, they are tilted into the abyss and swallowed there. One guyot lies on the flank of the Aleutian Trench, its flat top tilted eight degrees; another nearby is far deeper than most seamounts, at 1,400 feet, and it lies on the axis of the trench. Both of these are in the process of being destroyed.

CLAY AND OOZE

But it is not only the weight of isolated volcanic mountain masses like the Hawaiian Islands that press down on the sea bottom. The bottom of the ocean is covered with layers of sediments of varying thickness and composition. In some areas the elemental igneous rock

[31]

is nearly bare, but certain regions of the shallow sea floor are buried beneath a crushing load as much as 40,000 feet thick. The average on the open Pacific Ocean floor is about 1,000 feet, on the Atlantic twice that.

Sediments near shore are ordinarily "terrigenous debris," bits of the earth torn away by eroding water and transported to the shallows around the land. The quantity of such material is immense. The Mississippi River dumps some 750 million tons of the heartland of the North American continent into the Gulf of Mexico each year. In all, 3 billion tons of the land are estimated to be carried to the sea by the streams of the world in a year. Some sediments are formed from material flung into the sea by volcanoes, both those on land and those under water.

Far from shore, covering the rocky base of the deep-sea floor, there are three principal kinds of sediments. They are the red clays and two biological "oozes": calcareous and siliceous.

The inorganic components of the sediments, which constitute the clays, consist of minute particles of rock debris that were small enough to drift far from land before they sank to the bottom. Drifting in suspension, they may be borne by the currents for years or decades before they are finally dropped onto the sea floor. The organic sediments, the oozes, have their origin in the bits and pieces of the floating plants and animals that live in the sea. Except in relatively shallow areas, nothing survives the slow journey down except the inedible and insoluble shells; soft tissue and any other part that could possibly supply nourishment to a roving sea animal is engulfed or destroyed by bacteria long before it can reach the bottom.

FIGURE 2–8. *The material of ocean-bottom sediments. Terrigenous sediments are fragments of rock carried from the land; they are usually deposited in shallow water. Red clay consists mainly of fine particles of eroded rock; buried in it are ancient shark teeth*

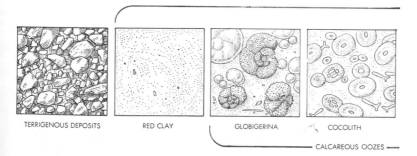

TERRIGENOUS DEPOSITS   RED CLAY   GLOBIGERINA   COCOLITH

CALCAREOUS OOZES

This stately drift of sediment to the sea bottom, although never-ending, is extraordinarily slow. Some abyssal red clays build a layer only one twenty-fifth of an inch thick in 1,000 years, others half that. Some of the calcareous oozes may grow faster; in 1,000 years they add as much as half an inch.

About half of the sea floor is covered with calcareous ooze, much the commonest kind of sediment. Calcareous oozes are made up principally of the shells of uncounted billions of billions of microscopically small one-celled animals called foraminifera. The shells, often delicately and marvelously arranged in graceful chambers, are the source of all the world's great deposits of limestone. The shells of pteropods (small mid-water snails) and the fragments of the shells of coccolithophores (very small plants) are also of calcareous origin and contribute to this common class of deep-sea sediments.

The siliceous ooze sediments are composed largely of the skeletons of diatoms, microscopic plants whose shells are tiny boxes of transparent silica. They flourish in fantastic numbers in cool seas and form the food of a great many small floating animals. Whether they are eaten or die unconsumed, diatoms leave a very concrete heritage behind them. Since the glassy shells are virtually indestructible, it might be said that nearly all the diatoms that ever lived are still around in body if not in spirit. And there have been incredibly large swarms of diatoms for hundreds of millions of years. Deposits up to 1,000 feet thick occur in the sea and on areas of the land once under the sea, and the deposits cover hundreds of square miles of surface—with several million diatoms per cubic inch.

These deposits have a commercial use. In the United States about

and whale earbones. Calcareous and siliceous oozes consist chiefly of the shells of microscopic one-celled animals; about half of deep-sea sediment is calcareous ooze. These pictures were not drawn to the same scale.

Scientific American

PELAGIC DEPOSITS

| DIATOM | RADIOLARIA | WHALE EAR BONE AND SHARK TEETH | NICKEL AND IRON SPHERULES | MANGANESE NODULES |

SILICEOUS OOZES

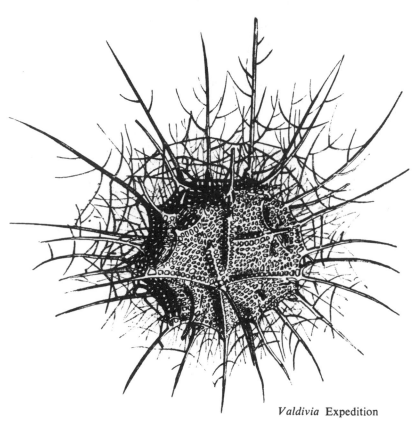

Valdivia Expedition

FIGURE 2 – 9. *This beautiful microscopic animal, whose shell forms siliceous ooze, is called radiolarian because of its radiating spikes. Its scientific name is* Oroscena regalis *var.* oroplegmoides. *Billions upon billions of such shells sink like snowflakes to the bottom of the sea, especially in the warmer regions.*

280,000 tons of diatomaceous earth are produced each year. The earth is used as a fine abrasive, for filters, as a base for toothpaste and powder, and for fine car polishes. It is a good insulating material, especially where great heat resistance is important.

Radiolarian oozes, which occur in warmer seas than those containing diatom oozes and which are considerably less common, are another siliceous type.

Radiolarians are small planktonic animals. The radiolarian shells that sink to form an ooze are among the loveliest of earthly forms, like snowflakes in their almost infinite variety. Some are basi-

[34]

cally spheres; many have tiny spikes radiating from a central structure (this is the origin of their group name).*

Intermediate in abundance between the two kinds of sediments produced by living creatures is the inorganic material called red clay, the finely divided fragments of eroded and transformed rock. The name is applied to sediments containing less than 30 per cent of organic materials. These sediments may be actually red, but very often they are some other color. For example, in the Pacific much of the "red clay" is brown in color, and many near-shore deposits are other shades. Buried in the red clay deposits are such interesting things as sharks' teeth, whales' earbones, and fragments of meteorites. Teeth and earbones are the hardest parts of animals, and often the

* In the appendix beginning on page 407 the genera (scientific names) of the animals mentioned in the book are listed according to the way biologists classify them. This will assist the reader in putting the animals in their proper positions in relation to the rest of the animal kingdom.

FIGURE 2−10. *"Shooting-star" rake. This device, used by the scientists of the Danish* Galathea *Expedition, is studded with small but powerful magnets. Dragged through deep-sea oozes, the magnets picked up tiny metallic pieces of meteorites that had accumulated over thousands of years in the sediments.*

Galathea Expedition

only parts to survive the decay that destroys the rest of living creatures. Teeth of sharks extinct for millions of years are sometimes dredged up from the bottom. Quite commonly the teeth, as well as whale and fish earbones, form the nucleus of concretions of iron or manganese oxides, which are deposited on their surfaces, eventually covering them and taking their shape.

The red clay also includes material from outer space. An amazing amount enters the earth's atmosphere—perhaps at least 5 million tons per year. Most of it is in the form of cosmic dust or other matter that burns up as it enters the atmosphere, so the amounts that reach the surface of the earth are small. But the dust has been falling a long time, and it is found in appreciable quantities in the slowly accumulated red clay sediments. Interestingly, examination of different levels of sediments shows that the amount of cosmic dust fall has varied considerably in geological history.

Since the sea-bottom sediments have been laid down layer by thin layer over the millennia, the strata are piled in neat sequence. The age of the sediments that have their origin in living creatures can be estimated by identifying the kinds of animals and plants present. The slow process of evolution has caused some creatures that once flourished in ancient seas to disappear and to be replaced by slightly different forms. The paleontologist who has learned to discriminate among hundreds of thousands of kinds of organisms can tell in what period of the earth's past history a particular layer of sediment was laid down. He can do much better than this, in fact. Through the techniques of modern chemistry he can actually tell the temperature of the ancient seas in which minute creatures, now long dead, had their vital being.

This clever calculation depends upon the existence of isotopes of oxygen. Isotopes are variations of a chemical element, identical in every respect except in their masses. Their nuclei contain the same number of protons and electrons but a different number of neutrons. There are three isotopes of oxygen: common oxygen 16, oxygen 17 (very rare), and oxygen 18. It happens that the relative amounts of these isotopes of oxygen stored in the calcium carbonate shells of the foraminifera, one kind of microscopic animal in the sediments, vary depending on the temperature of the water in which the animals lived. By precise measurements of these amounts, using a complicated machine called a mass spectrometer, it is possible to calculate the temperature of the waters at the time the forms lived, and so trace the climatic history of the world 200 million years ago.

3　The Deep-Sea Environment

The living space of air-breathing creatures like ourselves includes many different climates. There is the temperate climate of the middle latitudes near the sea, the hot and sticky climate of tropical rain forests, the dry heat of deserts, and the lifeless cold of high latitudes and mountains. And the sea? It all looks watery. The steady stirring of waves implies complete homogenization—a one-climate environment for the entire, vast world ocean.

The facts are different. Conditions vary much more widely in the ocean than on the land. "Climates" in the sea, as on land, are created by differences in temperature, light, and pressure. The circulation of the water by tides and currents corresponds to the circulation of the air, and in many ways changes the water environment more than air circulation affects the land environment. Factors such as light and pressure show considerably wider variations in the sea. The extremes of temperature, however, are held within a smaller range in the sea than on land. Moreover, temperature changes take place much more slowly in the sea than on land because water has a high "heat capacity." That is to say, more heat energy is required to raise the temperature of a given amount of water than of an equal amount of other substances; water does not undergo sudden rises in temperature such as occur in metals and many other materials. Conversely, water gives up its heat slowly—and it has large quantities to give—when its temperature is above that of its surroundings. This is why large bodies of water have profound moderating effects on the climate of areas nearby.

Scoop up a handful of blue sea water; it is colorless. Quite plainly, a large quantity of water does something to light. What it does varies with the location: there are the clear, deep blues of offshore waters, the varying shades of green inshore, and brownish or other tints in isolated shallow areas. The blueness of the sky is produced by the scattering of light from the sun by particles of gases and dust in the atmosphere. The blue of the open sea is partly the result of the same effect, in this case the light being scattered by the particles of the water itself and by very small pieces of suspended material. In addition, the blue end of the spectrum is absorbed least by water.

In general, the blue of clear offshore waters shades into green toward the land, where the water carries a much greater load of dissolved and suspended materials. This load consists of sediments, an enormous variety of dissolved chemicals carried from the land, and—more important—innumerable billions of minute plants and animals: the flora and fauna of the plankton ("floaters"). Sometimes the green of shallow waters derives directly from the presence of a thick soup of little planktonic plants with spots of greenish chlorophyll. More commonly, however, the reason is a mixing of blue and yellow to make green: the blue of the open sea is transmuted to green by the addition of yellowish pigments that occur near shore in the form of soluble substances apparently produced as decomposition products of the plant plankton. Since plankton is much thicker near shore, there is more of the yellow material in the water, and this results in the characteristic green of the shallows.

When the plankton is especially thick, the sea may take on the color of its inhabitants. The color of any object depends on the kind of light it reflects: a red book bounces the red fraction of the light back to the eye, a yellow dress reflects the yellow rays; in each case the other colors of the spectrum are absorbed. Sometimes the floating inhabitants of the sea are concentrated enough and distinctively colored enough to make the water yellow or brown or even reddish. The notorious "red tide" that causes havoc among the fish on the west coasts of Florida and Peru from time to time is a famous example of such discoloration, and is produced by countless billions of microscopic colored plants. The Red Sea gets its name because at times it is stained by swarming populations of another tiny plant, the blue-green alga *Trichodesium erythraeum*. The Gulf of California is sometimes called the Vermilion Sea because of such blooms of colored plants. The state of the sky also has something to do with the color of the sea, making it appear steely and dull on heavily overcast days and bright and sparkling on sunny days. However, the basic color is derived from the suspended and dissolved materials in the water.

We detect another influence of the sea on light as soon as we submerge. The surroundings become noticeably dimmer if we merely dunk our heads below the surface, and the light fails rapidly as we swim downward. It is hard to believe that half of the light that enters the sea is extinguished by the time it has passed through as little as 6 feet of water in some inshore areas; at about 14 feet, three quarters of the light is gone; and at 25 feet, 90 per cent has been blacked out. As we proceed offshore, away from the influence of the rivers and estuaries that throw silt and other debris into the water, the light penetrates deeper. In the Sargasso Sea, the clearest water in the world, light may penetrate five times deeper than near shore. The most crystalline fresh-water lakes are a good deal less clear than offshore ocean waters. For comparison we might note the depths at which 99 per cent of the light is extinguished in various waters. This occurs at the comparatively shallow depth of 26 feet in Woods Hole Harbor in Massachusetts, where the water is heavily burdened with sediments. In the Gulf of Maine the same 99 per cent extinction depth is at about 105 feet; the depth is increased to 490 feet in the transparent Sargasso Sea, and is about the same in parts of the Mediterranean. The clearest body of fresh water measured, Crystal Lake in Wisconsin, is murky by Sargasso or Mediterranean standards, with the 99 per cent extinction depth being about 27 feet, nearly the same as at Woods Hole.

While the light is extinguished rapidly as it passes through the water, there are still minute fractions of it to be detected even at very great depths. There comes, however, a point at which the human eye can no longer be stimulated; in order to set some arbitrary criterion for light penetration, we will define the depth to which "no light" penetrates in these terms. According to physiologists, including John F. Fulton of Yale and George Wald of Harvard, the human eye when adapted to darkness requires at least 5 quanta of light to react. In other terms, the eye, after being adjusted to darkness for a suitable length of time, can "see" light that is about one ten-billionth of full sunlight.

Measurements of the amount of light under the sea were made by N. G. Jerlov and Fritz Koczy during the round-the-world oceanographic cruise of the Swedish *Albatross*. They lowered photographic plates to various depths in the tropical ocean and recorded the amount of sunlight remaining. Light corresponding to the 5 quanta necessary to stimulate the human eye was detected down to about 1,800 feet. From the human point of view, then, the sea is totally lacking in light from the sun from this point down. (In effect, of course, it is completely "dark" by ordinary human standards at much shallower levels than this.) Whether fish and other deep-sea animals can see still dimmer light is another question.

Light from the sun is composed of different wavelengths. When they are mixed together they appear as "white." (In addition to the visible spectrum, light consists of ultraviolet radiation with wavelengths too short to be detected by our eyes, and other energy—infrared— with wavelengths too long.) Passing through a prism, the various wavelengths are bent differently and "unmixed," or spread out into a spectrum. The visible light with the longest wavelength appears as red, followed in order by orange, yellow, green, blue, and violet, the shortest.

The extinction of light by water is due to two processes, scattering and absorption. Both affect the different colors in different degrees. The longest rays are absorbed much more quickly than the shortest, so that in water red light is extinguished and red colors disappear first.

FIGURE 3 – 1 . *Dr. William Beebe* (right) *and Otis Barton beside the bathysphere, before their record dive of 2,510 feet on August 11, 1930.*

New York Zoological Society

William Beebe found that about half the red light disappeared just below the surface of the sea off Bermuda, and that it was nearly all gone at about 20 feet. Luis Marden, the veteran diver on the staff of the National Geographic Society, relates that he has been able to detect tones of red—"degraded to a deep maroon or brown"—to depths of 50 feet, although the color camera loses its ability to record reds before this.

One of the most striking effects of the disappearance of red light rays under water is the change that takes place in the color of blood. Many skin divers have been astounded to see an eerie green cloud pour from the side of a wounded fish. The green cloud is ordinary scarlet blood—transmuted to a ghostly new color by light that the water has robbed of all the warm tones.

The other colors of the spectrum follow red into obscurity. Diving in his bathysphere, Beebe noted that orange, dominant at a depth of 50 feet, had disappeared at 150 feet. Yellow was almost gone at 300 feet; half the violet had disappeared at 350, along with three quarters of the green. At 800 feet "I saw only the deepest, blackest-blue imaginable." The water "was of an indefinable deep blue quite unlike anything I have ever seen in the upper world, and it excited our optic nerves in a most confusing manner. We kept thinking and calling it brilliant, and again and again I picked up a book to read the type, only to find that I could not tell the difference between the blank page and a colored plate. . . . the repetition of our insistence upon the brilliance which yet was not brilliance was almost absurd." (From *Half Mile Down.*)

A small quantity of sea water, then, may appear to be transparent, but it is actually nothing of the kind. It absorbs, scatters, and reflects light. If sea water allowed the light to pass through it undiminished and unaltered, plants could carry out photosynthesis—the essential life process—all the way to the bottom. As it is, photosynthesis is halted for lack of light at depths ranging from about 100 to 325 feet.

TEMPERATURE OF THE SEA

The failure of light to penetrate beneath the outer skin of the ocean has an effect on the temperature of the oceans, since it is the radiant heat of the sun that warms the waters. Surface waters of the ocean in the middle latitudes can be very warm, sometimes reaching 90° F., or even higher. The temperature gradually decreases with increasing latitude until surface waters at the poles of the earth are below their freezing point and ice forms.

In the open sea the water temperature also declines from the surface downward, but at an uneven rate. Typically, there is a surface layer of about 65 to 650 feet of fairly uniform temperature. This is followed by a narrow layer, the "discontinuity layer," or "thermocline," in which the temperature and density change rapidly. At the bottom of the thermocline the water may be about 60° F., having dropped 20° to 30° in 500 feet or so; in the next 1,000 to 1,200 feet the temperature drops more slowly, to about 50° F. The rate of decrease is still more gradual in deeper water, so that at around 5,000 feet the temperature may be about 40° F. The deepest waters in the abyss range from about 38° F. to below the freezing point of fresh water.

It will be worth-while at this point to leave the ocean for a moment and consider the properties of fresh water. The colder substances get, the denser they become—that is, the greater their weight per unit volume. This is a universal rule in nature, applying to virtually every substance—almost including water. Water follows the rule up to a point and then breaks it. Fresh water increases in density continuously as it cools until it reaches a temperature of about 39° F. Then, for a reason related to its molecular structure, it reverses and becomes less dense at lower temperatures. The result is that water at the freezing point is lighter than water at other temperatures, and therefore floats at the surface as ice.

Consider the situation if water did not possess this remarkable and rare quality. When water cools at the surface under the influence of chilling air temperatures, it sinks, forcing warmer and lighter water to the top. This in turn is cooled and sinks. If the temperature continues to fall so that the water drops to 39° F., the bottom layer and then progressively all the layers above the bottom become a uniform 39°. Water colder than this is lighter; it rises to the surface and freezes at 32° F. If water continued to become denser right down to the freezing point, the ice on lakes would form on the bottom instead of at the surface. This icebound bottom layer would never melt, since warmth from the sun is absorbed by the surface layers, and sooner or later every body of fresh water in high latitudes would be solid ice from bottom to top. Life there would be impossible. It might be said, therefore, that our existence over much of the earth hangs on one curious rule of the physics of water.

Because of the dissolved materials in sea water, its temperature-

FIGURE 3–2. *An oceanographer aboard the* Crawford *attaches a Nansen bottle. By placing a number of these instruments at various depths on the same cable, scientists can obtain water samples and compare temperature, salinity, oxygen content, and other properties.*

density relationship is different from that of fresh water, and sea water keeps getting heavier as it gets colder. The result is that the coldest sea water is near the bottom. Fortunately, another factor comes into play to keep ice from forming from the bottom of the sea upward: the freezing point of sea water is depressed by the dissolved salts and by the great pressures at the bottom of the ocean. The very low temperatures required to freeze deep-sea water are rarely reached at the bottom of the oceans, where the coldest water is usually above 27° F. If on occasion deep-sea water does get cold enough to freeze, it still does not do so from the bottom up because as the particles of water solidify, they squeeze out the salt, producing nearly fresh-water ice. These light particles of ice float to the surface of the sea and form a crust of ice there.

The temperature of the sea, then, ranges from around 28° F. to something like 90° F. This is well within the range of temperatures that can be tolerated by living things, and is a considerably smaller range than the temperatures encountered on land.

PRESSURE

If the temperature of the ocean exhibits smaller extremes than are found on land, the same cannot be said of pressure. Pressure in the ocean increases steadily and rapidly with depth. This is because the pressure is produced by the weight of the column of water above any point, and of course the higher the column, the greater the weight. A cubic foot of water weighs 62.4 pounds, so that objects do not have to sink very far in the sea to have a great deal of weight pressing on them. At a depth of 3,000 feet, for example, the pressure is around 1,500 pounds per square inch—or about 100 times normal surface atmospheric pressure. In the deepest parts of the ocean it is a crushing 3½ tons per square inch!

A rare firsthand encounter with the tremendous force of deepwater pressure was experienced by William Beebe. On one of his expeditions he undertook to modify his bathysphere by fitting a quartz window into the space that had been provided for it but had previously been closed by a solid plug. The other two windows in the bathysphere had been fitted by expert workmen at the factory, but Beebe took the chance that his crew on the expedition would be skillful enough to do the job. As it turned out, this was a dangerous chance to take.

To test the new window the bathysphere, unoccupied, was lowered to 3,000 feet. When the great steel ball was hauled up, Beebe wrote in *Half Mile Down,* "it was apparent that something was very wrong, and as the bathysphere swung clear I saw a needle of water shooting across the face of the port window.

"Weighing much more than she should have, she came over the side and was lowered to the deck. Looking through one of the good windows I could see that she was almost full of water. There were curious ripples on the top of the water, and I knew that the space above was filled with air, but such air as no human being could tolerate for a moment. Unceasingly the thin stream of water and air drove obliquely across the outer face of the quartz. I began to unscrew the giant wingbolt in the center of the door and after the first few turns, a strange high singing came forth, then a fine mist, steam-like in consistency, shot out, a needle of steam, then another and another. This warned me what I should have sensed when I looked through the window, that the contents of the bathysphere were under terrific pressure. I cleared the deck in front of the door of everyone, staff and crew. One motion picture camera was placed on the upper deck and a second one close to, but well to one side of the bathysphere. Carefully, little by little, two of us turned the brass handles, soaked with the spray, and I listened as the high, musical tone of impatient confined elements gradually descended the scale, a quarter tone or less at each slight turn. Realizing what might happen, we leaned back as far as possible from the line of fire.

"Suddenly, without the slightest warning, the bolt was torn from our hands, and the mass of heavy metal shot across the deck like a shell from a gun. The trajectory was almost straight, and the brass bolt hurtled into the steel winch thirty feet across the deck and sheared a half inch notch gouged out by the harder metal. This was followed by a solid cylinder of water, which slackened after a while to a cataract, pouring out of the hole in the door, some air mingled with the water, looking like hot steam, instead of compressed air shooting through ice-cold water. If I had been in the way, I would have been decapitated." Such are the pressures of the deep sea.

SALT IN THE SEA

A small cousin of mine, raised on the Canadian prairies, emerged from his first contact with the ocean, spat out a mouthful of the Pacific, and blurted indignantly, "Someone put salt in this water!"

"Someone"—or something—had indeed put salt in the sea. When the seas were new they may have been fresh or somewhat salty—geologists disagree. In any case, through the eons of time, dissolved materials have been washed into the ocean in huge quantities, until now sea water contains about 3.5 per cent of dissolved substances, mostly salts. This process is still going on as vigorously as ever, with the rivers of the world washing about 273 billion tons of dissolved chemicals into the ocean every year. At present there is approximately a quarter of a pound of salt in every gallon of sea water, so that the sea contains about 4½ million cubic miles of salt, or enough to deposit a layer 500 feet thick over the entire land area of the earth.

The presence of salt in water increases its density, and sea water is therefore more buoyant than fresh water, as anyone who has swum in both lakes and the sea can attest. Marine animals are buoyed up to a considerable extent by the sea salts, and for many of them life is possible only because of this fact. The salt content has another and even greater significance to the life of the creatures of the ocean: it supplies the inorganic building blocks from which plants manufacture living substance, and is therefore the basic food on which is nourished the whole complex of animals in the sea.

The blood, or the fluid that serves the same function in animals that lack blood, is salty. The concentration of dissolved substances in the blood of sea animals is delicately adjusted to the salinity of sea water. If this were not so—if the concentration of salts in the blood were markedly above or below that of the ocean—the animals would either absorb water and swell up like balloons or become desiccated by having the water sucked from their blood and tissues. The reason is that two solutions on opposite sides of a membrane such as the cell membranes of an animal or plant undergo the process called osmosis: if the concentration of dissolved substances is different on the two sides of such a membrance, water passes from the side of lesser to the side of greater concentration. While the saltiness of sea animals' blood approximates that of their environment, the concentration of blood salts in human beings is less than that of sea water of the present age. It has been suggested that it may approximate the concentration of the less salty seas of many eons ago, when the first land creatures developed from animals that had crawled out of the ocean and started their pioneer existence on land. If this is so, we carry in our veins a reminder of the ancient seas that were the ancestral home of life on earth.

The least salty of open ocean waters are those near the poles, where melting ice reduces the salinity to 34 parts per 1,000. North Atlantic surface waters are about 35 parts per 1,000; waters within about 10° to 15° north and south of the equator are the highest, being approximately 37 parts per 1,000. The higher values in tropic waters result from the increased evaporation caused by high temperatures. The salinity decreases to a small extent with depth, but the differences are not great; the lowest concentrations in the abyss are just under 35 parts per 1,000. While these values are not very far apart numerically —only 2 or 3 parts per 1,000—the differences can easily be detected by swimmers, who are quick to note the extra saltiness of subtropical waters off Florida compared with those of northern beaches.

Over three quarters of the dissolved material in the sea is "table salt," sodium chloride, as shown in the accompanying table. The next

Elements present in solution in sea water, not including dissolved gases

ELEMENT	PARTS PER MILLION	PER CENT	ELEMENT	PARTS PER MILLION
Chlorine	18,980	55.2	Copper	0.001–0.01
Sodium	10,561	30.4	Zinc	0.01
Magnesium	1,272	3.7	Lead	0.005
Sulfur	884	2.6	Selenium	0.004
Calcium	400	1.2	Cesium	0.004
Potassium	380	1.1	Uranium	0.002
Bromine	65	0.19	Molybdenum	0.0015
Carbon	28	0.09	Thorium	0.0005
Strontium	13	0.04	Cerium	0.0005
Boron	4.6	0.01	Silver	0.0004
Silicon	0.02 –4.0		Vanadium	0.0003
Fluorine	1.4		Lanthanum	0.0003
Nitrogen	0.006–0.7		Yttrium	0.0003
Aluminum	0.5		Nickel	0.0001
Rubidium	0.2		Scandium	0.00004
Lithium	0.1		Mercury	0.00003
Phosphorus	0.001–0.10		Gold	0.000006
Barium	0.05		Radium	$0.2–3 \times 10^{-10}$
Iodine	0.05		Cadmium	Present
Arsenic	0.01 –0.02		Cobalt	Present
Manganese	0.002–0.02		Tin	Present

most common chemicals, in order, are magnesium chloride, magnesium sulfate, calcium sulfate, potassium sulfate, calcium carbonate, and magnesium bromide. The relative quantities of these in the sea diminish rapidly as we go down the list, so that there is only 0.2

per cent of magnesium bromide. A great many other substances occur in the ocean in still smaller concentrations and in most cases in extremely minute proportions. Silver is there, and so is gold. It is a commentary on the eagerness with which men strain for riches to learn that many schemes, some fake and others serious, have been devised to extract gold from the sea, even though the amount in a ton of sea water is about 0.00001 pounds.

There are other scarce substances in the sea of far greater value than gold: the nutrient minerals. But the problem of their extraction is left to the plants, which are much more clever at this particular game than humans. Plants manufacture living substance by combining water, carbon dioxide, and certain minerals, with the help of the energy of the sun and the green substance, chlorophyll. Water and carbon dioxide are in plentiful supply in the ocean; the main reason plants do not thrive equally well in all areas is the uneven distribution of the vital minerals. The most productive waters are in the inshore shallows where light can penetrate, and in the middle and high latitudes in places where vertical currents bring minerals that have sunk into the depths back up to lighted levels where they can be used by plants. These minerals are chiefly the phosphates and the nitrates, although a number of others are also involved, including iron, copper, and manganese, and recent work suggests that some more complicated substances such as vitamin B_{12} are also necessary for plant growth in the sea.

Large amounts of vital minerals are carried to the deep parts of the ocean in the bodies of dead or dying animals and plants. Once below the region where light can penetrate, the minerals are of no use to the oceanic plants, which require the energy of the sun to carry on photosynthesis. The energy cycle of the ocean depends, therefore, on the mechanisms operating to return the phosphates and nitrates and important trace elements back to the lighted zone. One of the most important of these mechanisms is convection. When the top layer of water is cooled and sinks, the water underneath is forced to the top, carrying with it fresh supplies of mineral nutrients. This mechanism operates in the high latitudes, where the water at the surface is cooled sufficiently to lower the surface density below that of water beneath; we find little or no convection in tropical and subtropical oceans. In areas like the Grand Banks and the North Sea, however, the oceans are overturned from top to bottom because of the sharp cooling of the surface waters. This is the key to the enormously abundant fish populations of these areas. The cold far southern seas bloom with tremendous vigor, supporting teeming new life and making the antarctic waters among the richest in the ocean.

A second important way in which deep waters are brought to the surface is through the action of two currents flowing away from each other. Where this happens, the space left as the currents part must be filled, and it can only be from below. So water rises from the depths of the sea, again bringing trapped nutrients to the surface. A deep current sweeping against an island or a sunken reef often deflects water upward too, and some areas of the sea are rich in life as a result of this mechanism.

The last important method of upwelling of deep waters is through the action of strong winds blowing offshore. In areas where the prevailing winds are steady and strong enough they can blow very large masses of water away from the land. Again, this must be replaced; cold, mineral-laden waters well up to fill in for the water blown away. This happens on the western coasts of the continents, where the winds blow strongly offshore. The waters of Peru and Chile and off the Atlantic coast of Africa are good examples.

Even the deepest of the upwelling currents like those off the west coast of South America come from moderate depths; the waters of what we have called the deep sea are not subject to such vertical movements. Once water has moved to the great depths of the ocean, it is brought back to the surface only by the sluggish, creeping currents described in the next chapter. Water from very deep areas may have taken centuries to get there and may require more centuries to return to the lighted areas of the ocean through the ordinary processes of current motion. Whereas moderately deep water off the coast of Peru is regularly and quickly returned to the surface by the action of the strong offshore winds, the deep abyss off the same coast may contain the "oldest" water in the world, a result of the depths involved plus the lack of deep currents in this area.

So now we may summarize what the most common environment in the world, the deep sea, is. It is pitch-black, without the least glimmer of the sun's rays to give it cheer; it is cold, only a little above freezing; it is under enormous pressure, with power to crush to a shapeless mass any body not constructed to combat it; it is salty and laden with nutrient minerals, but these are useless since the energy of the light is missing; it is virtually still, with only the most languid currents moving. Yet this unlikely living space is inhabited by a huge variety of fishes and squids and other curious animals that are lucky enough not to realize that their home is "unlivable."

4 The Circulation in the Abyss

Motion—whether the fast flow of the tide rip or the Gulf Stream or the barely measurable drift of currents in the abyss—is a quality as characteristic of the sea as its saltiness. We saw in the last chapter that some currents in the deep sea have their birth when surface currents become colder and therefore denser and slide slowly into the depths, leaving a space that is filled by other water nearby. In turn this motion, aided by the force of the winds, sets up other water movements until enormously complicated chains of currents are induced. The unending movements of the oceans are interconnected like a series of gears. It is difficult to decide where a cycle begins, but we can start our discussion of deep-sea currents at the surface.

The principal current-inducing force is wind. The most constant surface winds are the trade winds of the middle latitudes, which blow from the northeast in the Northern Hemisphere and from the southeast in the Southern Hemisphere. Blowing steadily and powerfully across the sea, they urge along the surface water through friction. Waves and ripples form; these present small vertical faces against which the wind pushes. Spray picked up from the water and flung across the surface with slanting blows has the same effect. The cumulative force is great enough to move prodigious amounts of water across the face of the earth.

The direction of ocean currents does not coincide for long with that of the wind that produces them. The currents are twisted by the "Coriolis force," named for the French scientist Gaspard Gustave de Coriolis, who first described it in 1835. The Coriolis force is not really a force but a phenomenon resulting from the spin of the earth on its axis. Its effect is to turn currents—of air as well as of water—to the right in the Northern Hemisphere and to the left in the Southern Hemisphere.

In the North Atlantic the great stream set in motion by the trade winds and deflected by the spin of the earth is called the North Equatorial Current. This current flings water toward the Western Hemisphere, where it piles up against the land barrier of the Americas (with the result, incidentally, that the surface of the sea has a small but distinct slope downward from west to east). When the North Equatorial Current reaches the Antilles Islands, part of it is shunted northward as the Bahamas or Antilles Current, while the remainder streams between the islands into the Caribbean Sea and onward toward the Gulf of Mexico. The islands and the Yucatán Peninsula of Mexico team up with the water already filling the basin of the Gulf of Mexico to keep the current from penetrating deeply into the Gulf. Instead, the stream turns sharply and moves with tremendous force and volume out of the cul-de-sac back into the Atlantic Ocean. Pouring between Cuba and the peninsula of Florida, it moves past Miami as the powerful Florida Current. As it proceeds up the east coast of the United States, at a point just north of the Bahamas it joins with the Antilles Current that had split off earlier, and the two make up the famous Gulf Stream.

In common with other currents on the western sides of the oceans, the Gulf Stream is swift and narrow. It flows north at a rate that sometimes reaches 6 knots (a little under 7 statute miles an hour). Fifty miles wide and 1,500 feet deep opposite Miami, it carries enormous quantities of water—about 1,040 million cubic feet per second—past that city. This is 1,200 times the volume of the Mississippi River at New Orleans.

The Gulf Stream comes closest to the shore at Palm Beach, where it is only about 100 yards offshore. North of that city the Stream begins gradually to move away from land. At Cape Hatteras, North Carolina, it is about 100 miles from shore.

In the reaches of the North Atlantic Ocean off the Grand Banks and Newfoundland the warm waters of the Gulf Stream meet head on the frigid waters of the Labrador Current and the East Greenland Current coming out of the north. Fog, which bedevils shipping in this busy area of the ocean, is one result of this marriage of waters. While the bulk of the Labrador Current wedges between the warm current and the shore, giving Newfoundland and Labrador bitter climates, the Gulf Stream turns to the east, to carry its tropic-warmed

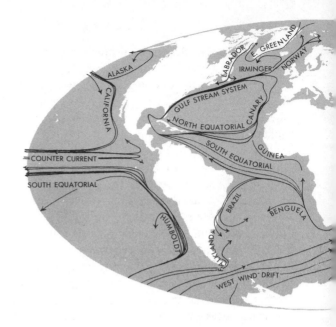

waters to the coast of Europe. Now called the North Atlantic Current, the water is given its easterly direction partly by the prevailing winds, which in this latitude blow across the ocean from America to Europe. Before it reaches the British Isles the current divides. One major branch laps the shores of northern Europe, presenting to countries like Norway and Sweden, lying in the same latitudes as Labrador, the gift of an equable climate.

The other main branch of the Gulf Stream, having by now given up its warmth, turns right to flow south past the European continent in a cool, broad, and sluggish current variously named the Portugal, Azores, and Canaries current. Off North Africa it turns west again to complete the circle, and is urged across the Atlantic Ocean by the trade winds, picking up heat once more from the tropic sun.

This simplified picture of the current system of the North Atlantic Ocean is that of an immense wheel of water whose rim is turning steadily and tidily around the edges of the ocean basin. We have, for example, described the Gulf Stream as comparable to a river contained in an unchanging channel. Recent studies of the Gulf Stream show, however, that, instead of a well-behaved stream flowing north in a fixed channel, it is actually a series of narrow, swift ribbons separated by countercurrents. It is as though the Mississippi

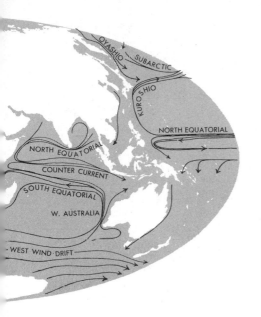

FIGURE 4-1. *Surface currents of the world oceans. A great circular current system exists in each of the North and South Atlantic oceans, the North and South Pacific, and the Indian Ocean. Coriolis force, produced by the turning earth, bends currents right in the Northern Hemisphere and left in the Southern Hemisphere. The currents of the Indian Ocean are particularly variable, changing with the shifting winds of the different seasons.*

River were split into many segments by water currents forcing their way upstream against the flow of the river. The countercurrents of the Atlantic that knife into the northern part of the Gulf Stream and separate it into ribbons are composed of water from the Labrador and East Greenland currents from the arctic region. The segments of the Gulf Stream wander over the face of the Atlantic in an erratic manner, forming loops and eddies and re-forming as narrow ribbons when they slide east again. Scientists of the Woods Hole Oceanographic Institution detected in the Gulf Stream one erratic meander forming a loop some 250 miles long outside the main current. Two days after it formed it broke off as an independent eddy. This playful excursion of the supposedly staid and steady current is credited with dumping 10 million million tons of subarctic water from the North Atlantic into the subtropical Atlantic. The volume of flow varies greatly; on any given day it may be only half what it was the day before.

The erratic behavior of ocean currents causes Dr. Walter Munk of the Scripps Institution of Oceanography to draw a distinction between the "climate" (the average, "normal" behavior) of currents and the day-to-day "weather"—the short-term departures from normal. Details of variable behavior can be discovered only by accurate observations, preferably at many places simultaneously. Says Dr.

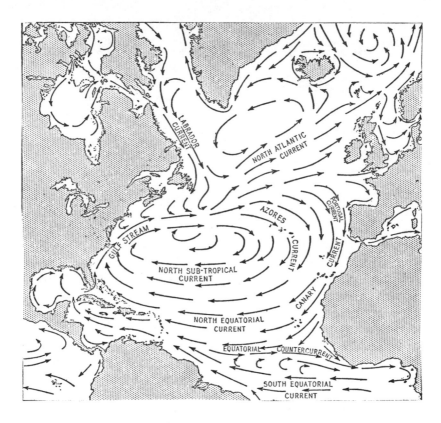

FIGURE 4-2. *The surface current system of the North Atlantic Ocean. A great circular current (North Equatorial Current) is set in motion when prevailing winds blow water toward the North American continent. The water is turned to the right by Coriolis force and sweeps up the coast of the United States as the Gulf Stream. Then, curving east, it divides to form the North Atlantic Current (warming northern Europe) and the Azores, Portugal, or Canaries current, to complete the circle. The Labrador Current brings frigid water from the north, and the Equatorial Countercurrent sweeps east between the North and South Equatorial currents.*

Sir Alister Hardy, *The Open Sea: Its Natural History.*
Part 1. The World of Plankton

Munk in *Scientific American,* "If 10 vessels strategically placed in the Gulf Stream were to measure the currents and make a 'weather map' of the Stream next Tuesday, the map would differ from the one for Friday. Not long ago we watched a freighter carefully holding a course which according to the climate chart should have speeded it on its way to Europe by taking advantage of the Gulf Stream. Actually the ship was bucking a two-knot counter-current; the Gulf Stream was 100 miles off its usual path!"

The Coriolis force has another curious effect besides turning the water to the right in the Northern Hemisphere. This is to tilt the water so that on the right side of a current the water is piled higher. This phenomenon has the result, for example, that on the Miami side of the Gulf Stream the water is actually 2 feet lower than it is at Gun Cay on the Bahamas side, some 50 miles away. The Sargasso Sea is the off-center "middle" of the huge North Atlantic gyre or current circle, and its waters stand appreciably higher than those on the circumference of the wheel, so that ships go "downhill" as they leave the Sargasso Sea in any direction. Water may "seek its own level," but in this case it fails to find it.

Long before seamen understood the circular pattern of the currents of the North Atlantic, let alone the details of variations in this

FIGURE 4 – 3. *The wandering Gulf Stream. From June 8 to 12 it followed the path shown by the dark lines. From June 19 to 22 it shifted to the route shown by the cross-hatched lines, and a huge eddy broke off, pushing enormous quantities of subarctic water into the subtropical Atlantic Ocean.*

Woods Hole Oceanographic Institution

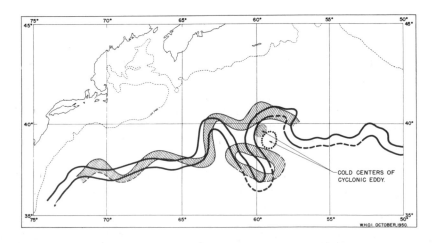

pattern, they made use of the push of the water to hurry them on their voyages. At first this was accidental. Columbus rode the North Equatorial Current on his epic voyage of discovery; if it had not been for the current it is quite possible that he would have failed to reach America. On their subsequent trips west the eager ships of Spain took the same general course to the gold lands as had Columbus. Going home, the Spanish seamen learned not to buck the Equatorial Current but to ride the Gulf Stream between Cuba and Florida and to make their course back to Spain north of the westerly setting currents.

To the new colonies in America, sea commerce was of critical importance; the sea was the basis of much of their wealth. The Yankee whalers learned early to take advantage of the south-flowing countercurrents in their travels along the North American coastline. They also knew that a massive current flowed from America toward Europe, and that it was to be used in a west-to-east crossing but must be skirted to the south by ships returning to America from Europe. English mail-packet skippers strangely lacked this knowledge—perhaps because they had none of the sharp profit motive of the whalers—and their boats were often delayed two weeks in bringing the mail to the colonies. When the Boston Board of Customs complained about this to the Lords of Treasury in London, Benjamin Franklin, then Postmaster General for the colonies, was told to see what might be done about the situation. A Nantucket whaling captain, Timothy Folger, had the answer: ". . . the difference was owing to this, that the [American] captains were acquainted with the Gulf Stream, while those of the English packets were not. We are well acquainted with that stream, because in our pursuit of whales, which keep near the sides of it but are not met within it, we run along the side and frequently cross it to change our side, and in crossing it have sometimes met and spoke with those packets who were in the middle of it and stemming it. We have informed them that they were stemming a current that was against them to the value of three miles an hour and advised them to cross it, but they were too wise to be counselled by simple American fishermen." With the help of Captain Folger, Franklin drew a chart of the currents of the North Atlantic for the General Post Office, and ships that used this chart improved their performance significantly.

Another American, Matthew Fontaine Maury, performed much the same service to water-borne commerce, but on a much wider

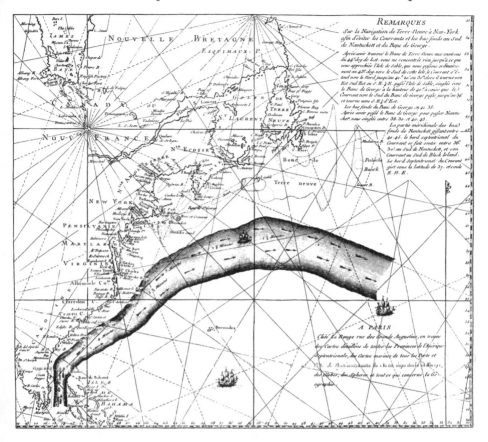

FIGURE 4 − 4 . *Benjamin Franklin's Gulf Stream. Franklin made this chart, when he was Postmaster General, to instruct mail packets from London how to avoid adverse currents during their voyage to the New World. Sailing directions, given in French, were compiled from information given to Franklin by Yankee whalers.*

James M. Snodgrass, Scripps Institution of Oceanography

scale and with far-reaching and permanent results. Maury turned his high talents and energies to the study of the currents of the oceans. As Superintendent of Charts and Instruments, he set about improving navigation charts. By searching old ships' logs stored in neglected profusion in Navy files, he was able to detect patterns in the innumerable records of wind and current observations by ship captains, and predicted that, by paying proper attention to these patterns, the sailing time between the ports of the world could be greatly reduced. Unsatisfied with the old records, he organized a world-wide system whereby more than 1,000 ships made observations on currents and other phenomena. From these Maury constructed "Sailing Directions" whose influence on shipping was profound. With the help of Maury's charts, ships all over the world took advantage of the currents and the other vagaries of the ocean's surface. The voyage from London to San Francisco was reduced from an average of 180 days to much less, the spectacular *Flying Cloud* setting a record of 90 days. The time from New England to Australia was reduced by 30 days. An Englishman estimated that British shipping would save 1 or 2 million dollars in the Indian Ocean alone if research like Maury's were conducted there. Today's "Pilot Charts" of the United States Navy Hydrographic Office are direct descendants of those constructed by Maury, and in recognition of his genius they still bear the legend, "Founded upon the researches made by Matthew Fontaine Maury, while serving as a Lieutenant in the U. S. Navy."

A great wheel of water similar to the Gulf Stream is set in motion in the Atlantic below the equator by the trade winds, but it rotates, of course, in a counterclockwise direction. As the trades push the water against the land mass of South America, the bulge of Brazil deflects part of this current to the north into the North Atlantic. This pirating of the southern current accounts in part for the greater strength of the North Atlantic current system as compared with the systems of other oceans, including the South Atlantic.

Between the two belts of trade winds in the Atlantic Ocean there is an area of light and variable winds or of calm called the doldrums. In this calm area no wind from the east pushes the water along, but the strong trades to its north and south actually mound the water against the land masses of the Americas, so that the water runs "downhill" to the east between the west-flowing currents as the shallow Equatorial Counter Current.

One more current in the Atlantic deserves mention: the musically named West Wind Drift that circles the Antarctic Continent. It is driven by the steady winds blowing from the west in the region of latitudes 40° to 60° South—the "Roaring Forties" of the days of sail. There is no land in this region, so the wind has nothing to stop its continuous influence on the waters, which consequently circle Antarctica without cease. This great wheel of water around the Antarctic connects with other wheels in the Atlantic, Pacific, and Indian oceans. In the Atlantic this is the Brazil Current, which has already been described.

In the South Pacific the great circle of currents starts at the shore of South America under the influence of the strong offshore winds. (Some ethnologists believe that the far-flung islands of the Pacific were peopled by men from South America, carried on frail craft by these currents. Thor Heyerdahl set out to test this proposition in 1947 when he and his companions rode the raft *Kon-Tiki* from Peru to the South Sea archipelago of Tuamotu.) With no land barrier comparable to the Americas to direct or impede the flow of the water, the South Pacific currents are less well defined than those of the Atlantic. Coriolis force bends the stream south, however, and then east to join the West Wind Drift and flow toward South America again, helped along by the prevailing westerlies. The circle is completed by the Peru, or Humboldt Current, the latter name given in honor of the pioneering German oceanographer who explored this region.

The area occupied by the Peru Current is one of the richest and most interesting in the ocean. The water is cold, and it was logical to believe, as early oceanographers did, that it was from the Antarctic regions. Now we know that this is only partly true: the current does include Antarctic water, but most of the chill is from the deep sea. Moving north as part of the great Pacific counterclockwise gyre, the water of the Peru Current is blown away from the shore by the prevailing southerly and southeasterly winds. As the surface waters are swept seaward they are replaced by upwelling from the ocean depths.

These waters, as we learned in the preceding chapter, are not only cold but also rich in minerals, and they set in motion the luxuriantly rich food chain leading from planktonic plants through small crustaceans to fish to birds and to man. Here the quantities of fish are enormous—a fact that is only now being realized by man, but which some marine birds of South America have known for a long time. The so-called guano birds, primarily cormorants, boobies, and pelicans, at their peak are numbered in the tens of millions off the coast of Chile and Peru. They are able to exist in these prodigious numbers because the rich waters of the area support billions of anchovetas, little fish akin to sardines. Guano, the basis of an industry that flourished before the conquistadores arrived in Peru and which still supplies rich fertilizer, is the dung of the greedy birds. As many as twenty-six anchovetas, each 5 to 6 inches long, have been found in the stomach of a cormorant. The birds, as many as 30 million in some years, are estimated to consume an average of 2.5 million tons of fish a year. Humans as well as birds are predators on the anchovetas. These small fish are not used for human food directly but are reduced to fish meal, for which there is a heavy demand in the poultry feed industry, and to oil used chiefly in the manufacture of margarine. Since 1950, when the first fish-meal reduction plant was built in Peru, the new industry expanded enormously, from a few thousand tons of anchoveta a year to 12.3 million metric tons in 1970. When fishing was at its peak the big purse seine boats could fill their holds in a few hours, and some vessels made two or three trips a day, coming back with full loads each time. From twenty-sixth rank among the fishing nations of the world in 1957, Peru rose to first place in 1963, almost entirely on the strength of the huge anchoveta stocks.

The tens of millions of birds, billions of fish, and millions of billions of other organisms in the great living complex depend on the replenishment of nutrients from the deep sea into the Peru Current. When the winds blow in their usual pattern, things go well. The surface waters of the ocean are swept offshore, the cold, deep water wells up, minerals are renewed, and the sea life flourishes. But in some years—about every 7—something happens to the nice balance, and havoc results. The wind currents shift, the surface waters are not skimmed off the top of the Peru Current, and the year of "El Niño" arrives. (It is called "The Child" because the shift comes around Christmastime.) Warmer water from the north wedges between the shore and the Peru Current and pushes the latter offshore. The Peru surface water is not swept away, and nutrient-rich water from the deep sea is not brought to the surface. Without this renewed supply of minerals, which provide the spin to the wheel of life in the sea, the food of the fish, the fish themselves, and finally the birds go hungry, sicken, and die. A severe Niño struck in 1972, and this, perhaps combined with overfishing, caused a collapse of the Peruvian anchovy industry. The catch in 1972 fell to 5 million tons, and plummeted to a little more than one million in 1973. Since then production has increased a little, but it was still not certain in 1975 whether the anchovy population will

FIGURE 4 – 5. *Peruvian guanaye birds* (Phalacrocorax bougainvillii) *sitting in craterlike nests made of their own dung. Immense numbers of these birds feed on still vaster swarms of anchovetas, little fishes whose food is derived from mineral-rich water swept upward from the deep sea.*

ever recover to its former level.

Along with El Niño the area off the Peruvian coast is sometimes visited by other unpleasant phenomena with the colorful names of red tide and Callao Painter. The red tide is caused, under certain water conditions, by the explosive production of immense numbers of one-celled plants called dinoflagellates. The Callao Painter (so-called since it occurs conspicuously in the harbor of Callao, Peru) results from the putrefaction of enormous numbers of bodies of marine animals killed by El Niño and the red tide. These release hydrogen sulfide, and the water turns curious shades of pink, salmon, red, and other colors. The hydrogen sulfide combines with the lead of the paint on ships in the area, blackening them.

In the Peru Current area we have an excellent illustration of the interrelation of surface currents and those of the deep sea. We have, in addition, a dramatic example of the fact that animals in the sea adjust themselves to a particular set of conditions, and that when these conditions are altered and the delicate balance of nature disturbed, catastrophe can be the result.

The circular system in the North Pacific starts with the North Equatorial Current, a permanent east-west flow kept in motion by the trade winds. It turns north at the Philippine Islands and flows past Taiwan and east of the main islands of Japan. Here it is known as the Kuroshio Current, the "Asiatic Gulf Stream." Carried across the North Pacific to the coasts of North America, it gives coastal southeast Alaska a mild climate, contrasting strongly with the frigid weather experienced by the inland areas of that state. Coastal British Columbia, Washington, and Oregon are an "Evergreen Playground" because of the heat carried by this current from the tropical Pacific on a long journey around the rim of the ocean. Flowing south down the American coast as the California Current, the water turns west again to join the North Equatorial Current off the coast of Lower California.

The current system of the Indian Ocean also has a circular pattern, but the circles are not as well defined as those of the Pacific or the Atlantic. Seasonal differences in the direction and strength of the currents are much greater in the Indian Ocean than in the other two major oceans. Seasonal shifts in direction set up complex eddy systems that make the Indian Ocean currents much more difficult to understand. This is one of the principal reasons why the massive Indian Ocean Expedition was launched in 1962.

DEEP-SEA CURRENTS

We have seen that the surface currents of the ocean owe their motion principally to the energy supplied by the wind. Some deep-sea currents, on the other hand, are set in motion when surface waters become cooler and denser than the water beneath them and begin to sink. Any swimmer is familiar with the way water forms in layers according to its temperature; it is a common experience to tread water with the upper part of the body in warm water and the legs in a cold layer.

Sea water becomes denser in two ways: through an increase in the amount of dissolved salts and through a decrease in temperature. When salt water freezes, the salt migrates out of the ice crystals, leaving the ice nearly fresh. In a commercial ice plant, where water is frozen in metal cans, the ice forms from the edges of the can toward the center. By the time most of the water is frozen, a pocket of unfrozen water in the center of the block contains nearly all the salts or impurities that were in the water, having been concentrated there by the freezing action. This water is siphoned off in the process of ice manufacture, and fresh water is poured into the hole before the final freezing takes place. In the sea the salts are likewise squeezed out of the water as it is frozen, leaving the unfrozen water saltier than before and therefore considerably denser.

Under the influence of this increased saltiness and the bitter cold of the extremes of the earth, two deep-sea currents are set in motion at the edge of Antarctica—currents that inch their way with extreme slowness toward the north underneath the surface currents. These two deep currents are the Antarctic Bottom Water, generated in the Weddell Sea, and the Antarctic Intermediate Water, which has its origin at the edge of the Antarctic Ice Cap.

The Weddell Sea is the great bay of the Antarctic Continent that forms the most southerly part of the Atlantic Ocean. It is here, in the grip of the southern winter, that the salt is squeezed out of

FIGURE 4 – 6 . *A section through the Atlantic Ocean showing variations in saltiness and the main water currents. The Antarctic Bottom Water and Intermediate Water currents flow from the Antarctic Ice Cap into the Northern Hemisphere. These enormous masses of water are replaced by currents from the North Atlantic and Mediterranean.*

Sir Alister Hardy, *The Open Sea: Its Natural History.*
Part 1. The World of Plankton

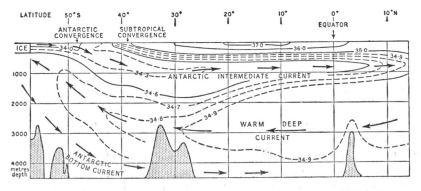

[63]

the sea water as ice is formed. The icy-cold (about 29° F.), salty water that remains begins a slow, silent slide down the continental shelf and over the ocean bottom all the way to the abyss. This is the densest, coldest water of the ocean, and it occupies the deepest pockets of the ocean floor. It can be traced, by its characteristic temperature and salinity, across the equator and north as far as the Bahamas, far from its place of origin.

The Antarctic Intermediate Water is formed when the sea is cooled near the southern continent, at about 50° South latitude. It is not as cold as Antarctic Bottom Water, nor is it as salty, since melting ice and precipitation add fresh water to the mixture. As a consequence, it does not sink all the way to the bottom of the ocean, but forms a layer above the Antarctic Bottom Water. Somewhere in mid-depths it begins a slow, horizontal spread northward. In the Atlantic this water can be traced to 20° (1,300 miles) north of the equator; in the Pacific it does not spread so far. The enormous masses of water leaving the vicinity of the Antarctic Continent must be replaced by a southbound stream. Water from the North Atlantic wedges between the two antarctic currents to balance the account.

Turning to the North Atlantic, we find that the North Atlantic Bottom Water is formed in the Labrador Sea and in the ocean between Spitsbergen and Greenland. The process is the same as in the Antarctic: temperatures fall to about 28° F. and the salinity is raised as ice forms. In addition, water in this region is made saltier by the Gulf Stream. Coming from the hot middle latitudes, surface water of the Stream is evaporated by the tropical sun, leaving it saltier than before.

Dense water from the northern region of the Atlantic spreads slowly over the deepest areas of the North Atlantic and creeps as far south as the equator. This frigid water, only a little above the freezing point, fills all the deepest canyons of this part of the ocean. There is plenty of water to do this, since an average of 4 million tons of it sinks every *second* the year around.

An important layer of deep water in the Atlantic that does not have its origin in the polar or subpolar regions is the one that comes out of the Mediterranean Sea. In this nearly landlocked arm of the Atlantic, rapid evaporation takes place from the surface. Mediterranean water is saltier than that of the Atlantic because the amount of water flowing into the Mediterranean from rivers, plus the rainfall, is less than the surface evaporation, which lifts off an average of 100,000 tons of water a second. This extra-salty water sinks below the surface, pours out over the sill at the entrance to the Mediterranean at the Strait of Gibraltar, and spreads under the Antarctic Intermediate Water and above the Antarctic Bottom Water. It can be traced west to the Bahamas, and it reaches far north and south. As this subsurface current flows out of the Mediterranean Sea it must be replaced, and this replacement comes as a surface current flowing from the Atlantic Ocean.

The Gulf Stream transports far more water north than the Canaries Current carries south. The water balance in the Atlantic is maintained by massive deep-current movements of water to the south.

The slow spread of waters over the oceans of the world at great depths thus completes the circulation at the surface, which we examined earlier. Until recent years it was considered that the deep Atlantic current was an enormously broad one, encompassing the whole extent of the Atlantic Ocean to join the antarctic current and flow into the Indian Ocean and thence into the Pacific. Henry Stommel of the Woods Hole Oceanographic Institution believes that this theory must be modified. He believes that the flow of deep water should act like the flow of the atmosphere. Air does not move in a broad band over the earth from an area of high pressure to one of low pressure; it circulates around the highs and lows because of the Coriolis force. Stommel believes that the cold water originating near the poles should be carried in a narrow, relatively strong stream along the western side of the ocean basins, but in a direction opposite to that of the surface currents. Hence there should be a massive deep current flowing down the western North Atlantic and up the western South Atlantic; these should merge and flow east toward Europe. A German oceanographer, Georg Wüst, analyzed pressure distributions to show theoretically that deep-water currents in the South Atlantic were indeed confined to relatively narrow streams near the continental slopes of South America. But the direct checking of Stommel's theory has been difficult.

Now a new tool invented by Dr. John C. Swallow, a British oceanographer, has lent support. The buoyancy of the Swallow float can be adjusted delicately so that it is made to sink to a predetermined level and remain there, floating in water of the same density as itself. The float has a small ultrasonic transmitter that announces its position to tracking vessels. An American-British oceanographic expedition in 1957 released Swallow floats in the deep waters below the Gulf Stream off Charleston, South Carolina. At a depth of 6,500 feet a current was discovered flowing southward at speeds ranging from 2 to 8 miles a day. The Swallow float off the coast of Portugal recorded currents from zero to two miles per day. Other estimates place the speeds of abyssal currents at about a mile a day.

The Pacific Ocean borrows all its deep water from the Atlantic. The vast trenches and basins of this biggest of oceans are filled with cold water flowing from the South Atlantic between Australia and Antarctica. Most of this water sank originally from the surface into the depths in the Weddell Sea; the rest is from the Mediterranean and the North Atlantic. The "secondhand" character of the Pacific deep water may account in part for its excessive sluggishness. Deep currents are not set in motion as strongly by the slow flow between Antarctica and Australia as they are by the surface sinking that occurs in the Atlantic.

We are nearly ignorant of the motion of the deep currents in the Pacific. This ignorance was strikingly illustrated by the discovery as recently as 1951 of an immense subsurface current stretching at least halfway across the Pacific Ocean. This current, perhaps the second largest in the Pacific after the Kuroshio Current, carries over half as much water as the Gulf Stream, at nearly as fast a rate. It seems incredible that a current transporting 40 million tons of water a second—as massive as thousands of Mississippi Rivers—could escape the notice of men for so long. The current came to light in a curious manner. Scientists of the United States Fish and Wildlife Service, probing waters for tuna at the equator south of the Hawaiian Islands, had set their gear for a depth of about 165 feet in an area where the surface waters flowed westward under the influence of the South Equatorial Current. They were astounded to see the buoys head unmistakably to the east! Apparently there was an easterly current only 100 feet or so below the surface. The next year the Fish and Wildlife Service sent one of its oceanographers, Townsend Cromwell, to the equatorial Pacific to make a careful study of this unexpected current. His work brought to light its remarkable characteristics.

Subsequent work by a group of scientists from the Scripps Institution of Oceanography headed by Dr. John A. Knauss has shown the Cromwell Current to be one of the major components of the Pacific current system. It is at least 3,500 miles long. It extends eastward to the Galapagos Islands; in the west it has been identified at 150° West (about the longitude of Tahiti). There are hints that it exists at longitude 160° East, near the Solomon Islands, and Japanese oceanographers have turned up currents on the far side of the ocean that may well be the other end of the Cromwell Current, giving it a total length of 8,000 miles.

A remarkable feature of the Cromwell Current is its shallowness. As Dr. Knauss has written, if a scale model of the current were made 5 inches wide, its thickness would be only that of a heavy piece of paper. At 140° West longitude, the top of the current is about 65 feet below the surface and its core 325 feet deep. The current slows down below the core; by 1,640 feet what movement remains is in the opposite direction. This thin ribbon of water is about 250 miles wide—1,500 times as wide as deep—extending 2° on each side of the equator.

Most remarkable of all, perhaps, is the constancy of the current. Over the whole length that has so far been charted, it extends a steady 125 miles on either side of the equator, with its center exactly on the equator. Its speed of about 2½ to 3 knots and its volume apparently vary surprisingly little. There seems to be little doubt that this constancy of behavior is associated with the position of the current precisely on the equator. Current action is a combination of horizontal forces supplied by the wind or by density gradients and by height differences, plus the Coriolis force. The Coriolis force varies with latitude, being greatest at the poles and zero at the equator: at the equator only the horizontal forces act on the water, with no twisting action to modify them. Although this concept may seem fairly simple as stated, it is by no means fully understood, and the theoretical problems presented by the actions of the Cromwell Current are very intriguing to oceanographers.

MEASURING THE CURRENTS

An indirect method of measuring deep-sea currents is through precise determination of their oxygen content. Water starts at the surface with a relatively rich supply of oxygen, and as it sinks into the depths this is gradually used up by the animals living in it and by processes of oxidation of dead organic material, so that the extent to which the oxygen is depleted is a measure of the time since it left the surface layers of the sea. During the *Discovery* Expedition of the early 1930's, measurements were made by Dr. G. E. R. Deacon, the British oceanographer, of the reduction of the oxygen content of the Antarctic Intermediate Water. At the same time the salt content increased, since the salty ocean water added part of its contents to the relatively fresh antarctic water originating from the melting ice mass. However, the salt content did not increase steadily as the oxygen content of this water decreased, as would be expected. Instead, Dr. Deacon found that these changes took place in a series of waves, with the top of the salt-content wave corresponding to the bottom of the oxygen-content wave. He interpreted the curious behavior of these changes to represent the progression of seasons: In the antarctic summer, when relatively high temperatures hasten the melting of ice and encourage the violent growth of planktonic plants, water masses of especially low salinity and high oxygen are produced, while in winter this action is much reduced. Since there were seven peaks and troughs in the measurements, Dr. Deacon decided that it took the Antarctic Intermediate Water at least 7 years to flow from the place of its origin to the Northern Hemisphere.

Thus the "age" of waters in the deep sea (meaning the elapsed time since they left the surface) can be deduced from the pattern of their oxygen content. Those with the highest oxygen content— the youngest waters—occur in two areas. These are around the Antarctic Continent and in the western North Atlantic. All the oceans of the world eventually receive their deep water from these two areas, the currents carrying it first into the Indian Ocean and eventually into the Pacific. The oldest water, with the smallest load of oxygen, occurs in the abyss off the coast of Peru.

Particular "parcels" of water commonly contain quantities of salt that differ from the amount contained in other water that originated in different areas. The saltiness of the water can be measured with great precision. In the same manner, slight differences in temperature can be detected in different parts of the ocean, whether on the surface or in the deep sea. A particular parcel of water will have its own combination of temperature and salinity, and this combination will usually be different from the values for any other water mass. Hence, water of common origin can often be traced over very great stretches of the ocean by its telltale "T-S" diagram.

CARBON-14 DATING

A second and more accurate method of aging sea water is by carbon-14 dating. This technique, a product of the most recent era of science, depends on the fact that carbon has six forms, or isotopes, with identical chemical characteristics but different atomic weights. Of the two main forms of carbon, one, with an atomic weight of 14, is extremely rare, and the other, with an atomic weight of 12, is common. Carbon 14 decays to carbon 12 at a fixed rate, so that after a certain interval a very exact ratio of the two is present as a mixture. Measuring the ratio determines the age of any substance containing carbon. This is an extremely useful technique, since all living material, and all material that was ever living in the past, contains carbon. Carbon dating has given the precise age of a great many substances, including the fires of prehistoric man and the Dead Sea Scrolls found in caves in Palestine.

Using carbon-14 dating in deep waters in the eastern South Pacific, California oceanographers estimated the ages to be from 1,350 to 1,910 years, implying current velocities of about 7½ miles per year. Estimates by other means are usually lower than this, however, giving an age of less than 1,000 years for deep Pacific water. It is ordinarily estimated that deep Atlantic waters are not as old as those in the Pacific; estimates of the age of Atlantic waters deeper than 6,500 feet vary all the way from 50 to 1,000 years. The deep water in nearly enclosed basins like the Mediterranean and the Black Sea takes much longer to be replenished than that in the open oceans. The deep waters of the Black Sea apparently remain isolated for very long periods—at least 2,500 years and perhaps as much as 5,600 years.

We are still groping for trustworthy information about deep-sea currents. This is of great theoretical interest, but it also has immense practical significance, since countries operating nuclear power reactors find themselves with radioactive wastes, which are difficult to dispose of safely. The ocean is beginning to be used as a garbage dump for these wastes, and will be to a greatly increased extent in the future unless oceanographers advise us that currents in the deep sea are likely to deposit the dangerous wastes back on our shores again. The conclusion so far is that it will probably take many years for the wastes to find their way back to shallow waters if they are dumped in the deepest sea, and that this will permit their radioactivity to decay to a safe level. But we are still none too sure of the exact figures or about the wisdom of creating great atomic waste dumps in the sea. In this situation the scientists and officials responsible for such decisions have shown a laudable tendency to be conservative.

5 Meadows of the Sea

Man is inclined to judge other living things in terms of their size, assuming that evolution has passed by the small, one-celled creatures in a steady climb to the multicelled animals and plants. But evolution has operated in another way too. In addition to encouraging a division of labor in the many-celled creatures, it has improved the structure and efficiency of the plants and animals that have remained one-celled. Many of these are complex, and it would be hard to name any more successful organism than the microscopic floating plants of the sea. They do not build cities and destroy them with atomic bombs like humans; they do not flex great muscles and spout water vapor in high plumes over the antarctic sea like whales. What they do may be regarded as far more important: they carry the whole of the vast assemblage of living creatures of the oceans on their shoulders.

According to Isaiah, "All flesh is grass, and all the goodliness thereof is as the flower of the field: surely the people is grass."

In other words, the animals of the land depend finally on plants for their life. Some animals eat plants directly; others feed on the flesh of plant eaters. Man does both; with all the other animals, he is created by the biblical "grass." The pastures of the land are familiar, and so is the mechanism by which grass is transformed into beef, corn into pork, grain into chicken. Presumably all animals in the ocean must also depend on plants; but where are the pastures of the sea?

Floating pastures do exist, and they are actually far vaster than those on land. All the vegetation on all the continents is estimated to produce 40 billion tons of carbohydrates per year, while the plants of the sea produce from 80 to 120 billion tons. But the meadows of the sea are not so conspicuous as those of the land—first, because inhabitants of the sea are rarely as easy to observe as those on land, and, second, because with few exceptions the plants of the sea are too small to be seen without a microscope. These uncounted myriads of little plants, together with the great numbers of tiny animals that feed on them and on each other, form the plankton. "Plankton" is the name applied in 1887 by Victor Hensen, a German professor of zoology, to the great company of marine creatures that drift at the mercy of the currents, as distinguished from the "nekton," animals like fish and whales that are able to swim against the moving waters, and from the "benthos," the plants and animals attached to the bottom or crawling upon it. There are many degrees of helplessness among the planktonic creatures. All the plants and some of the animals float passively, while other animals are able to make their way feebly or with some vigor against the currents. There is no clear-cut dividing line between planktonic and nektonic creatures; some of the fish and other marine animals belong to the plankton during the early stages of their lives and to the nekton later.

PLANKTON PLANTS

Consider first the plants of the plankton. They are so minute that most people are not aware of their existence. This small size even kept them from the attention of most scientists until a little over 100 years ago, when European naturalists (including Charles Darwin) began to tow small conical bags of muslin or fine silk through the water to strain out planktonic plants and animals.

A few marine plants that grow in very shallow water bear flowers and seeds, but the great majority are members of a simpler plant group, the algae. These include the seaweeds like kelp and sea lettuce and a large number of other forms, both attached and floating. Most of the algae large enough to be seen easily are not members of the plankton, but a notable exception is the sargassum weed, the attractive brown plant with little bladder floats that collects in great aggregations in the vast eddy of the Sargasso Sea. Like other large-sized plants of the sea, sargassum weed grows initially attached to the bottom. It breaks off from shores in Central America and the Caribbean islands during storms and is borne on currents into the open sea, where it grows and divides, although it does not reproduce.

But the really important plants of the sea are far less conspicuous than the sargassum weed or the kelp that writhes in great shiny ropes up to 150 feet long on the shores of the Pacific and elsewhere. The microscopic plants of the plankton are never seen by the mariner or the bather unless they "bloom" in such prodigious numbers that they discolor the water, as in the case of the red tide or the pungent-smelling "Dutchman's baccy juice" which sometimes stains the North Sea a slimy brown and drives away the herring and the herring fishermen.

The little plants that repel the herring are diatoms, while the

FIGURE 5 – 1. *Two principal groups of the plant plankton, shown here photographed alive, are diatoms* (Biddulphia sinensis, Melosira borreri, Coscinodiscus conicinus) *and dinoflagellates (anchor-shaped* Ceratium tripos). *Plankton are the floating "pastures of the sea" that feed sea animals—and, eventually, many land animals.*

Douglas P. Wilson

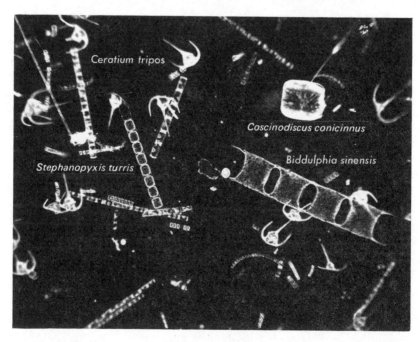

red tide is caused by one of the dinoflagellates; these are two of the principal groups of the plant plankton. Both convert dissolved materials into food by the process of photosynthesis, and do it within the confines of a single cell. From the simplest of chemicals (carbon dioxide and water) they synthesize carbohydrates (sugars and starches). These in turn form the basis of nearly all the food of the multitudes of animals in the sea, and eventually of many animals on land.

THE DIATOMS

There are perhaps 15,000 to 20,000 species of diatoms with an enormous variety of structure (Figure 2-8). Basically, they are all in the form of tiny pillboxes of silica (the main ingredient of glass) containing shining masses of protoplasm, the living matter of cells. Imbedded in the protoplasm are small colored bodies—grass green, pale emerald green, or brownish. These are the chloroplasts, containing chlorophyll, the green chemical necessary for the process of photosynthesis. The pillboxes consist of discs or oblongs, rectangles or rods, with an apparently endless variation of exquisite adornment. Little spines or horns extend from the corners of some kinds of diatoms, while the surface is pitted, pierced, and sculptured with slits and pores in geometric tracery. Since the glasslike silicon walls of the diatom are impervious, the contact of the living material of the cell with its environment is through these minute openings. Some of the diatoms live singly, but very commonly several cells of the same kind are strung together in chains or bundles.

Reproduction in the diatoms consists of simple division of the cell, one half taking the lid of the pillbox and the other taking the smaller lower piece. This latter section becomes the new lid of the daughter cell, so one of the divided diatoms is smaller than its parent. There is consequently a good deal of difference in size among individuals of the same species, and the average size of individuals in the population gradually declines as successive divisions take place. There is a limit to the dwarfing, however; it is halted eventually when some of the diatoms produce large swellings of protoplasm that bulge out between the valves, producing an auxospore, which in a bound restores the original size of the plant. But even after this increase in size the giants among the diatoms are still barely visible to the naked eye—about one one-hundredth of an inch in diameter —and most species are smaller than this.

It is useless to try to give an accurate account of the numbers of diatoms in the sea. Before we get beyond the consideration of a tiny fraction of the ocean we are dealing with numbers so large that they become meaningless. In times of abundance there are diatoms in every drop of sea water in the temperate zones, and a single diatom may have 100 million descendants in a month. When the ocean meadows become as lush as that, the animals that feed on the diatoms flourish and become fat, and the creatures on the next step up the food ladder wax abundant too. Eventually fishermen benefit, bringing in bigger catches.

The amount of plant life produced under a given area of the sea surface is often about as much as grows on a similar area of tropical rain forest. In the Gulf of Maine, for example, the water beneath one square meter (not much more than a square yard) was found to contain 7 to 8 billion diatoms during one period of abundance. Copepods, tiny animals so small that they can barely be seen, have been found to consume 120,000 diatoms in one day, and a single copepod probably consumes many millions of them in its lifetime. It has been estimated that 4 million tons of copepods exist at one time in the 36,000 square miles of the Gulf of Maine, and over a season this quantity is many times more. Consider, then, the number of diatoms required to keep even the copepods in bread and butter, apart from the vast numbers eaten by other creatures!

The growth and the reproduction of diatoms are very rapid. When conditions are favorable, there can be two or even three new generations in a day. This extremely efficient use of the environment by diatoms is made possible by their structure and notably by their tiny size. In order to manufacture carbohydrates a plant must absorb carbon dioxide, water, and dissolved minerals. The amount of absorption that can take place depends on the amount of surface the plant presents. The surface area of a plant increases in relation to its bulk as its size decreases: whatever the size of any object, if it is split into two pieces the amount of surface is increased—and the more times the same bulk is cut up, the greater is the surface exposed. This is the basis of the tremendous advantage the microscopic plants have over larger plants floating in the water, and the reason why the tiny diatom is the most efficient of plants.

[74]

PHOTOSYNTHESIS

Photosynthesis is basically the same whether carried on by land plants as big as the giant redwood or by microscopic plankton in the sea. The crucial role is played by chlorophyll, the green substance in plants that acts as a "catalyst," or helper, in performing a near-miraculous chemical manufacturing job. With chlorophyll as the catalyst, a plant uses light—solar energy—to convert simple chemical substances (carbon dioxide and water) into more complicated ones (carbohydrates). These complex "organic compounds" have in common the possession of carbon atoms derived from the carbon dioxide. As more complicated substances are gradually manufactured, the plant rearranges the carbon atoms into rings of six; attached to these rings are other atoms in an infinite variety of arrangements. Thus, in addition to many kinds of carbohydrates (sugars and starches), the plant synthesizes proteins, fats, and vitamins. For example, the vitamin for which cod liver oil is famous is manufactured by diatoms, and is stored in the liver of the cod when the cod eats little fish that have previously eaten the diatoms.

Something of immense importance is involved in the process of photosynthesis: energy is stored. The radiant energy of the sun is converted into chemical energy in the form of the bonds that cement together the atoms of the compounds fashioned by the plant. The stored energy is released when the process is reversed as the substances created by the plant are broken down again. This can take place in the digestive tract and tissues of an animal that grazes on the plant or as the result of bacterial action should the plant die. As the organic compounds are broken down again into simpler substances, the energy stored during their synthesis is set free. Here, then, is the ultimate source of the boundless energy of a healthy child—of the energy, indeed, of every living thing on earth. The sun was its father, the plants its mother, and the chlorophyll the attendant midwife.

Food production by the plants of the sea does not take place much below about 100 to 130 feet in the murky inshore waters of northern latitudes, because light strong enough to energize photosynthesis can penetrate no deeper than this. In the clear waters of the tropics, especially offshore, this distance increases to about 325 feet. It is only in the top skin of the ocean, then, that photosynthesis can take place. Living diatoms and other plants are found below these depths, but they have drifted there through the pull of gravity, and have presumably ceased to carry on food manufacture or reproduction. There may be a small exception to this generalization. In the 1920's, German biologists on the *Meteor* Expedition filtered some puzzling tiny plants, light olive green in color, from waters all the way from the surface down to 16,000 feet. The greatest concentration was at about 1,000 feet below the maximum depth where photosynthesis has been thought to take place. These tiny blobs of life may be blue-green algae, members of the same group of simple plants as the green scum on the surface of ponds and ditches. If their green color means that they carry out photosynthesis in the great depths where they are common, they are certainly the most efficient light absorbers known. Perhaps, though, they do not synthesize, but merely use the small amounts of organic material dissolved in the water; until we know for sure, they present a fascinating problem.

THE FLAGELLATES

The diatoms are the first of the two dominant groups of planktonic plants; the other group comprises the flagellates or whip bearers. There are many different kinds of flagellates, some naked and others covered with armor; they all have in common the possession of one or two whiplike appendages that propel them jerkily through the water. A group with two whips, called dinoflagellates, are the most numerous and best known of the flagellates. The unarmored forms are delicate and easily damaged in collecting nets. The others have skeletons of cellulose, tiny mosaics arranged in striking and multiform designs. From each tiny point of the mosaic project spines and horns, knobs and bumps, creating all manner of fanciful shapes. The various species of dinoflagellates resemble Chinese hats, carnival masks, children's tops, urns, pots and vessels of many kinds, the spiky knobs of medieval war clubs, balloons on strings, hand grenades, or lances. One of the whips lashes transversely from a groove girdling the middle of the body, and the other works from the rear. The latter serves to drive the little plant forward, while the transverse whip sets it waltzing through the water in an erratic motion. Unlike the diatoms, which often join into colonial strings, the dinoflagellates are nearly always solitary.

It may have occurred to some readers that we are ascribing to plants the power of motion, a characteristic supposedly the sole possession of animals. The fact is that we are dealing here with creatures so low on the evolutionary scale that they have not struck out positively along the path toward either the animal or the plant kingdom. Zoologists claim them as animals, and botanists regard them as plants—and both are right. The flagellates are animal in their ability to propel themselves and to ingest solid food. Most of them also have the green, yellow, or brownish pigments of chlorophyll or other photosynthetic materials, and this makes them plants without quibble.

Like all other plants, the dinoflagellates are dependent on light as the source of energy for the process of food manufacture. To help them stay in the lighted zone, many have a minute red eyespot, or stigma, which is sensitive to light and tells them whether it is getting stronger or weaker. The whips come into play to propel the dinoflagellate toward the light when the stigma gives the proper signal. In some forms the stigma is highly developed, rather like an eye. This is all the more confusing if we are trying to place these creatures in one or the other of the great life kingdoms, for here is a typical animal organ, the eye, being used to assist in an exclusively vegetable function, photosynthesis! But it is not necessary or useful to try to make this separation; we must simply accept the curious fact that here are organisms that are simultaneously animal and plant.

Until very recent years textbooks on the plankton gave first place in importance in the economy of the sea to the diatoms. Now the flagellates are generally regarded as being even more important. This change has occurred as better collecting gear has revealed the prodigious quantities of flagellates in the sea, their swarming numbers having previously escaped full notice because of the minuteness of many of the individuals, and because flagellates are typically creatures of warm tropical and subtropical waters, which were less thoroughly studied in past years than the temperate seas, where diatoms are most common. Some concept of their vast numbers can be obtained from the fact that catches were made in the Gulf of Mexico off Florida as high as 60 million to the liter (a little over a quart) of *Gymnodinium brevis*, the red-tide organism.

Among the flagellates that failed to receive proper attention from biologists because of their small size are the coccolithophores. They are members of the "nannoplankton" or "dwarf plankton." These tiny spots of life are so small that they pass through the finest nets and can only be isolated by filters and centrifuges, whirling devices for separating the heavier from the lighter components of a mixture. Diminutive as they are, the coccolithophores have a skeleton made of many sections in the form of calcareous plates fitting neatly together. The plates have a knob or a spike springing from their center, or sometimes a projection of more elaborate shape, like a trumpet. The minute plates were known long before the living plants were discovered, since the broken skeletons form part of the deep-sea sediments (Figure 2-8). Geologists saw them through their microscopes and named them "coccoliths," or seed-stones. Sir John Murray, brilliant co-leader of the *Challenger* Expedition, discovered the true identity of the coccoliths, and since then biologists have come to realize their importance to the rest of the creatures of the sea. Like the dinoflagellates, the coccolithophores have a pair of vibratile whips, and like them too, they sometimes occur in vast numbers. At such times they give the water a milky appearance, the result of light reflected from the calcareous skeletons. An English herring-boat captain, Ronald Balls, has described the appearance of this white water as "a fairy glow," and says that it gives "the queer impression of whiteness coming upwards: as if the light was below the sea instead of above it." At such times as many as 5 million coccolithophores per quart have been counted. Fishermen welcome the appearance of white water since the coccolithophores seem to attract herring. This more than makes up for the small nuisance caused when the dried shells of the little plants leave a chalky deposit on the gill nets.

Recently it has been discovered that the coccolithophores are of special interest in relation to life in the deep sea, since many of them live at remarkably great depths—at least down to 13,000 feet. Nor do they occur there merely by chance: counts made from re-peated hauls have yielded as many as 240,000 individuals per quart. It appears certain that the coccolithophores are not manufacturing food at this depth but exist by assimilating dissolved organic matter from the water and incorporating it into their body tissue. They may not reproduce here either, but depend on recruitment of fresh indi-viduals from the water above for maintaining the population; this point has not been settled. These deep-living coccolithophores and the mysterious olive-green cells described earlier supply important amounts of food to deep-sea animals.

Much of the phosphorescence seen at the surface of the sea

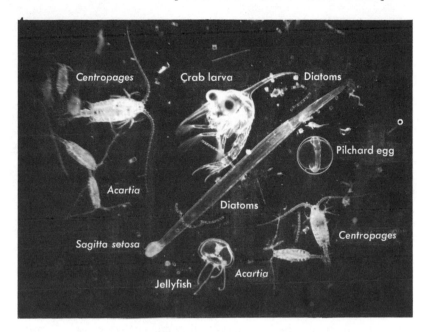

FIGURE 5–2. *Living animals of the plankton: the transparent arrowworm* Sagitta setosa, *the zoea (early larva) of a crab, a round gadoid egg with a developing fish inside, and a small jellyfish. The larger copepods are* Centropages *and the smaller ones* Acartia. *Less easily visible are the copepods' bread and butter, the diatoms.*

Douglas P. Wilson

is due to the presence of dinoflagellates. Sometimes this shows as a general area of ghostly light, sometimes as separate sparks of light in the water. When the wake of a boat flames behind in a night sea, or when the oars form pockets of liquid fire, the light is usually caused by luminescent dinoflagellates being stimulated to show their colors. Two of the best known of these are *Noctiluca*, the "night light" (Figure 14-1), and *Pyrodinium*, the "fire whirler."

PLANKTON ANIMALS

Depending for their food on the substance manufactured by the planktonic plants are gigantic swarms of animals of bewildering variety. The zooplankton (that is, the animals of the floating community) includes representatives of nearly every one of the numerous groups inhabiting the sea, from fish on down. Some of them, especially the fish, are represented in the plankton only temporarily as eggs or larvae, but many groups are permanently planktonic (Figure 15-2).

FORAMINIFERA AND RADIOLARIA

The first of the planktonic animals to be considered, the foraminiferans and the radiolarians, are like the flagellates in being one-celled, but they are unmistakably animal. They do not ever manufacture any of their food; instead, they capture it by snaring it in sticky protoplasm. In both groups the protoplasm (living substance of the cell) is supported by a skeleton. Few of these animals are big enough to be seen without a microscope, and some require considerable magnification. They live mostly in the upper 600 feet of the sea, although they have been captured as deep as 16,000 feet. They are largely warm-water forms.

The skeletons of most foraminiferans are made of calcium carbonate, and are pierced by innumerable tiny holes (Figure 2-8). Through the holes streams the protoplasm in sticky strings that combine again on the outside of the shell. The result is a branching network of filaments—a sort of living spider web, except that, as Sir Alister Hardy points out, the spider and the web are one. Little animals and plants, smaller even than the forams, are ensnared in this device and digested outside the shell.

In some living foraminiferans, needlelike projections from the shell form further support for the living stuff of the cell, but they are broken off when the animal dies. The skeletons are in the form of miniature, chambered spirals, much like the shells of very small snails. *Globigerina*, a genus with a large number of species, is one of the most ancient and common of this group, its shells having accumulated on the ocean floor for millions of years (Figure 2-8). About 45 per cent of the deep-sea sediments are once-living members of the order Foraminifera, with *Globigerina* remains predominating. The white cliffs of Dover, the storied symbol of England that gave it its ancient name of Albion, are composed of incredible numbers of shells of foraminiferans. Miles in length, these cliffs are over 500 feet high in places. One teaspoonful of the chalky material contains half a million little shells.

In the radiolarian, radiating threads of protoplasm spring from a perforated central skeleton (Figure 2-8). This skeleton is often of great beauty, a delicate basketwork tracery symmetrical in design. One species, *Oroscena regalis*, is shown in Figure 2-9. Fashioned of glasslike silica, the skeletons assume shapes resembling urns, hats, helmets, and even ornate carved spheres within spheres, like those intriguing Chinese ivory balls. Radiolarian ooze consisting of the shells of long-dead animals forms a wide band straddling the equator, but is restricted almost completely to the Pacific and Indian oceans, and covers thousands of square miles of the bottom of the sea.

WORMS AND ARROWWORMS

Of the many kinds of worms in the sea, the most interesting is a beautiful creature called *Tomopteris*. This graceful worm swims in the water with an undulating sidewise motion, employing a row of winglike feet on each side of its body. *Tomopteris* is a voracious carnivore, devouring uncounted planktonic creatures. It is perfectly transparent when it lives near the surface of the sea, but there is a deep-sea form that is a brilliant crimson color.

The next group of animals are worms in name only. The chaetognaths ("bristle jaws") are called arrowworms because of their long, slender shape (Figures 5-2 and 15-2), but they are not related to the worms. As a matter of fact, zoologists have been unable satisfactorily to link them with any other group of animals, although they are of great interest here since they are dominant members of the plankton. The arrowworms are like fine threads of glass, ranging from less than a quarter of an inch in length to several inches; the commonest size is about three quarters of an inch. They are so completely transparent that they usually cannot be seen when alive unless they have food in their bellies or unless their tiny black eyes show up. Their mouths are ringed with hooked teeth; paired fins and a horizontal aileronlike tail equip them efficiently for swift pursuit of prey. They feed hugely on copepods and on small fish, including young herring.

SEA BUTTERFLIES

If one set out to pick two animals that offer the widest possible contrast in appearance and behavior, a dramatic pair of "contraries" might well be the snail and the butterfly. The snail drags a bulky

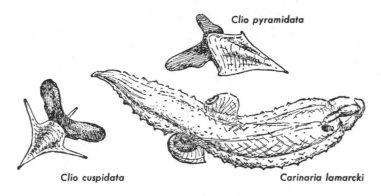

Clio pyramidata

Clio cuspidata

Carinaria lamarcki

FIGURE 5 – 3. *Planktonic pteropods* (Clio cuspidata *and* Clio pyramidata) *and a heteropod* (Carinaria lamarcki). *The latter is drawn natural size, while the pteropods are magnified one and a half times. Called "sea butterflies," pteropods propel themselves through the water with extensions of the foot that resemble wings.*

Sir Alister Hardy, *The Open Sea: Its Natural History. Part 1. The World of Plankton*

shell around on a slimy foot, moving ponderously and with great deliberation so that it scarcely seems to move at all; the butterfly, *ne plus ultra* of lightness and delicacy, floats and flirts through the air on fragile wings. In the face of this, it is a surprise to find snails in the plankton that look and behave enough like butterflies to be called "sea butterflies." These pteropods (meaning "winged feet") are transparent, delicate creatures that are among the loveliest of the planktonic animals. They "fly" through the water with rapid beating of wings that are lateral extensions of the foot. Some have what no terrestrial butterfly possesses, a shell. Unlike the bulky shell of most snails of the land or of the sandy shallows of the ocean, it is a fragile structure, variously in the form of a delicate cone, a triangle, or a tightly coiled spiral (Figure 2-8). Some sea butterflies have no shell and, as a consequence, work less hard to keep themselves afloat. An example of such forms is the common *Clione*, which resembles a miniature slug with wing extensions. As an adult it may reach an inch in length but is usually smaller. *Clione* is pale gray in color, with shades of delicate salmon pink on the wings and tentacles.

A deep-sea species, *Diacra trispinosa,* is a rich chestnut brown color. The shell-less forms live mostly in the temperate zones, whereas their shelled cousins are tropical species. This is why the pteropod ooze, one of the three main kinds of plankton shell deposits on the sea floor, is found mostly near the equator. These deposits of sea-butterfly shells are not as extensive as those formed by the diatoms or the dinoflagellates, but are often encountered around coral islands and on isolated sea mounts.

THE SALPS

The next group of planktonic creatures, the salps, are usually regarded as being more closely related to the vertebrates (animals with backbones, including ourselves) than to the invertebrate creatures we have discussed so far. This kinship is apparent only to the professional zoologist, who sees in the larval stages of some of these forms a tadpolelike creature with a notochord, regarded commonly as the precursor of the spine. To the ordinary person few things on earth look less like fish or birds or other backboned animals than the sea squirts (clinging like inert lumps of brownish or gray-black clay to pilings or rocks) or their relatives the salps, transparent tube-like plankters. Salps look like little barrels, 1 to 6 inches long. There is an opening in each end through which water flows from front to back, and eight colorless or bluish bands circle the salp like the hoops of a barrel. The bands are muscle fibers that cause the barrel to pulsate, opening and closing the fore and aft apertures in proper sequence and thus passing a stream of water through the animal. This flow of water has three functions: it brings food (diatoms and other small plants and animals) into the body of the salp to be filtered out by a curtain of mucus; it supplies oxygen; and it propels the salp slowly through the water. Salps keep mostly to warm waters, and their capture in the northern Atlantic Ocean indicates the presence of the Gulf Stream.

Related to the salps is the interesting colonial form, *Pyrosoma* ("fire body") (Figure 5-4), a genus of small salplike creatures that join forces to form an elongated luminescent cylinder. The mouths of the individuals of the colony point outward, drawing water in and passing it out through a common orifice. Colonies range from the size of a thimble to 4 feet or more in length. The luminescence shows up when the animals are touched, each individual reacting on its own. When a finger or other object is brushed across the surface of the colony, a glowing line marks its path. Members of the staff of the *Challenger* Expedition amused themselves by writing their names on the flanks of *Pyrosoma* colonies and watching the glow persist for several seconds. A planktonic shrimp, *Phronima,* has a grimmer use for *Pyrosoma* colonies. It eats the animals from the surface of the colony and then moves into the transparent shell. It may be then seen, with a brood of young, navigating its stolen home through the water.

Salpa democratica

Pyrosoma

FIGURE 5 – 4 . *Salp and a* Pyrosoma *colony. In their larval stages some salps have a rudimentary spine called a notochord. An adult salp has a mouth at each end of its body, and water is pumped through by encircling muscle bands. A* Pyrosoma *consists of colonial animals that join to form a cylinder. The individuals light up if touched.*

N. B. Marshall, *Aspects of Deep Sea Biology*

THE PLANKTONIC CRUSTACEANS

After the true "forage" species—the undoubted plants like the diatoms and the plant-or-animal flagellates—one group of planktonic animals plays so large a role at the base of the food-chain pyramid that one might wonder whether any other group deserves to be called important at all. These are the crustaceans. The group includes the familiar lobsters, crabs, and shrimps, all of which pour their young into the plankton, but we will concern ourselves in this chapter with only one group of the crustaceans, the copepods. Most of these are permanently in the plankton, and not only in the larval stages, like nearly all the other members of the group.

The crustaceans are members of the same great group of animals as the insects and outdo them in their most famous attribute: their multitude. Says the *Encyclopaedia Britannica*: "Entomologists have described more than 625,000 kinds of insects, and estimates of the total living forms vary from 2,000,000 to 4,000,000. These are kinds or species and are not estimates of individuals. No scientists familiar with insects have attempted to estimate individual numbers beyond acres or a few square miles in extent. Figures soon get beyond the grasp of the mind."

There may not be as many different kinds of crustaceans as there are insects, but there certainly are vastly more individuals. Sir Alister Hardy thinks it likely that one kind of crustacean alone—the copepods—exceeds in abundance all other multicelled animals combined. This means that there are more copepods than there are insects, plus the herrings (so numerous that half a billion individuals are landed in the United Kingdom alone each year, leaving more billions for other fishermen to catch and still more billions to swim in the sea), plus all the other kinds of fishes, plus all the oysters and snails and shrimps, all the snakes and bears and apes, all the humans— *all* the other animals on earth with more than one cell to their bodies.

Off Bermuda over 91 per cent of the animals captured in the plankton have been found to be crustaceans, and over 75 per cent were copepods. The next most numerous group, the arrowworms, made up only 3.1 per cent, leaving to all the other animals combined only about 5 per cent of the numbers. Of course, not all plankton hauls contain 75 per cent copepods—some have even more and some have fewer—but it would be hard to make a haul anywhere

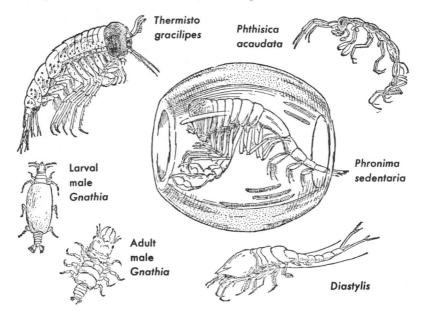

Thermisto gracilipes

Phthisica acaudata

Larval male Gnathia

Phronima sedentaria

Adult male Gnathia

Diastylis

FIGURE 5 – 5 . *Odd crustaceans of the plankton.* Phronima sedentaria, *a little shrimp, literally eats the* Pyrosoma *out of house and home—and moves in.*

Sir Alister Hardy, *The Open Sea: Its Natural History.*
Part 1. The World of Plankton

in the ocean without catching a few. Copepods figure prominently in the photograph of plankton in Figures 5-2 and 15-2.

The comparison of the copepods with the insects is fully valid insofar as their structure is concerned, for both are members of the phylum Arthropoda ("jointed legs"). But as a vital link in the food chain of the sea, the copepods are more like the pigs and cattle of the land than they are like the insects, for the copepods are grazers, devouring the plants of the sea and converting their substance into animal tissue, to be eaten in turn by fish and sea mammals and birds. These grazers of the sea are diminutive by comparison with those of the land because the grass of the sea—the diatoms and the flagellates—comes in such small bundles.

The commonest of the copepods, *Calanus finmarchicus* (called "brit" by the fishermen), is one of the most important animals in the world despite its tiny size—about that of a grain of rice—and its name may someday be a household word as increasing knowledge of marine biology expands our use of marine resources.

At about three eighths of an inch, *Calanus* is of average size among the copepods. The smallest copepods are about as big as the dot made by the point of a sharp pencil—just visible without magnification—and the largest are about half an inch long, but few species attain this size. Into its small size the copepod packs an intricate anatomy. The body consists of three parts, but the head and thorax appear as one long, oval shape. The tail or abdomen is shorter and narrower, and ends in a fork. Several pairs of appendages are attached to the head, the most prominent of which are the great antennae, often exceeding the length of the body; these stick out like caricatures of a Victorian handle-bar mustache. When it is not swimming, the copepod hangs head uppermost in the water, suspended partly by the antennae and the thick fringes of hairs attached to them. In motion the copepod uses its mustaches to help row it along. However, movement is chiefly the responsibility of other appendages on the head and of the five pairs of paddlelike legs, flattened and joined at their base, that scull the animal along in a series of jerky hops. In a jar of freshly caught plankton the copepods can be picked out quickly by their agitated dance in the water.

Like so many of the surface plankton, the copepods are nearly transparent. Specks of color show in the body of many species, with reds and yellows predominating. Often there is a faint iridescence, the colors shifting and changing with every turn of the animal. These spots of color are for the most part oil droplets. In times of rich forage, the oil stores may be large in individual copepods, and the numbers of animals may be so enormous that the sea is colored. At such times the fishermen refer to the copepods as "red feed" and expect good catches of herring. Some species, especially those living in the warm waters of the tropics, are more strikingly colored than temperate-water species, with scarlet or bronze fringes on their antennae and other parts of the body. The color of the appendages frequently contrasts sportily with the color of the body: some copepods have blue bodies with red or brown spots, and wave orange or reddish plumes on their appendages; others have orange or brown bodies with blue and green plumes. Deep-water species of copepods assume vivid colors, such as the clear rose-red of *Euchaeta*. Some very deep species, like *Candacia,* are black.

Copepods occur in all ocean waters but are most abundant in the colder areas of the temperate zones. Like most other creatures of the sea, they become more thinly scattered the deeper the water, but this reduction in numbers does not take place gradually. Instead, there appear to be two centers of abundance, one near the surface and another between 1,500 and 4,000 feet. The young live at shallower depths than the adults of the same species; for example, individuals of the genus *Eudeuchaeta* live at about 300 feet as juveniles

FIGURE 5 – 6. *Three deep-sea copepods, which are a vital link in the food chain. Greedier than pigs, they eat half their own weight in a day and are in turn eaten by fish, sea mammals, and birds.*

N. B. Marshall, *Aspects of Deep Sea Biology*

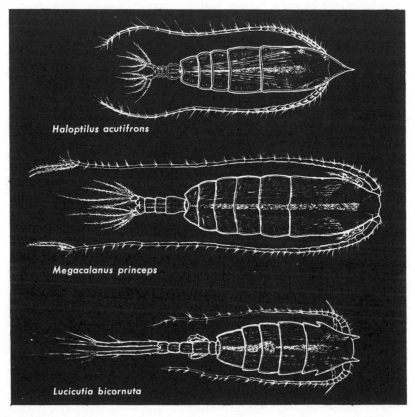

and about three times as deep as adults; *Bradycalanus*, an abyssal species, lives at 4,600 feet in its early life and at 7,700 feet as an adult. Three other species of deep-sea copepods are shown in Figure 5-6.

In foraging for the little plants of the sea, most copepods utilize an efficient filtering apparatus composed of feathery, curved setae, or bristles, borne on appendages that curve over the mouth. The movement of these and other appendages sets up water currents that move the copepod slowly forward through the water, and at the same time creates swirls of water that bring diatoms, coccolitho- phores, dinoflagellates, and other forage creatures into the path of the filter. Copepods are big eaters, consuming half their own weight of diatoms in a day. No cow can eat that much, nor can even that symbol of gluttony, the pig. Other species of copepods are carnivorous.

The next link in the food chain is very frequently the herring. Copepods—*Calanus* in particular—are the principal food of this fish, which pursues and snaps them up in great numbers; as many as 60,000 may be in a herring's belly at a time. Many other fish eat copepods, and this includes the largest of all, the whale sharks (attaining a size of 50 feet and several tons) and the basking sharks (40 feet or more). The latter open their great mouths and swim slowly through the water, straining out the copepods with a fine network of tissue that covers the gill openings. Whales use this filtering method too. The sei whale is especially fond of copepods and has a filtering apparatus fine enough to capture these small creatures. Among the birds that eat copepods, the stormy petrels might be mentioned. These birds of the open ocean exist largely on plankton, scooping it from the water as they fly over the surface.

In this chapter we have examined the meadows of the sea and the creatures that crop them. We have described, too, some of the animals on the next few levels of the food chain above the plants that form its base. To land animals like ourselves, familiar with the rooted grass and cereal crops of the soil, the invisible, floating herb- age of the sea seems strange. Strange, too, are the herbivores of the sea: salps and copepods and other creatures so unlike the cows, sheep, and pigs of our farms. In the next several chapters we will describe some of the other creatures of the ocean that depend directly or indirectly on the floating pastures, and are as curious and unlikely in their way as the plants and animals of the plankton.

6   *Stick-in-the-Muds*

It was once considered absurd to expect life to exist in the deep sea —cold, perpetually dark, and subject to crushing pressures. Over the years, however, as collecting apparatus has probed deeper and deeper, it has become clear that no depth is without its animals; none even lacks plants, for there are bacteria and diatoms even in the abyss.

A little reflection on the unpromising marginal environments that land animals have conquered should have prepared biologists to find that the deep sea was long ago colonized by marine creatures. The kangaroo rat lives in the desert of Death Valley, where ground temperatures may soar to a sizzling 190° F. in summer and plunge far below freezing in winter. What deep-sea animal is likely to toil harder for an existence than an arctic snowshoe rabbit or a desert lizard? The stress and teem of burgeoning populations have driven many animals, including men, from the kindlier areas of the earth; the same pressures have driven representatives of all the sea creatures from the lighted, richly endowed shallows of the sea. Like some hermits of the land, human and otherwise, the creatures that seek the lonelier areas of the deep sea are often grotesque in form. One might be anthropocentric and think of these animals as having fled to the depths because of their appearance. The fact is, of course, that each curious adaptation is of specific help to the animal in facing its environment; these creatures appear grotesque to us merely because they are unfamiliar.

It is little wonder that we have been slow to gather specimens from the depths of the ocean: the difficulties of catching them are formidable. It has been said that man's fumbling essays at sampling the bottom of the ocean are comparable to those of a blind giant riding a cloud 2 miles above the earth and dragging a big butterfly net over the surface of the land. He might capture an occasional bird or insect in mid-air and scrape up a few slow-footed muskrats or dogs. He might even catch a man once in a while, but his chances of getting a fair sample of the living creatures of the earth would be slight until he had made a very great number of drags over a very wide area. It is certain that we are far from having collected examples of all the animals of the deep sea, and even farther from an accurate measure of their comparative abundance.

But let us examine those that have so far been encountered. For convenience we will group the creatures in a number of loose categories based on their structures and habitats. Beginning with the invertebrate animals (those without backbones) living on or close to the deep-sea floor, we will go on to the invertebrates that swim in mid-water, the fishes living on or near the bottom of the deep sea, and then the fishes of the mid-water depths. Such divisions are of course arbitrary, for there is much overlapping. Some of the creatures sometimes live close to the deep-sea floor and at other times swim far from the bottom. Then, too, different branches of the same large group of animals—for example, the crustaceans, whose members include the shrimps and crabs—may occupy different habitats, so we will be dealing with the same group more than once.

Since all of these creatures derive from shallow-sea animals, it will be necessary in describing the principal inhabitants of the abyss to tell something of the form and behavior of the more familiar shallow-water representatives of some groups.

The ancestors of the deep-sea animals lived in shallow seas at the edges of continents eons ago. Because a few ancient types of animals are found in the deep sea, it has sometimes been suggested that life appeared there first and then moved on to other habitats. But this is clearly far from the mark. The uncompromisingly harsh environment of the deep sea could never have cradled the fragile cluster of molecules that held the first spark of life. Only the warm, dimly lighted, dilute broth of ancient shallow seas could have nurtured these primeval living things until the centuries, the millennia, the eons of time, could mold and change them into the vast assemblage of plants and animals that eventually swarmed in the ocean's margins. In time, as the press of population made emigration an easier way of survival for some individuals than competition in the middle of the crowd, plants and then animals moved onto the land. Later, others moved into the deep sea.

THE BEGINNINGS

The origin of life has been one of the most intriguing and hotly contested of human questions. In centuries past, its answer seemed easier than now: religion taught that creatures sprang full blown into the world, placed there by an omnipotent divinity. For a long time there was no compelling reason to question this dictum, but as knowledge accumulated concerning the mutability of living things, the first easy explanation lost certainty for many people. Then there was substituted the theory of spontaneous generation. This seemed a more rational viewpoint, since it apparently tallied with observation. The horsehair worm appeared out of nowhere in the rain barrel; maggots sprang from decaying meat; worms appeared from mud; mice issued from piles of refuse: life came from nonlife, and one had better believe one's observations, for this was the dictate of the new force, Science. The belief in spontaneous generation was made easier for most men because many of the most respected and able minds accepted it. Aristotle, one of the giant intellects of the world, believed in spontaneous generation; so did William Harvey, whose epochal descriptions of the circulation of the blood did so much to advance medicine; so did Sir Isaac Newton, the British physicist and philosopher who formulated and proved the law of gravitation; so did René Descartes, the French mathematician who helped to lay the foundation of the modern scientific method.

But then careful experimentation began to replace too hasty observation, and the belief in spontaneous generation was gradually eroded away. In the seventeenth century, Francesco Redi, an Italian, showed that meat protected from flies never developed maggots. In the eighteenth century, another Italian, Lazzaro Spallanzani, boiled nutritive broth sealed from the air and showed that this broth never spoiled. Spallanzani's experiments were attacked with the claim that the boiling had destroyed the capacity of the broth and the air in the flask to sustain life. He could show that this was not true, since when he broke the seals of the flasks, microorganisms entered and multiplied and the broth rotted. But he could devise no experiments to show that the air had not been vitiated; this remained for the brilliant experiments of Louis Pasteur in the latter part of the nineteenth

[92]

century. One of these experiments was to allow air to pass freely in and out of the flask through a long S-shaped tube that prevented dust and microorganisms from entering and contaminating the nutrient mixture. After boiling, the broth remained uncorrupted. With this and a long series of other convincing experiments, Pasteur finally destroyed the theory of spontaneous generation.

Science had thus thrust aside the idea of divine creation of life in its present forms and had then brought crashing down an apparently rational alternative. There seemed to be no third way; nothing was left. But now we find science returning to a version of spontaneous generation. The modern theory was first outlined by the Soviet biologist A. I. Oparin in a book entitled *Origin of Life,* published in 1936. The modern theory, unlike the one destroyed by Redi and Pasteur, does not assume that animals and plants were created full blown in their present forms. It assumes instead that the first living thing, neither plant nor animal, was exceedingly simple, possessing only those indispensable gifts of all organisms: the power to reproduce itself and the power to respond to the press of environment by adapting. Furthermore, the modern theory recognizes that creation of life as postulated could not take place in the world as it is constituted now, but that conditions on the young earth were once favorable for this solemn and portentous accident.

The substances needed to create living organism are water, carbon compounds, and salts containing sulfur, nitrogen, phosphorus, potassium, and calcium. As it happens, these are precisely the substances found in the sea. The right conditions included a rich solution of these substances in constant circulation to permit, eventually, the chance encounter that in one dramatic moment produced a living substance.

But more special conditions were necessary, this time to avoid the opposite of spontaneous generation: spontaneous degeneration. Any delicately poised aggregation of molecules created in the shallow seas today would have a brief existence. If bacteria did not destroy it, oxygen would. But in the ancient seas there were no bacteria, since life had not yet been created. And there was very little oxygen, since the air contained virtually none. Most of the earth's oxygen was bound in the water molecule and in metal oxides, only to be released later by living things. The creation of life was possible, then, because no life existed. Even in the absence of bacteria and oxygen the chance of the new living aggregation surviving seems

unbelievably small. Add this tiny chance to the still more unbelievably small chance of its formation in the first place—the im-
probability of the right substances in the right proportions being
present under the right conditions—and there is an exceedingly
slight likelihood of the event occurring. But regardless of the in-
finitesimally small likelihood of this occurrence, out of billions of
billions of chances it only had to take place once. And time was on
its side—billions of years of time.

So here we sit, all of us, living proof that the improbable event
did occur. Before our presence on land was possible, however, the
primal organism had to evolve, include chlorophyll in its make-up,
and create more organic substance. Animals had to differentiate
from plants; one-celled creatures had to clump into aggregations;
some aggregations had to form tissues and then organs to produce the
multicelled creatures. Life had to diversify and ramify until the
shallow regions of the sea sustained a rich variety and a teeming
wealth of creatures. Sponges were there, and jellyfish, worms, starfish, clams, and crabs. Five hundred million years ago, in the early
Cambrian Period, the fossil record starts, when algae left impressions
in the mud and animals had skeletons hard enough to be preserved
in the rock.

But there was still no life outside the water. The dry land presented a barren, harsh aspect, noiseless except for the wind and rain,
bleak and uninhabited. There were no plants and therefore no soil.
Before the animals could come, the plants had to precede them to
prepare a place. This happened about 350 million years ago, in the
Silurian Period. An ancient relative of the scorpion, a jointed-legged
arthropod, crept out of a tide pool left by the receding sea and found
existence out of water tenable—not merely tenable, indeed, but
offering advantages in food supply and the absence of enemies. Later,
other creatures, including fishes that had learned over the millennia
to breathe air when they were stranded in the mud by retreating
tides, discovered the joys of land existence, and the migration from
the sea gained momentum.

The movement in the opposite direction, from the shallow edges
of the sea to the deep sea, did not take place until much later—
millions of years later—than the migration to the land. Pushed by
burgeoning numbers to the edges of the areas where they were born,
some pioneers among these animals found existence easier in the
open waters of the ocean or on the bottom of the sea in deeper

water. The unrelenting pressure of population pushed them farther and farther from their original homes. Gradually, with creeping slowness, the whole of the ocean was inhabited.

The deep sea was populated from two directions. Some of its creatures crept down the slopes into the abyss, clinging to the bottom all the while. These are the demersal animals of the deep-sea floor—the stick-in-the-muds. The others struck off boldly into mid-water, to occupy the great reaches of the sea between the bottom and the surface. These are the pelagic animals—the swimmers.

This chapter will deal with the animals without backbones that live on the ocean bottom—attached to it, burrowing in it, or moving only short distances from it.

SPONGES

The first to be considered are creatures low on the evolutionary scale, the sponges. The common conception of a sponge is a curiously textured piece of material full of holes, more like a plant than an animal. This is only the skeleton of the sponge, however, and of a relatively uncommon kind at that. It took mankind a long time to accept the fact that sponges are animals—despite Aristotle's statement that they were; like many another truth discovered by that exceedingly sagacious observer, it was denied by everyone for centuries thereafter. John Ellis appears to be the first after Aristotle to rediscover the truth; in 1765 he watched living sponges closely, noting the water currents passing through them and the slow but unmistakable movements of parts of their bodies.

It is easy to see how, without this kind of careful observation, sponges could be misclassified as plants. They sit firmly rooted to the sea bottom. They usually have irregular shapes, encrusting the rocks like moss or growing up in various haphazard forms, or in regular shapes that are occasionally strikingly beautiful: vases, urns, baskets, or cups. In size they range from a tenth of an inch to 6 feet across. The usual colors are drab duns or grays, but some sponges are brilliant coral red, scarlet, yellow, blue, violet, fawn brown, or snow white; some kinds of commercial sponges are a glistening jet black when alive.

The sponges are the most primitive of the common multicelled animals. They are loose aggregations of protozoa-like cells; the cells of the internal cavity closely resemble certain of the flagellates. The sponge does not possess separate organs to carry on the various processes of living. There is no digestive system, the individual cells absorbing their own nutriment; no excretory system, the cells throwing off their wastes, which are swept out of the sponge by the water current flowing through its interior; no nervous system; no appendages.

The vital being of the sponge is possible because of the water current flowing through it. This is set up by the beating of myriads of tiny whips—the flagella—of the cells lining the inner cavity of the sponge. These beat constantly (but not together), sweeping water into the body through thousands of tiny pores in the outer layer. The water then passes through a complex of canals in the body, bathing the cells, bringing food and oxygen to them, and carrying off waste. Eventually the water collects in the center cavity of the sponge and is expelled through a sizable hole, the osculum. The food is necessarily very small, since it must pass through the pores. It consists of bacteria, the smallest diatoms and flagellates, plant spores, eggs and sperms of marine creatures, and bits of organic material.

Between the thin outer layer of cells forming the skin and the inner layer of whip-bearing cells there is a layer of jellylike substance supported by a skeleton. The skeleton is sometimes of a fibrous, horny material called spongin (as in the case of the familiar bath sponge of commerce) or more often of complicated spicules of calcium carbonate or of glasslike silicon. Sometimes spongin and spicules occur in the same sponge. The spicules are of a bewildering variety, ranging from straight needles keenly sharpened at the ends to tridents, snowflakelike patterns, and stars with three, four, or six points. The rays of the stars may be straight or end in knobs, mushrooms, whisk brooms, or a great variety of graceful forms. These needlelike, brittle skeletons make most sponges harsh and sharp— quite unsuitable for bathing or washing the car. The glassy and brittle siliceous skeletons have tiny spikes that can penetrate fingers painfully.

Like plants again, the sponges can reproduce by the simple method of budding. But they also reproduce sexually, releasing into the water sperms and eggs that combine to form a swimming larva that comes to rest in time on some firm surface and grows into another sponge. Sponges possess very considerable powers of regeneration. Cut in two, a sponge will grow into two new individuals; in fact, a sponge cut into a large number of separate pieces will make as many new sponges. In some experiments sponges have been squeezed through fine silk; the dissociated cells, placed in sea water, reunited themselves and formed a new animal. In the Bahamas and elsewhere, commercial varieties have been cultured by cutting large sponges into pieces about 2 by 4 by ½ inches and fastening these to a cement disc. Each piece grows into a well-shaped individual big enough to market in about 4 years. Difficulties in protecting sponge "farms" from pirates, together with the ravages of a fungus disease, have halted the culture of sponges in the Bahamas and British Honduras.

Sponges are in little danger of their lives unless man covets them for their porous skeletons. They are unattractive to predators because of their spiky and sharp spicules and their unpleasant odor and taste. In addition, some produce massive quantities of slime; for example, a kind called *Desmascedon* will make the water in a bucket thick and slimy with its mucus. Consequently, sponges found in the stomachs of fish probably got there by accident as the fish greedily snatched at some more delectable animal attached to the sponge or living inside it. Many sponges shelter an amazing complement of boarders: shrimps, crabs, worms, clams, and even fish. The immense loggerhead sponges of Florida

FIGURE 6 – 1. *Watering-can sponge, or Venus' flower-basket. About 5¼ inches high, its scientific name is* Euplectella aspergillum. *Often the permanent home of a pair of small crustaceans, and thus the symbol of permanent and happy marriage, it is sometimes given to newlyweds in Japan.*
Valdivia Expedition

FIGURE 6-2. *Glass sponge* Pheronema carpenteri, *which is anchored to the mud bottom by strong glass ropes.*

N. B. Marshall, *Aspects of Deep Sea Biology*

and the West Indies, growing to the size of big washtubs, always harbor a great variety of creatures. In one sponge Dr. A. S. Pearse counted 17,128 animals, mostly little shrimps but also other creatures up to fish 5 inches long. It seems unlikely that the sponge derives any benefit from this arrangement, but the other animals obtain shelter as well as food brought in by the currents of the sponge.

Some of the inhabitants of sponges are permanent tenants. One of the most attractive of the deep-sea sponges, the watering-can sponge or Venus's flower-basket, found at depths of 650 to 1,000 feet, usually is host to a pair of crustaceans. The sponge is formed of an intricate network of lacy glass spicules, and the little crustaceans squeeze through the interstices of this network when they are young. As they grow they become unable to get out again, and spend the rest of their lives inside the sponge. The permanent marriage of the two crustaceans inside the sponge so struck the imagination of the Japanese that they often present newlyweds in that country with a Venus's flower-basket and its pair of inmates as a wedding gift symbolic of permanent wedded happiness.

The sponges that occur in the deep sea, some to depths of 3½ miles, are more likely to be drab in color than the shallow-water species. They are also more likely to be regular in shape, rather than having the amorphous forms of those living where currents are strong. The absence of any but feeble currents often has another effect: the deep-sea sponges may be bent so as to point down-current. In the absence of brisk water currents to carry away waste products, the sponge would foul itself unless the osculum were located so as to expel wastes into the small currents that do exist.

Most deep-sea sponges have silica skeletons built up of complexly interlacing needles and hairs of spun glass. A net laden with sponges from the deep sea looks "as though it had fouled a sunken haystack—with here and there a bird's nest attached." The deep-sea glass sponge *Pheronema* is an oval basket woven of fine needles with a great cushion of matted glass hair at one end. Sponges living on mud bottoms develop stout ropes of glass that anchor the animal in the bottom. The Japanese unhook the glass-root sponges from the bottom and sell the anchors for curios: they have been mistaken for the delicate work of skilled oriental artists. *Monorrhaphis chuni*, found in the Indian Ocean off Africa, is called the one-needle sponge, since its root is a single stout needle of glass that protrudes from the bottom of the sponge and presumably spikes it to the sea floor. *Geodia*, another deep-sea sponge of curious appearance, forms great rounded masses on rocks in deep water. These masses may be many feet across and are dead white in color, resembling the whitened bones of some prehistoric monster.

FLOWERS AND JELLY

On a branch above the sponges on the evolutionary tree rest the members of the phylum Coelenterata. The jellyfish are one kind of creature included in this large and diverse group of animals. Corals, sea anemones, and sea fans are others. At first it may seem unjustified to group together such different-appearing creatures as the gossamer jellyfish, the corals with heavy skeletons, flowerlike sea anemones, and sea fans, whose purple or pink lacework makes them look far more like plants than animals. But all these creatures have in common a simple, basic body plan: a hollow pouch opening at one end, the margin of the opening being fringed with tentacles.

Aristotle thought that the coelenterates were poised on the edge of the plant and animal kingdoms, sharing some of the characteristics of each. It was not until the eighteenth century that their identity as animals was finally established. Then it was recognized that the tiny zooids that are the individuals of the colonial corals and sea fans are essentially similar in structure to the bell-shaped jellyfishes and the brightly colored sea anemones.

The true identity of sea anemones as animals is most clearly revealed when food comes within reach. Shrimps or small fishes are in danger of being engulfed if they approach close enough to be seized by the tentacles surrounding the anemone's mouth. Abundant in the shallow areas of the sea where they are attached to rocks, these wide-ranging polyps lend variety and color to both tropical coral reefs and the rocky areas of northern oceans. Sea anemones are also among the commonest deep-sea invertebrate animals. Underwater photographs have shown them to be the most abundant creature in some areas off the east coast of the United States in depths of 2,300 feet. Anton Bruun, the late famed leader of the Danish *Galathea* deep-sea expedition, described the excitement of trawling in the very deepest parts of the abyss and realizing that his net had brought living sea anemones from depths of over 6¼ miles.

Other polyps are also common in the deep sea. Sea pens and gorgonians, or sea fans, are two kinds of colonial animals in this group. Instead of existing as single individuals, these humble creatures share their existence with a great many others of their kind, forming branching colonies that wave in the water in a most unanimallike manner. The colonies arise when some pioneering individual settles on a rock as a young larva. Growing to adulthood, it may bud off progeny that, instead of leaving home, stay close to the parents—so close in fact that the two are joined, sharing a common skeleton and usually gaining something in ability to get food, in protection, and perhaps in other ways. Over a period of time the original animal may build a considerable population of offspring, representing many generations. Corals are the best known of these colonial animals, and some of them, too, inhabit the deep sea.

Sea pens and sea fans, common in shallow waters, are also sometimes caught in nets dragging the bottom of the abyss. A sea pen has a long stalk as a central supporting axis to which are attached numerous fleshy colonies, yellow, red, or brown in color and as much as 6 feet in length. The common name comes from the striking resemblance of some species of the group to old-fashioned quill pens.

Sea fans are also called horny corals. A "stem" supports a flattened network of skeleton to which are attached very large numbers of individuals of the colony. Sometimes the network is like an intricate and delicate lace, brightly or softly colored. To the layman the sea fans are among the best known of this group, since they may be seen tacked to the walls in seaside restaurants in Florida or in collections of tropical shells. Unfortunately, in these circumstances they look sad and bedraggled, their lovely colors faded, their fans cracked and broken. In life they form fairy forests, waving gently in the currents, showing pastels of pink, violet, and red, and sheltering such interesting tenants as shrimps and beautiful little pink and white snails called flamingo's tongues.

Some of the sea pens are intensely luminescent. One species, *Pennatula phosphorea*, from the mud off Scotland and Norway, lights up in the darkness when it is touched. From the point of contact, the light spreads from branch to branch until the whole colony is aglow. In the big sea pen *Funiculina quadrangularis* the main stem is especially sensitive, and, if it is touched, light flickers up and down it like a flame. Waving fields of sea pens in the deep-sea mud, where the animals are often abundant, must present an eerie and astonishing appearance at times. One sea pen of the depths, with the graceful name of *Umbellula*, shines with a delicate bluish light; others emit an entrancing pale violet luminescence. It is not hard to imagine the path of a fish being traced in brightness as it moves through a bed of these strange animals, setting aglow the stygian gloom of the sea bottom.

WORMS

The many worms of the deep-sea bottom constitute an important part of the food of many of the creatures dwelling there. The bristle-footed worms, relatives of the common earthworm of our gardens and bait cans, and of those that occur in great variety in shallow waters, include some of the most irritating animals alive: the silky bristle worms. These creatures, sometimes as long as a foot but usually smaller, have rows of bristles along the sides of the body, often white, red, or green in color. When the worm is touched, the bristles blossom out into tufts; they can sting fiercely if they penetrate the fingers of an unwary person, and the wounds may be painful for a considerable period because the bristles are like the spines of a cactus, sharp and difficult to remove.

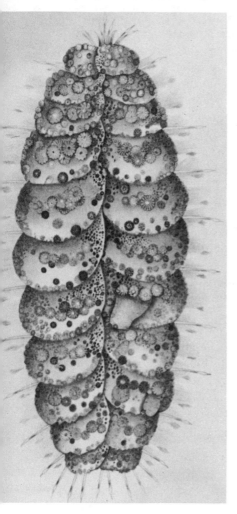

FIGURE 6−3. *Deep-sea bristle-footed worm* Euphione elizabethae, *from 900 feet of water south of the Cape of Good Hope. It is beautifully patterned with colored rosettes. The bristles, also colored, blossom out into tufts when touched.*

Challenger Expedition

Another marine worm of shallow waters is the palolo worm, whose reproductive habits are particularly intriguing. Rejoicing in the feminine name of *Eunice,* this worm lives in the coral reefs of the South Seas. Its spawning is controlled in a mysterious and remarkably precise way by the periodicity of the moon: spawning swarms appear at the third quarter of the moon in November, and natives of the islands always know exactly when to expect them. The precise relationship between spawning and the phase of the moon is not the end of the strange story. When zoologists first collected these animals, they were bewildered to discover that although the worms were alive and were swimming vigorously, they had no heads! It was some time before it was discovered that the worms, stirring in their hiding places in the coral reefs under the stimulus of maturity, cast off the hind part of their bodies. These broken remnants of the animals, containing the eggs and sperms for the next generation, swim to the surface of the sea, leaving the forepart below. The palolo worms are esteemed as food by the island peoples, and ritual feasting follows their annual appearance.

The deep-sea relatives of these worms have neither the tendency to sting nor the showy sexual behavior of the shallow-water members of the clan. Nonetheless, they are closely related to species living in the lighted areas of the sea and are probably derived from them.

SPINY AND OTHERWISE

Very few broad groups of animals belong exclusively to the sea. One that does is the phylum Echinodermata, containing the starfish and their allies: sea urchins, brittle stars, sea cucumbers, and sea lilies. The group name is from the Greeks, who called both the sea urchin and the hedgehog *echinos* for the armature of spines they carry. Not all the echinoderms are truly spiny-skinned; the sea cucumbers, for example, are extremely soft-skinned, their "spines" being merely microscopic bits of calcareous material in the integument. The outstanding common characteristic of the echinoderms is their radial symmetry, exemplified by generally star-shaped bodies. This is perfectly obvious in the starfish and some others of the group, but much less so in the sea cucumbers. Radial symmetry here seems to be a secondary adaptation from bilateral symmetry, in which the two sides of the animal are roughly mirror images, since the larvae have this more common shape. The change was presumably a result of the sedentary habits of these animals, which are either attached permanently to the bottom of the sea or move sluggishly over it. The most efficient adaptation to this mode of life is to do without a "head" end and present all edges of the body equally to the outside world. As an example, the starfish moves indifferently in the direction that any of its arms points.

The starfish has numerous slender "tube feet" extending from the underside of the arms. The tube feet are equipped with tiny suction discs that can grip tightly to a rock or an oyster shell. The starfish moves by attaching the sucking discs of its tube feet to a rock, elevating the ends of the arms slightly for the purpose. Then the tube feet are retracted, pulling the animal slowly along. A starfish can even right itself with the tube feet after being turned on its back: it extends the feet greatly, attaching them to a surface on one side of its body, and then by slow stages peels itself over into a normal position. The suckers of the tube feet can also seize small prey, including such active animals as small shrimps and even little fish.

Some echinoderms are notable for their tremendous depth ranges. For example, the scarlet sea star *Henricia sanguinolenta* can be found from the shallows out to the abyss; the brittle star *Ophiacantha bidentata* ranges from about 15 to 14,600 feet. The most familiar of the five classes of echinoderms is the starfish, so well known as to be practically a symbol of the seashore. The most familiar types are five-armed and red in color. Some species have six arms and others as many as fifty, and a few, particularly in the deep sea, are five-sided but lack arms. Besides the red species there are others in various vivid colors: yellow, blue, purple. No starfish is exceptionally small; the biggest, *Pycnopodia heliathoides,* attains about 32 inches.

The sea stars fed chiefly on shellfish, shrimps and other crustaceans, and tube worms. In attacking an oyster the starfish envelops it and applies pulling force against the two sides of the shell with its tube feet. Anyone who has tried to pry a live oyster apart with his fingers knows how tightly the two valves of the shell are held together. But the muscle of the oyster is no match for the persistent strength of the starfish. The unrelenting pull eventually tires the shellfish, and the shell gapes slightly. Then it is thought that the starfish injects a poisonous substance that makes the oyster open all

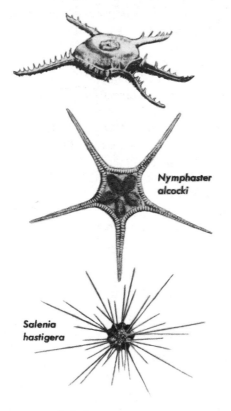

Nymphaster
alcocki

Salenia
hastigera

FIGURE 6–4. *Two abyssal starfishes* (Styrcaster *and* Nymphaster alcocki) *and a deep-sea urchin* (Salenia hastigera). *Many deep-sea starfishes search for food by running over the mud on the tips of their arms.*

N. B. Marshall, *Aspects of Deep Sea Biology*

the way, and the battle is lost. At this point a remarkable ability of some starfishes comes into use: it everts its stomach to surround the edible part of the oyster and digests it outside its own body. The inside-out stomach produces mucus and digestive enzymes that reduce the oyster flesh to fluid, whereupon the stomach is withdrawn again into its owner. In captivity, starfish have been observed to devour fifty small clams in 6 days. They do immense damage to clam and oyster beds, and commercial growers spend large amounts of money and energy to "mop up" the starfish or to destroy them in other ways. "Mopping up" is not a figure of speech in this case, since huge shaggy mops are actually dragged over the oyster beds; the pests become entangled and are dragged to the surface and destroyed. Before they learned better, some oystermen wasted their efforts. They cut up the starfish they had captured with their mops and threw them overboard. Since starfish have great powers of regeneration and can lose considerable portions of their bodies and grow them back again, most if not all of the mutilated pests lived to eat oysters again.

Besides eating useful animals, starfish may do harm indirectly by devouring the food of fish. In one area of the North Sea it was once calculated that starfish ate about as much food in a year as was required to support the plaice caught there. There are few natural predators to keep starfish populations in check, although a few of them are eaten by other starfish.

Starfish reproduce sexually, the eggs and sperm being cast into the water, and the resulting larvae swim freely for a time before settling down to existence on the bottom. In this scheme of things parental care is not the rule, so the case of *Henricia*, the scarlet sea star mentioned earlier, is worth comment. This creature is a careful parent, protecting its young during the early stages. To spawn it stands on the tips of its five arms, thereby creating a cavity under its body into which it releases its eggs. For the three weeks or so during which the eggs develop, the starfish stands guard over them. Only after they hatch and swim off as larvae does she begin to feed again.

Starfishes of the deep sea differ in some important respects from many of the shallow-water representatives of the group. Since starfish locate their food principally by touch, their success in finding food is related to the amount of ground they cover while hunting. In the shallows, where food is abundant, they make out well enough by averaging only about 15 feet per hour, with the fastest species doing perhaps twice that. In the deep sea, on the other hand, food is sparser and more ground must be covered if starfishes there are to survive. Many abyssal species are believed to be capable of speeds of 100 feet per hour; they achieve this by actually running over the bottom on the tips of their arms instead of dragging themselves with their tube feet. Many of the deep-sea species, indeed, have lost their sucking discs—which are in any case useless in fine sand or mud where there is no hard surface to cling to. Deep-sea species also lack the strange power to evert their stomachs. This skill, so useful in helping a toothless creature devour an animal too big to stuff into its mouth, has been lost in the deeper species, which find few prey of such size.

The second of the five classes of echinoderms are the brittle stars. Their bodies consist of small round discs to which are attached five long, snaky arms. These arms, jointed and highly flexible, give the group its alternate name of serpent stars. Some members of the group have arms that divide and divide again, ending in fine tendrillike tips. These are called basket stars, and they are among the most delicate and curious of animals. One, called *Gorgonocephalus,* has such a complicated mass of repeatedly subdivided arms that it forms a writhing mass resembling the head of the mythical Medusa and her sister Gorgons, whose hair consisted of snakes.

The brittle stars are creatures of the night, hiding under stones or seaweed during the day. For the most part they are ungainly creatures, whipping their arms about and moving clumsily, albeit more rapidly than the starfish. Upon occasion they swim, using their arms much as a human does in moving through the water. Their limbs break off easily, but are almost as easily regenerated. Brittle stars must be among the most numerous of bottom-living creatures, if indeed they do not take the first place in this regard. They occur in enormous numbers in many places in the ocean. Deep-water photographs show masses of these brittle stars, their reptilian arms intertwined in a dense tangle so thick that it covers the bottom.

On one square meter of the bottom (an area a little larger than a square yard) 500 brittle stars have been counted. Because of their fragility, the numbers of brittle stars appearing in net hauls from the bottom of the sea do not represent their abundance. Many individuals are smashed and their fragments flushed through the meshes of the net. Yet trawls fishing in the very deep sea have brought up

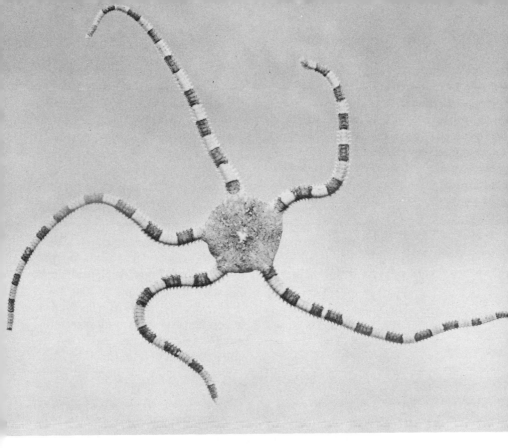

FIGURE 6–5. *Brittle star* (Ophiolepis elegans), *often called serpent star because of its jointed and very flexible arms. Although they break off easily, the arms quickly grow back.*

Marineland of Florida

several hundred brittle stars per haul. This surprising abundance of a single kind of animal rejects forever the idea that the deep-sea bottom is a desert.

Unlike the starfish, the brittle stars are eaten by many kinds of fishes. In turn they eat little crustaceans (shrimps and others), clams, and similar creatures; they also ingest a good deal of detritus from which they extract whatever nutrient is available. These animals detect food by its odor, and a piece of meat dropped near a crowd of brittle stars will cause a great stir and commotion among them.

Seen alive, the members of the next group of echinoderms, the sea urchins, do not seem to resemble the starfish or the serpent stars too closely. The five-armed pattern is revealed, however, when the spines covering the body drop off following the death of the animal; the "arms" curl under and help to form the rounded shell that encloses the body. The shape of a sea urchin is roughly like that of a tomato, with the stem underneath; this is the position of the mouth of the animal. In life the appearance is more like a pincushion, with long spines in almost constant motion over the surface. These are attached by ball and socket joints to the plates of the shell. Between the spines there are numerous three-jawed pincers whose function is to kill small larvae or other creatures that settle on the delicate skin of the animal, and to throw off larger foreign objects.

The sea urchin travels like the starfish, by means of long tube feet, but it also uses its spines and even its five teeth to help inch it along. The teeth are enclosed in a complicated apparatus surrounding the mouth. This structure, called "Aristotle's lantern" (for its shape and for the Greek philosopher, who first described it), enables the urchin to chew up the sea grasses that are its main food.

Some sea urchins use their spines to excavate lairs in hard sand or soft rock, where they lurk and in some cases create a real menace to bathers. One villainous species of the shallow waters of Florida and the West Indies is the black sea urchin *Diadema,* whose sharp spines may reach a length of 9 inches. They can easily penetrate the sole of a rubber shoe, and if they puncture the skin the wound may be exceedingly painful for days.

Some deep-sea species are said to be poisonous even to touch. One kind inflicts a terrible sting from glands at the base of the spines, and this is reported to have been fatal on occasion. Dr. Henning Lemche, one of the scientists aboard the *Galathea* during her round-the-world cruise in 1950–52, describes the great care the biologists took to handle certain deep-sea species with rubber gloves. These sea urchins were of a curious sort, lacking the hard shell of most of the group. They collapsed like pancakes when they were dredged up, only the buoyancy of the water having maintained their globular shape. The largest of the sea urchins occur in the deep; *Hygrosoma* is a foot thick, compared to shallow-water forms of only a few inches. In the deep sea the urchins, like the starfish and some other echinoderms, feed on the oozes covering the bottom, extracting organic material from them. Perhaps it is to facilitate plowing through the abyssal mud that some deep-sea urchins are curiously elongated, with narrow necks (Figure 6–4).

In shallow water, fish eat sea urchins if they can get past their spines. Even *Diadema,* with its formidable black armor, is the victim of some marine animals. The helmet shell, *Cassis,* a big snail of tropic seas, eats them, as do some of the stronger and braver fishes. Sea urchins are the primary food of sea otters off the coast of California. Other urchins are eaten by starfish, which surround smaller individuals with their everted stomachs; if the sea urchin is too big to permit this, the starfish forces its stomach down the mouth of the unfortunate victim, digesting it from the inside.

In the United States, where abundance has made people finicky about their food, the idea of cracking open a sea urchin and spooning out the raw eggs is looked on with horror. In other parts of the world they are a delicacy. In Mediterranean countries sea urchins are peddled in the streets. Vendors in Marseilles, Naples, and other ports supply vinegar with the urchins for the buyer to sprinkle over the eggs, which he sucks raw from the shell. In Barbados, urchins are called sea eggs and their roe is sold in the streets in little cones fashioned from leaves. Hotels serve them for breakfast cooked like scrambled eggs.

The name "echinoderm" is not very appropriate for the sea cucumbers, which have surfaces as far removed from spininess as could be conceived. They are usually markedly soft and yielding, with a slimy feel, and several of their habits are as repugnant as their appearance. First of all, one of the sea cucumber's methods of self-protection is to eject a good part of its viscera when disturbed. This presumably confuses and repels the would-be predator, and the sea urchin placidly grows a new set of interior plumbing. Another defense mechanism is the discharge of sticky white threads from the anus; these swell in the water and prevent a predator from seizing the sea cucumber. This trick gives the name of cotton spinner to the animal. The sea cucumber carries its respiratory gills inside the anus too, and, to cap the story, several small animals, including the little fish *Carapus,* commonly make their home inside this strange gill chamber.

Looking like a great slug or worm, the sea cucumber plows slowly through the bottom of the sea, swallowing mud and the little animals living in it. In the deep sea, bacteria in the mud are its principal nutrient. To avoid suffocation in the ooze, many deep-sea cucumbers have developed a large tail appendage that is apparently carried aloft like a periscope to bring oxygen to the burrowing animal as it plows through the mud.

There is plenty of mud in the deep sea, and sea cucumbers constitute a major segment of the deep-sea fauna. In the North Atlantic the *Michael Sars* caught them as deep as 10,000 feet in many localities. The *Galathea* caught a strange species, *Scotoplanes globosa,* in 21,850 feet of water in the Kermadec Trench in the South Pacific Ocean. They were about 3 inches long and looked rather like little pigs with strange protrusions growing from their backs. *Scotoplanes* and another sea cucumber, *Myriotrochus,* were captured by trawls in the Philippines Trench at 32,800 feet. A haul by the *Galathea* from 23,400 feet in the Sunda Trench south of Java yielded the greatest catch ever made at this depth, and it was principally sea cucumbers. Over 3,000 individuals of *Elpidia glacialis* were caught, as well as 144 larger sea cucumbers called *Periamma naresi.*

These bottom dwellers from such enormous depths were exciting enough, but the most notable sea-cucumber catches by the *Galathea* resulted in the identification of three swimming species.

FIGURE 6 – 6. *Abyssal sea cucumber* Scotoplanes globosa. *This curious creature has been caught in the Pacific Ocean—in the Kermadec Trench at 21,850 feet and in the Philippines Trench at 32,800 feet. It probably spends most of its time buried in the mud.*

Challenger Expedition

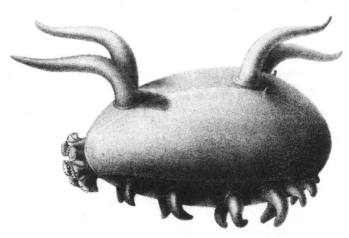

They are described in the next chapter, which deals with the swimmers of the deep sea.

Sea cucumbers are more commonly used as human food than are sea urchins. They are called trepang or *bêche-de-mer,* the French name being derived from a Portuguese phrase, *bicho da mar,* the sea worm. They are usually boiled, dried, and made into gelatinous soups that are gourmet luxuries among the Chinese, Japanese, Malayans, and other oriental peoples. The Japanese also eat the dried entrails. There is a fishery for sea cucumbers in California, primarily for the oriental trade.

The sea lilies, last of the echinoderms to be discussed here, should perhaps have been first, since most of them are very ancient and are represented only by fossils in rocks millions of years old. Those that remain are authentic living fossils.

The sea lily is strikingly flowerlike, some having a long stem supporting a "blossom" of considerable grace and beauty on its top (Figure 12-1). This blossom is in reality a cuplike disc corresponding to the round body of the sea urchin or the starfish; springing from it are five arms that divide immediately so that ten appendages wave on top of the disc. Sometimes the arms branch into feathery pinnules; the resemblance of these forms—which may be a brilliant yellow, orange, red, white, green, or brown—to flowers is striking. It is primarily the deep-sea forms that have the stalk, however; most members of the group living in the shallow seas usually lack any supporting stem as adults and crawl about by their basal tentacles. Until deep-sea expeditions began catching stalked sea lilies in considerable numbers they were great rarities, highly prized in zoological collections. Unstalked forms, which sometimes occur in the shallows, are known as feather stars, and they are among the loveliest marine animals, with their delicate colors, finely branched arms, and ability to swim gracefully and slowly through the water. In contrast to the starfish, the feather star swims and lies on its back, mouth upward. It feeds on microscopic plankton and detritus.

Having virtually no enemies, when the stalked species establish themselves they form extensive forests in clear water. But lack of predation has not resulted in the proliferation of great numbers of sea lilies as compared with the crowds of other creatures in the sea, and there are apparently far fewer today than in times past. There are about 2,100 fossil species but only about 800 living representatives. Occasionally, however, deep-sea trawls come up with big catches. In a single net load, the research ship *Albatross* brought up 3 tons of feather stars. The deepest have been found at about 27,000 feet.

CRUSTACEANS

In many respects the crustaceans—shrimps, crabs, lobsters, and related forms—are the most important and interesting of the many animals without backbones. They are the most numerous of all animals; they are among the handsomest and most varied in form; they include species of great economic value. Some crustaceans were described in the previous chapter, since among the plankton the copepods loom so large in importance; the shrimps and prawns will be described in the next chapter as members of the mid-water swimming community. There remain the barnacles, crabs, and lobsters, which spend most of their time on the bottom of the sea.

Barnacles are so unlike most of the other crustaceans, being encased in a hard conical castle of protective carbonate, that even biologists must sometimes pause to remind themselves that the barnacles are indeed related to the shrimps and crabs. Taking a little biological license and paraphrasing Dr. T. H. Huxley, the famous British biologist, we might say that a barnacle is a shrimp that has glued its head to a piling and built a marble cone around its body; it sticks its feet through the open end of this cone and kicks plankton and detritus into its mouth. The barnacle's true identity as a crustacean is revealed by the larval stages. The egg of a barnacle hatches into the same kind of larva, the nauplius, as that of the majority of other crustaceans. The nauplius soon changes into a shelled larva resembling in form the ostracod, a little crustacean that carries a bivalved shell. Because of this resemblance this stage of the barnacle bears the name of the well-known ostracod *Cypris*. The cypris larva swims briefly in the water and then settles down on some hard surface—a piling, a rock, the bottom of a ship, even the back of a whale or a turtle—and attaches itself forever by means of adhesive glands on its head. The bivalved shell is cast off, and calcareous plates are formed that fuse together into the familiar conical shape.

If barnacles settle on the hull of a ship, they contribute to many a headache for the shipowner and the captain. The water resistance created by barnacles can slow a ship by as much as 50 per cent. Delayed voyages, wear and tear on ship's machinery, and high fuel consumption are the result. To combat barnacles, expensive lay-ups in dry dock are necessary in order to scrape and repaint the hull of the vessel. A big boat may carry around 100 to 300 tons of this fouling cargo; the annual cost of combating barnacles in the United States alone is more than 100 million dollars.

The barnacles described so far are the familiar "acorn shells" of pilings and rocks. Another group of barnacles of different appearance, the goose barnacles, are found mostly on drifting logs or other flotsam as well as on ships. These differ from the acorn shells in the possession of a fleshy stalk, formed from the head, that attaches the animal to some firm surface. At the outer end is a shell consisting of five plates, from which protrude six pairs of feathery appendages, or "feet," resembling the tail feathers of a bird. This similarity to a bird's tail, together with the stalk shaped rather like the curved neck of a goose, led to a widespread belief among the ancients that the marine animal was the young stage of a goose. That is what gave the group the common name of goose barnacle and gave a principal species the scientific name *Lepas anatifera* ("goose-bearing barnacle"). The belief had official sanction too: as late as the nineteenth century, Catholics in some parts of Europe were still permitted to eat goose on Friday, in the face of the Church's proscription against flesh or fowl.

In Spain and Brittany one of the stalked barnacles, *Pollicipes,* is a gourmet item; a huge species, *Balanus psittacus,* is relished in Chile. The latter attains a size of 9 inches across the base and 3 inches high, comparing in size with a small lobster. Indians of the northwest coast of America also eat large barnacles, as do those at the opposite end of the Americas, in Patagonia. The Japanese, who wring from the sea every scrap of sustenance possible, cultivate barnacles on bamboo stakes set near shore. These barnacles are too small for food, but they are scraped off every 3 months or so and used as fertilizer on the hungry fields of Japan.

Some of the barnacles that attach themselves to other animals are exclusive in their choice of host. The turtle barnacle, *Chelonobia testudinaria,* is found only on loggerhead turtles. Some whale barnacles are restricted to one kind of whale and even to particular parts of the whale. *Coronula diadema* is found only on the humpback whale, and attaches to the throat and the corrugated belly; *Coronula reginae* is found only on the lips and the front edge of the flippers of whales. For a barnacle the next step after hitchhiking a ride on the exterior of another animal is to invade the tissues of the host and become a parasite. The best known of these parasites is *Sacculina,* which is a serious enemy of a fellow crustacean, the crab. After a short free-living larval existence, *Sacculina* invades a crab, bores through the tissues, and attaches to the abdomen, on the underside of its victim. Now it loses all crustacean characteristics and degenerates into a shapeless, saclike mass attached to the crab by a short stalk. The outer skeleton is lost, as are all internal organs except those for reproduction: there are no appendages, no alimentary canal, no mouth. From the parasite's body rootlike processes spread throughout the body of the unfortunate crab to draw nourishment from its tissues. The crab becomes stunted because the parasite keeps it from molting, and as a final insult the parasite renders the crab sterile by invading its reproductive organs.

CRABS

Perhaps it is partly the indignities imposed by *Sacculina* and other parasites that give crabs such disagreeable personalities that their name is a synonym for gruffness and irascibility. The reputation is well deserved, since many crabs are pugnacious. The velvet "fiddler" crab *Portunus puber,* a large swimming crab of the coast of France, has so low a boiling point that the French fishermen call it *le crabe enragé.* The other behavioral characteristic of crabs that has added a common phrase to the language is their habit of walking and running sidewise, a mode of locomotion that sometimes develops surprising speeds.

Several related groups of animals are called crabs, including the common crabs, whose reduced abdomens are tucked under their cephalothorax; the hermit crabs, which occupy the discarded shells of marine snails; and the spider crabs.

The color of crabs is extremely variable. Many crabs assume protective coloring that blends subtly into their backgrounds. *Ocypode ceratophthalma* is sometimes so similar to the background of sand on which it lives, scuttling along tropical shores at the water's edge, that it is only revealed by the shadows it casts as it holds its body high on long legs. If this ghost crab stops running and presses its body against the sand, the shadow disappears and the crab becomes virtually invisible.

The fighting fiddler crab of the Atlantic, *Uca pugilator,* changes color in rhythmical sequence with the time of day. At night it is pale and in daytime it assumes a darker color. The changes are related to the tides, with the darkest color occurring about 50 minutes later each day, always coinciding with the period about an hour before low tide. In this manner, maximum darkening is attained at the time when the crab is most active in foraging for food, and this helps to conceal it against the dark sand.

Since the time of low tide varies along a coast, crabs from various beaches are darkest at different times. They maintain their own unique schedule of color change even when taken to other places, including the darkness of laboratory cages. The crab evidently has a highly accurate internal clock that is independent of light. It is almost independent of temperature too: very low temperatures change its rhythm, but a return to normal temperatures restores the original timing. Even the massive changes in environment, time of daylight, and tidal rhythm involved in an experimental trip from the Atlantic to the Pacific coast failed to disturb this amazing internal clock of *Uca*. Some of these crabs were kept in darkness for 6 days in their new habitat and still maintained their normal rhythm, varying in color along with their fellows left behind on the Atlantic coast.

HERMIT CRABS

The abdomen of the hermit crabs is bigger than that of the true crabs, although small compared with the fleshy tail of a lobster. It is soft and unprotected, and its vulnerability causes the hermit to seek protection—often in a snail shell. The crab thrusts its tail into the shell and curls it into the spirals of its house. The appendages of the right side of the abdomen have been lost in most hermits, but hooked appendages on the left side grip the central spiral of the shell to permit the crab to drag its protective abode around with it and to prevent predators from pulling it out. One species of deep-sea hermit crab from Malaya lives only in joints of bamboo that have sunk to the bottom.

If what the "hermit" wants is to be alone in its shell, it is sometimes disappointed, since it may have to share its abode with a number of companions. In some cases a bristle-footed worm lives in the shell, popping its head out when the crab has caught some item of food and unceremoniously snatching fragments from the jaws of the bigger animal. The crab does not try to catch the worm, which it could do easily, but tolerates its presence. Copepods, amphipods, smaller crabs, and other animals sometimes share the hermit's house. The addition of parasites makes for a crowded community. Near the aperture of the shell house of some hermit crabs, hydroids and sea anemones grow beyond the edge of the shell and thus increase the size of the dwelling. This is especially useful where shells are scarce, as in the deep sea. There the hermit crab *Pagurus pilosimanus* welcomes the house-enlarging services of the anemone *Epizoanthus parasiticus*. Despite the name, the latter animal is not a parasite but a commensal, meaning that it contributes something of value to its host in exchange for the advantage it gains. The advantage to the anemone is the mobility provided by the crab, with a consequent increased chance of food; its contribution to the welfare of the crab is the enlargement of its dwelling and the protection afforded by the anemone's stinging cells. Some sea anemones living on the outer surfaces of the shells in which hermits live refuse any other station in life. One big hermit crab of the Philippines, *Pagurus deformis,* lives in empty conch or helmet shells, to the outer surface of which the anemone *Calliactis* attaches itself. Sometimes as many as seven or eight anemones ride the same shell. If the crab abandons a shell, as it does when it has grown too big for it, the anemones leave too. More than that, the crab literally captures the anemones and transfers them to its new abode. The anemone allows this to be done, not contracting when the crab grasps it in its claws, but detaching and allowing itself to be rolled to the new shell. It makes no attempt to sting the crab, although much less rough treatment by any other animal would set the stinging cells into violent action.

　　If the presence on its shell house of a sea anemone with formidable stinging cells is of value to a hermit crab, consider how much more impressive it would be if the anemone could be waved in the face of an enemy. And this is precisely what some hermits have learned to do. *Melia tessellata,* a hermit of the Seychelle Islands, and *Polydectes cupulifera,* from Hawaii, both habitually carry a small sea anemone in each claw. The claws are very slender, and their inner surfaces bear rows of sharp teeth like those of a saw. The anemone is carried lightly in the half-shut claw, but not so lightly that it can be detached without tearing it. The crab can release it spontaneously, but rarely does so, since he uses it as an offensive and defensive weapon. The mouth and tentacles of the anemone are turned upward, and when the crab is disturbed, it thrusts its claws toward the intruder, brandishing the anemones like weapons. Sometimes the anemones capture food, and when they do they must gulp it down fast, for if any part of it protrudes from the mouth of the anemone the crab will snatch it. Much of what the crab eats is stolen from its slaves. Unlike other crabs, these species carry food to their mouths with the second legs instead of the first legs, which are fully occupied in holding the captive sea anemones.

Some of the true crabs also use other animals for protection. The furry, clumsy sponge crab, *Dromia vulgaris,* has short last legs bearing claws with which it holds masses of live sponge on its back for concealment and to discourage enemies. Biologists of the *Galathea* Expedition captured deep-sea crabs of the genus *Ethusa* at 15,800 feet and found that their short legs curved over their backs like those of the shallow-water sponge crabs; the scientists assumed that this odd behavior had been carried to the abyss.

Their abode in very deep water had not allowed these crabs to escape from parasites, since *Sacculina*-like crustaceans were attached to their undersides. The same haul that captured the deep-sea sponge crabs also yielded some hermits, and these, too, resembled their shallow-water relatives, twelve of the thirteen individuals carrying a reddish-violet sea anemone.

Another deep-sea hermit crab, *Parapagurus pilosimanus,* carries a sea anemone on its abdomen. Some deep-sea crabs, on the other hand, are blind. Many of these are of special interest, since living representatives of their particular family, the Eryonidae, live only in the deep sea; the other known members of the family are fossils from the Jurassic Period, over 150 million years ago. When they were alive, in the shallow seas of that ancient time, they possessed efficient eyes; their modern relatives, which have retreated to the deep sea, are all blind and pallid. Some crabs of this and other families are found clinging to sponges and other "rooted" animals of the abyss. Others, with well-developed eyes, range actively over the bottom oozes. Since these oozes are less compact than the sandy bottoms of many shallow areas, the animals of the deep-sea bottom have difficulty in preventing themselves from sinking. Some of the crabs have solved this problem by developing excessively long legs on which they stilt over the sediments.

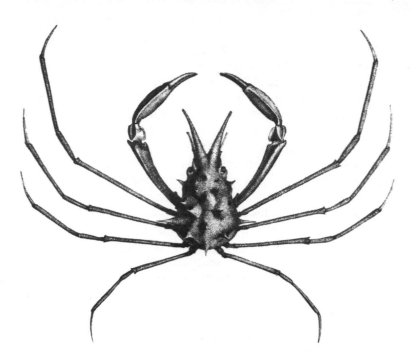

FIGURE 6 – 7. *A crab* (Anamathia pulchra) *from 2,250 feet of water off the Philippine Islands. Its long legs help it stride over the soft sediments of the deep sea.*

Challenger Expedition

SEA SPIDERS

The sea spiders may be said to be truly of the deep sea, although there are shallow-water species. These relatives of the crustaceans are remarkable for the great length of their legs in relation to the rest of the body. Many sea spiders have the body reduced so far that there is no longer enough room for all the organs, some of which have had to be tucked into the joints of the legs nearest the body. In certain species the legs are between three and four times the length of the body, and this gives the sea spider a great advantage in striding lightly over the soft oozes of the deep sea where it lives. Some sea spiders are very large, 2-foot individuals having been photographed on the sea bottom. They are common too, if we can judge from the numerous tracks seen in photographs of the soft deep-sea muds.

[118]

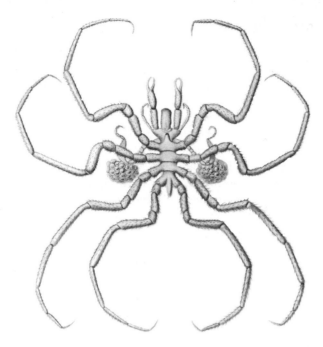

FIGURE 6–8. *Sea spider* Chaetonymphon altioculatum. *Its body is so reduced that some of its organs have to be carried in the joints of the legs nearest the body.*

Valdivia Expedition

LOBSTERS

Perhaps you have walked through a large hall in the Metropolitan Museum of Art in New York City and seen its display of medieval armor. How magnificent were the knights who wore those gleaming suits, and how graceful they must have been on their horses. But how helpless out of the saddle, sprawling immobile on the ground, their dignity gone, while they waited for their squires to rescue them.

A lobster out of water is not unlike an unhorsed knight. His steed is the sea. Without its salty buoyancy the lobster flounders helplessly on land, borne down by its dark blue-green armor. In the sea the lobster's body is just heavy enough to give its legs purchase on the bottom, and it can walk or run forward with such delicate tread that a 3- or 4-pound lobster can walk with pointed claws on flatfish without disturbing them.

[119]

The northern lobster (*Homarus*) lives on rocky or sandy-to-muddy bottom from near shore out to the edge of the continental shelf, becoming scarce beyond about 30 fathoms. It walks or runs forward, or when alarmed it can dart rapidly backward amid violent flappings of its broad and powerful tail.

The food of the lobster, to which it is attracted mostly by the odor, includes dead plant or animal material, but live food is preferred: fish, mussels, clams, and sea urchins. Lobsters are particularly fond of crustaceans, and other lobsters are favorite fare. The fact that lobsters are notorious cannibals is the cause of one of the greatest difficulties that lobstermen encounter. They are usually sent to market alive and are kept in pounds or holding tanks. Lobsters awaiting shipment in these tanks must have their claws immobilized by carefully inserted wooden pegs; otherwise any lobster that sheds while surrounded by its voracious fellows is doomed. A hundred lobsters left long enough together would eventually become one large lobster. If a lobster is lucky enough to find more food than even its greed can cope with at one sitting, it buries the excess, like a dog with a bone. It does this by bracing its rear legs and pushing sand or gravel over the food with its front legs and mouth appendages.

Probably all the very big individuals are males, since they grow faster and larger than the females. This is because the female sheds its shell only every second year after achieving maturity, while the male, a little more carefree about its reproductive duties, molts and therefore increases in size once a year.

The female reproduces every second summer, carrying the eggs one year, mating and spawning the second. As she grows, the number of eggs increases from 10,000 or so at 10 inches to double that

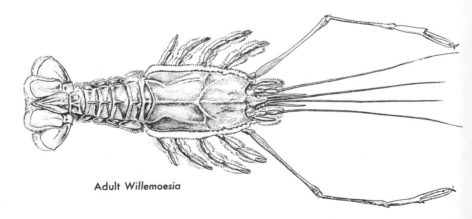

Adult *Willemoesia*

number at 12 inches; 2 more inches of length again doubles egg production to 40,000, and the biggest females, 15 to 16 inches in length, produce nearly 100,000 eggs.

Over the years the annual catch of lobsters has fluctuated widely. There are tales of unbelievable abundance in the "good old days." Dr. Waldo Schmitt, crustacean expert of the Smithsonian Institution, quotes a statement from the records of the Salem Plantation of colonial New England that "the least boy in the plantation may catch and eat what he will" of lobsters. After heavy storms lobsters were sometimes washed ashore in windrows, and their carcasses were carried by the wagonload to fertilize nearby fields. Lobster fishing was profitable in the early days, even though the price in colonial times was only "a penny each," apparently regardless of size. Landings reached their maximum in the United States in 1889, with over 30 million pounds. The catch fell rapidly thereafter, to a low of 9 million pounds in 1933.

Spiny lobsters lack the great claws of the northern lobster, their front legs being pointed like the rest of their limbs. Most of the meat is in the powerful tail. These creatures are brown, yellow, even purplish in color, with beautiful sculpturing; great curved spines protect the eyes, and smaller spines stud the carapace. Other common names for the spiny lobster are "crawfish" and "crayfish" (although in the United States the latter name is more often applied to freshwater members of this crustacean group). The French call them *langouste,* and over the whole world they are regarded as delicacies.

Spiny lobsters make surprisingly loud noises by rasping the base of their antennae against a corrugated surface on the carapace. Violent contractions of the abdomen produce this sharp sound when

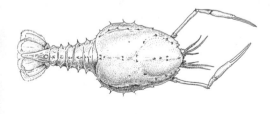

Poul H. Winther, *Galathea* Expedition

FIGURE 6–9. *Blind lobster* Willemoesia *and its offspring, caught in a trawl off the Pacific coast of Central America in 12,173 feet of water. The adult's body was about 4¾ inches long, and the young lobster was 1½ inches long.*

the animal is alarmed. In groups, when undisturbed, they make another sound: a slow rattle that must have some sort of community meaning.

The larva of the spiny lobster is flat and leaflike, with long, complex appendages. This phyllosoma ("leaf body") or glass lobster is almost transparent in the water. Some species go through eleven larval stages before they molt into the adult form.

The Norway lobster, *Nephrops norvegicus*, is a deep-dwelling crustacean compared with the spiny lobster or the northern lobster. *Nephrops* lives on mud bottoms at 300 feet or more (although it occurs in waters shallower than this too) in the seas around England and Scandinavia. It is common in the Irish Sea, for example, and has the name of Dublin prawn on the British market. It is a handsome creature, with an orange body bearing red and white markings and long, slender claws.

Several lobsters occur in the deep sea. One of these is the squat lobster *Galathea,* which resembles a small red crawfish. It lives at 400 to 500 feet or deeper. *Nephropsis* is a deep-sea lobster closely related to the Norway lobster but with vestiges of eyes. These consist of very short and slender eyestalks, showing at their tips the merest trace of what were eyes in ancestral lobsters. The large, squat, white lobster *Munidopsis,* which occurs in waters thousands of feet deep, is remarkable for its very large eggs, about a seventh of an inch in diameter.

MOLLUSCS

The molluscs are not especially well represented on the deep-sea floor, but one member of the group, *Neopilina,* deserves particular mention. It is neither a clam nor a snail but a relic of a class of molluscs believed to have disappeared from among the living some 300 million years ago. It was discovered first by the scientists of the *Galathea* on May 6, 1952, when the trawl brought up specimens from the bottom of the Pacific, in 10,000 feet of water (Figure 12-3). Here is a true living fossil, an example of an animal known previously only from ancient remains in rocks of the Cambro-Devonian times 350 to 550 million years ago.

7 Invertebrate Swimmers

of the Deep Sea

Man and most of his air-breathing companion creatures share a two-dimensional existence, ranging back and forth across the surface of the land. Bound to the earth by the fierce strength of gravity, a man can leap no higher than about the length of his body, and that only if he is extraordinarily athletic. If he penetrates the sea or the atmosphere at all deeply, it is with the aid of elaborate machines that carry a small part of his normal environment with him. Some burrowing land animals can penetrate a thin layer of the soil—a few feet at most. Insects and birds make larger penetrations of the atmosphere, but these are usually only a few hundred feet in extent.

Animals of the sea, on the other hand, move as easily up and down as they do to and fro, and this third dimension vastly increases their living space and their freedom. This is not to say that any single kind of creature lives successfully in every part of the ocean, for living conditions—notably temperature, light, and pressure—are greatly altered with distance from shore, and more particularly with depth. Nonetheless, some animals have become adapted to extreme conditions, and the whole vast bulk of the ocean has been penetrated by living animals. Some of these cling to the bottom, while others have struck out into the trackless wilderness of the open waters.

While the species found on the bottom, even in the deep ocean, represent most of the common kinds of animals of the shallows, not as many have succeeded in making the transition to the more specialized environment of the open waters, so that fewer kinds of creatures are represented there. The chief groups with representatives among the deep-sea pelagic animals are the coelenterates and the crustaceans, with which this chapter largely deals, and the molluscs (principally squids), and fish.

JELLYFISH

Among the most conspicuous inhabitants of the sea are the jellyfish, those discs of transparent substance that throb in the water like animated parasols and often drive swimmers onto the beach for fear of their stinging tentacles. Some jellyfishes are included in the plankton, since their swimming powers are so low that they qualify as drifters. But others swim purposefully, if feebly, and are therefore not plankton in the strictest sense. The discussion here is limited to the two important pelagic groups, the medusae and the siphonophores.

The umbrella-shaped jellyfishes represent only one of the two characteristic forms assumed by the coelenterates. Most members of this great phylum carry on a Jekyll and Hyde existence in which they are sometimes free-swimming creatures capable of sexual reproduction, and later plantlike growths attached firmly to the bottom, reproducing "vegetatively" like a strawberry runner. From the body of this phase, buds are pinched off that float away to grow into the pelagic phase again. The sexual, free-swimming jellyfish phase is called the medusa, named for the resemblance of the fringing tentacles to the writhing, snaky tresses of Medusa in Greek mythology. The individuals of the attached phase are the polyps (from the French *poulpe,* meaning "octopus," although there is no real relationship between a polyp and an octopus); these were discussed in the preceding chapter.

In *Aurelia,* the moon jelly whose plate-sized discs are familiar to many swimmers, the fertilized eggs of the jellyfish adult develop into tiny larvae covered with small beating hairs or cilia. The larva swims briefly and then settles onto a stone or other hard surface, where it develops into a polyp like a miniature sea anemone. The polyp grows taller, and a series of furrows develop in its trunk. These furrows bite deeper and eventually cut through to form a set of discs like a pile of saucers. These are the new medusae, which break loose and float off to grow into adults again.

Some medusae are so small or so transparent that they are only discernible by the faint shadows they cast, while others are big, solid

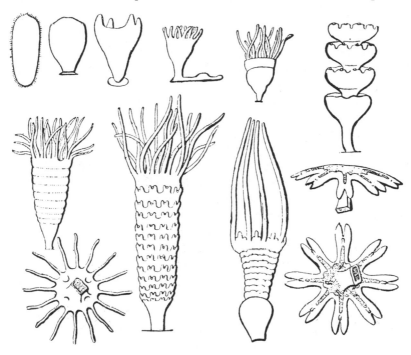

FIGURE 7 – 1 . *Development of the jellyfish* Aurelia aurita *from the egg. The upper series shows the larva changing into a flowerlike polyp. Furrows develop in the polyp's trunk, and these cut through to make a pile of discs, which break free and float off as medusae—named for their resemblance to Medusa's snaky hair. The medusae produce eggs to repeat the cycle.*
J. Murray and J. Hjort, *The Depths of the Ocean*

chunks of jelly with bold markings and brilliant colors. From individuals of microscopic size the jellyfishes range to animals like *Periphylla,* with a bell 2 to 3 feet in diameter, *Rhizostoma,* the size of a soccer ball, or the big blue plates called *Cyanea,* whose northern representatives are 6 or 7 feet across. The tentacles of these giants may trail 120 feet behind the animal, endangering small fish and other marine animals to which a brush with the stinging cells of *Cyanea* means death.

SIPHONOPHORES

The siphonophores are medusa-like, but differ in enough respects to be put in a separate group. They are colonial animals, with one or more swimming bells serving as the "head" of the colony, and long tentacles, sometimes many feet long, streaming down into the water. Some siphonophores are as lovely as lace, and in delicacy and color are among the most beautiful of marine creatures. One of the commonest of siphonophores, easily seen because it sails along at the surface driven by the wind, is *Velella,* the before-the-wind sailor. It consists of a small oblong raft about 2½ inches long and 1½ inches broad with a small triangular sail set diagonally across it. Beneath the raft are long tentacles bearing stinging cells, and the individual polyps of the colony, some specialized for digestion and others for reproduction.

The most famous of all the siphonophores is *Physalia,* the Portuguese man-of-war. *Physalia*'s fame rests both on its beauty and its reputation as one of the most dangerous of marine creatures. The visible part of *Physalia* is a highly colored float, in various shades of blue and pink, sometimes with purple, mauve, orange, and other colors intermixed. This float, up to a foot long, supports a veritable city of individuals, for this is a colony and not a single animal. The float catches the breeze and the colony is propelled through the water at an angle to the wind. At intervals the float is twisted by muscular action so that it dips into the water to wet its surface and then resumes its upright position to catch the wind again.

Trailing beneath the colony are long tentacles. These are capable of great contraction, and are sometimes pulled up so that they are not much more than stubs; at other times they stream out surprising distances behind the float, sometimes to 100 feet. The tentacles bear batteries of stinging cells whose function is to repel enemies and to shock prey into submission so that it can be consumed.

Stinging cells are possessed by all the jellyfishes, and their complex structure and efficiency in offense and defense make them among the most interesting of animal structures. They consist of a trigger hair that is sensitive to the lightest touch and an internal mechanism of great complexity, all neatly packed in a single cell. The mechanism consists of a very long, thin, hollow tube, sealed at one end and opening into a bulb at the end nearest the trigger. The tube is coiled inside the cell, and sharp spines are attached to it. When

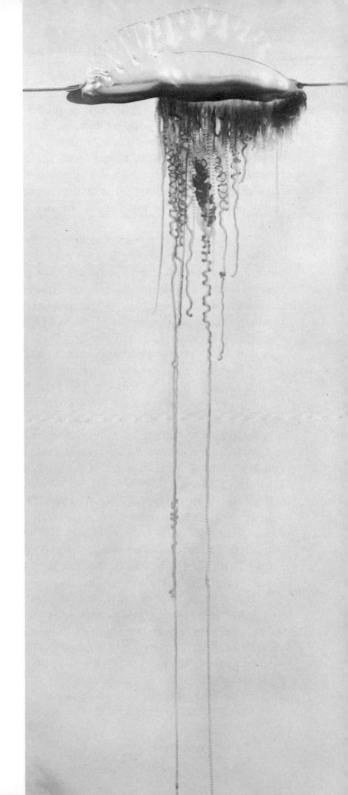

FIGURE 7–2. *Portuguese man-of-war,* Physalia. *This beautiful but dangerous colony of animals has a body as much as a foot long and stinging tentacles that sometimes stretch to 100 feet.*

Charles E. Lane

the trigger is touched, there is a sudden increase in the pressure inside the liquid-filled bulb, and this forces the tube to turn inside out with lightning rapidity, as though a rubber glove had been blown into with great force, popping out the inverted fingers. The razor-sharp little spines at the base of the tube lacerate the surface of whatever has touched the trigger, and the little tube, kept rigid by internal pressure, is shot into the opening. Now the poisonous contents of the bulb are injected into the wound, killing, paralyzing, or irritating the victim, depending on its size and its resistance to the sting.

Some idea of the power of stinging cells of various coelenterates can be obtained from their effect on man. In some cases they are too weak even to be felt, while in others they produce a nettlelike rash that is uncomfortable to painful, according to the susceptibility of the victim. In a few cases, larger jellyfish and siphonophores can produce exceedingly painful and occasionally dangerous stings. The Portuguese man-of-war packs a wallop like a bare electric light cord, and can lay a strong man low with a maddening pain lasting for hours. It can even affect breathing enough so that the victim has to be placed in an oxygen tent. In Miami a husky student of marine biology who had been a steel worker during summers scoffed at the ability of the frail-looking jellyfish to harm him. He boldly wrapped a long, streaming tentacle of a man-of-war around his arm—and was immediately convinced. The *Physalia* gave him a violent sting that laid him up for several days. Dr. Charles E. Lane, physiologist at the Institute of Marine Science, has discovered that the poison of the man-of-war is nearly as virulent as cobra venom. But in spite of its violent sting and the power of its poison, there is no record that *Physalia* has killed a human being. Neither has *Cyanea,* the big jellyfish of European waters described above, even though Sir Arthur Conan Doyle had a murderer use this jellyfish as the lethal weapon in one of the Sherlock Holmes stories.

But if these famous jellyfishes are not killers, there are others that are. *Lobenema* is credited with killing a person, and two Pacific species are certainly responsible for human deaths. These latter are variously called sea wasps, fire medusae, and box jellies in Australia, where they have killed at least eleven victims. The information about these dangerous jellyfishes comes mostly from Dr. R. V. Southcott, a physician and zoologist from Adelaide.

The fatalities took place in the waters of northern Australia, mostly near Darwin. The stories connected with these deaths are frightening. Most of the case histories record violent pain and include the phrase "died in a few minutes." Known attacks go back to the case of a "youth aged 17 years, Darwin baths, between 1882 and 1896. Died in a few minutes." A recent case concerns "L.M.S., female aged 11 years—December 13, 1957. Shortly before 11 A.M. —at North Mission Beach, Tully, N. Queensland. When in water about 2½ feet deep, about 15 yards from the shore, she gave a loud scream and appeared to be in pain. Her grandfather, who was in the water nearby, grasped her and carried her from the water in an unconscious state. As he did so he noticed a jellyfish about a foot away from her in the water, light blue in colour and about 8 inches across. When the girl was placed on the sand at the water's edge her eyes were closed, her lips and nose were blue, and numbers of jellyfish tentacles were seen clinging to her legs from thigh to ankle. A few minutes after being stung she made two successive convulsive movements, and became limp, apparently dead."

The jellyfish responsible for these rapid deaths is *Chironex fleckeri,* belonging to a group called the Cubomedusae from their squarish or cuboidal shape. They are about the size of a coconut and light blue, but nearly transparent. They swim just below the surface of the water with hardly a ripple, and hence are very hard to see. Another species, *Chiropsalmus quadrigatus,* so similar to *Chironex* that the two are difficult to tell apart, has killed swimmers in the Philippines, Borneo, and Malaya.

While the stinging cells usually serve as defensive weapons or to capture food for the jellyfish, they occasionally protect little fish or other animals that have learned to swim among the tentacles and still avoid them. Baby whiting use the big blue umbrellas of *Cyanea* for shelter, swimming under and around them to pick up food and darting beneath their canopies when alarmed. Young horse mackerel, butterfish, haddock, and cod also use this protective cover. The Portuguese man-of-war has a constant companion whose existence is so closely tied to that of the jellyfish that it is rarely encountered except swimming under the float and tentacles of its host. This is the Portuguese man-of-war fish, *Nomeus gronovii.* As a protective coloration it has adopted the brilliant blue of its host, in the form of vertical bands against a silvery background. These vertical dark stripes may resemble the feared tentacles of *Physalia,* and thereby cause predators to avoid the little fish. *Nomeus* darts in and out among the tentacles, apparently unconcerned at the prospect of sudden death. The fish has more resistance than most other creatures to the effects of the poison, but if it is forced against the tentacles in an aquarium, it is killed like other creatures of its size. In nature it seems to evade this fate, and sets up an unusual partnership with its dangerous companion.

No other group of animals has the same kind of stinging cells as those of the coelenterates. It was once thought that some of the sea slugs possessed them; but it turned out that the slugs eat coelenterates and that the stinging cells pass, intact and unused, through their gut, later to be arranged in definite patterns in their tissues and to serve the same purpose there as they did for their original owner.

One swimming octopus has learned the extraordinary trick of brandishing the stinging cells of the Portuguese man-of-war on its arms and using these as offensive and defensive weapons. The octopus, *Tremoctopus violaceus,* carefully detaches pieces of the tentacles of *Physalia* and fastens them to its arms. Everett C. Jones, a marine biologist with the Bureau of Commercial Fisheries Laboratory at Honolulu, caught a small female octopus in a dip net while fishing under a night light in the Pacific. He was surprised to get a severe sting as he lifted the animal from the net, and was even more taken aback to find that the cause of the pain was a series of fragments of *Physalia,* arranged in an orderly fashion along each row of suckers on the four dorsal arms of the octopus. About three quarters of the suckers on each arm carried bits of *Physalia.* The suckers seem to be adapted especially for the function of carrying these strange weapons, with a cavity shaped nicely to receive them. "It is interesting to speculate," says Jones, "on the method *Tremoctopus* might use in obtaining the tentacle fragments. Unless the octopod is immune to *Physalia* toxin, the approach to the coelenterate must be made quite cautiously, and one can only imagine the pickpocket type of dexterity which would be required to obtain enough tentacle fragments to cover the eight rows of suckers." (From a recent article in *Science.*)

Considering the presence of stinging cells, which can produce their effect even after the animal is dead, and the insubstantial character of the bodies of jellyfish, which are about 96 per cent water, it is not surprising that they are not much used as food. But the Chinese and Japanese do eat some jellyfishes. Two species of *Rhopilema,* big jellyfishes averaging about 1½ feet across the bell, are consumed as a pickle or an appetizer. They are soaked in a solution of alum and brine, wrapped in oak leaves and steamed, and then before being served are soaked in water and flavored with spices.

Some of the handsomest and largest jellyfishes live in the deep sea. In general, the species that occur near the surface are colorless or insubstantial shades of blue, green, pink, orange, or yellow, although some exceptions have already been noted. In the deep sea the colors of these coelenterates deepen. Dense shades of red or chocolate brown are common on their undersides, and Beebe once saw a 3-inch black jellyfish from his bathysphere. *Atolla,* one of those most common in deep waters, is colored red, cream, and purple "like exotic water lily flowers." *Agliscra ignea* is flaming red and several other species, including *Halicreas rotundatum,* have bright red markings. *Nausithoe rubra,* looking like the insubstantial creations that bridesmaids call hats, is colored shades of reddish purple, while *Periphylla hyacinthina,* whose tall cone resembles a wedding cake more than a bridesmaid's cap, is blue and purple.

Beebe expected to see many luminescent jellyfishes, but was surprised to find that only about an eighth of the coelenterates that he saw from the bathysphere were luminous. Those that did shine put on a lovely show. There were luminous bands on the umbrellas and pale spots on the edges, or the tentacles were lighted. One entire school at 1,650 feet was brilliantly illuminated with pale green lights. Some species appear as round balls of white fire in the sea, one of these being *Pelagia noctiluca.* Sir William Herdman, a distinguished English marine biologist of the early part of this century, describes the intense luminescence of *Pelagia* in his book *Founders of Oceanography and Their Work:* "A small tankful of them once gave us a magnificent display in the dark at the Port Erin Biological Station, and when taken out in a bucket they looked like balls of fire, or rather incandescent metal, as the light is white and very intense. It was difficult to believe it would not burn one's fingers when touched." Another time, when Herdman was anchored over the pearl oyster banks in the Gulf of Manaar "in an intensely dark night, I saw the black sea around us in all directions lit up by an innumerable assemblage of what looked like globes of fire, waxing and waning in brightness, all simultaneously glowing and then fading away into darkness, and after a few seconds lighting up once more. This periodic display continued for about an hour and then disappeared."

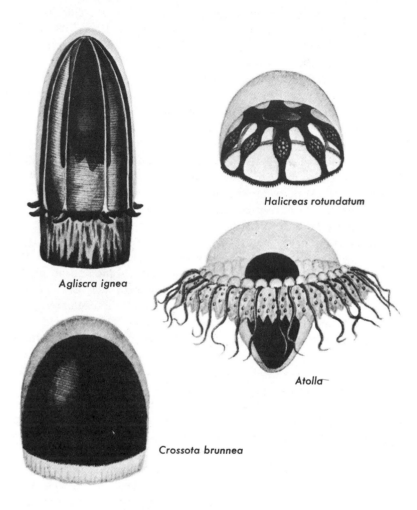

Halicreas rotundatum

Agliscra ignea

Atolla

Crossota brunnea

FIGURE 7 – 3. *Deep-sea jellyfishes are among the most striking inhabitants of the ocean.* Agliscra ignea, *about ¾ of an inch high, is flaming red;* Halicreas rotundatum, *¾ of an inch wide, has bright red markings.* Atolla, *the commonest, measures about 3 inches across and is red, cream, and purple.* Crossota brunnea *is dark brown and only about ⅓ of an inch wide.*

N. B. Marshall, *Aspects of Deep Sea Biology*

COMB JELLIES

There is another group of animals, the ctenophores, whose tissues are gelatinous and pellucid. They are therefore often called jellyfish, although they do not really belong to the group. They are the comb jellies, which have eight bands of beating comblike plates encircling them. The animal also has a pair of long tentacles whose function is to sweep the waters for copepods, larval fish, and other morsels from the plankton. This the comb jellies do with devastating efficiency, and swarms of them in the water wreak great havoc among schools of crustaceans or newly hatched fish. Comb jellies do not possess stinging cells, but capture their prey with sticky lasso cells that spring at right angles from the tentacles. When the questing tentacle has succeeded in entangling a copepod, an arrowworm, or a small fish, it brings the victim up to its mouth; if the prey has size and strength to resist, the comb jelly may have to play it like a salmon on a line before it can finally subdue it.

Among the handsomest of the ctenophores is *Beroë,* a finger of pink or soft lavender jelly. This lovely animal was named for the daughter of the Greek god Oceanus. In appetite *Beroë* is not very ladylike, engulfing its prey by rapid and wide extensions of its mouth, and sweeping in prodigious quantities of plankton rather like a vacuum cleaner. Venus's girdle, *Cestus veneris,* is another beautiful member of the group. It is untypical of the comb jellies because of its elongated shape, and is a giant among the ctenophores, reaching the length of a foot.

All the ctenophores are luminescent. Very commonly the light organs are aligned along the combs, and, when the animal is fluorescing, the eight bands are outlined like a Christmas tree decorated with colored lights.

SWIMMING CUCUMBERS

Few phyla of marine animals seem more truly bottom dwellers than the echinoderms—the starfish, sea urchins, and sea cucumbers described in the previous chapter. It was especially exciting, then, to discover sea cucumbers in deep water that were true mid-water swimmers. The first of these, *Pelagothuria natatrix,* a 4-inch rose-colored animal with a dark violet hind part, was discovered in 1891 off the west coast of Central America. It has been encountered sev-

FIGURE 7–4. *The sea cucumber* Galatheathuria, *unlike nearly all other sea cucumbers, does not grub on the bottom but swims in mid-water. It is 9 inches long and dark violet in color. Caught in the China Sea between 11,155 and 12,468 feet down, it was described as a "sensational find."*

Poul H. Winther, *Galathea* Expedition

eral times since by deep-fishing expeditions in tropical parts of the Pacific and Indian oceans. It lacks the spicules of most of the commoner sea cucumbers, and this reduces its weight, as does the thinness of its body wall. It swims mouth-upward and probably eats plankton.

Two other deep-sea swimming cucumbers of the genus *Enypniastes* have also been known for many years from such widely separated areas as the Bay of Bengal and the Great Australian Bight. These two creatures attain buoyancy by their possession of a thick, gelatinous body wall and their lack of spicules. The *Galathea* Expedition of 1950–52 found a third species of *Enypniastes* in the China Sea in very deep water, over 11,000 feet. It was a pale, transparent blue with light brown tentacles and was much bigger than the others, reaching 10 inches in length. In the same area the *Galathea* captured a fourth swimming cucumber, new to science. It was broad and oval in shape, about 9 inches long, 6 inches wide, and dark violet in color. Unlike all the other swimming sea cucumbers, the new species had numerous cruciform spicules. It was named *Galatheathuria*, after the vessel.

MORE CRUSTACEANS

The importance of the crustaceans to the economy of the sea is almost in proportion to their numbers. They are near the broad supporting base of the food pyramid: they consume the small plants of the ocean and the small animals that have fed on these little plants, and are in turn eaten by creatures farther up the steps of the pyramid. If the crustaceans were to disappear, starvation and extinction would face a great variety and number of animals in the sea. We have so far dealt with the copepods, which are permanently planktonic, and with the lobsters and crabs and some other crawlers. The present chapter will discuss the shrimps and other creatures that swim in mid-water.

Such a division—copepods as plankton, shrimps as swimmers, lobsters and crabs as crawlers—is arbitrary. Virtually all the crustaceans are planktonic for at least part of their lives; many of the shrimps and other "swimmers" move so feebly that they are practically planktonic; others included here as swimmers spend much of their existence on the bottom, rooting in the mud for food; and some crabs and other "bottom" forms swim vigorously. Nonetheless, if it is realized that the divisions between planktonic, pelagic, and demersal forms are blurred and overlapping, it will be useful to separate the crustaceans in this manner for our present purpose.

An external skeleton, a jointed body and limbs, and two pairs of antennae or "feelers" are the principal characteristics that set off the crustaceans from other animals. The skeleton, which surrounds the body with overlapping shelly rings, is composed of chitin, a substance similar in character to our fingernails; it remains soft and flexible in some animals, but becomes hard and rigid in others through the deposition of carbonates of lime. At the joints the rings of armor are linked by soft and flexible articular membranes.

The possession of the outer skeleton profoundly affects the mode of life of the crustacean. For example, growth is by fits and starts, taking place at times when the skeleton (or shell) is shed, instead of steadily as in most other animals, including ourselves. In this process the skeleton is split along definite lines, and the animal strains and struggles to force its way out through the split. This is a hazardous time, and often the animal dies in the attempt to cast its old shell or, if it does succeed, it may be eaten while the new shell is still soft and the animal relatively helpless. To avoid this fate, crustaceans like lobsters seek a hiding place until their shell has stiffened again. Meanwhile, an increase in size takes place, the animal swelling rapidly inside the soft new shell as it takes in water. Then the shell hardens again with the deposition of lime salts, a process aided by the nearly universal habit of crustaceans of eating their own cast-off shells and thus re-using the lime.

A great many crustaceans are luminous. In most cases the luminescence is produced by chemicals made in the body of the crustacean itself, but in a few it is caused by captive bacteria. A fresh-water shrimp, *Xiphocaridina compressa,* from Lake Suwa, Japan, glows in a captivating manner as a result of luminous bacteria. On hot summer nights the shrimps swarm to the surface of the lake and provide such a beautiful spectacle that they are protected as a national asset by order of the Japanese government. A small, bivalved crustacean, the ostracod *Cypridina,* is famous for the dependability of its light production.

Of all the luminous crustaceans the shrimplike euphausids are most remarkable for this characteristic, since all members of this big group produce light (the name "euphausid" can be translated roughly as "true shining light"). The euphausids are among the most important of the swimmers of the sea since they occur in such vast numbers, second only to the copepods in abundance. They swim actively and vigorously by means of a powerful abdomen and five pairs of swimming legs that drive the little animal forward with a rhythmic beat. The euphausids are considerably larger than the copepods, one of the principal species, *Euphausia superba,* being 1½ to 2 inches long.

In color the euphausids, or "krill," are transparent or translucent, with tints of pink or red. The light organs of the krill appear as seven pairs of red spots on their sides, the biggest of which are inside the base of the two eyestalks. The light organs flash brilliantly, beaming a bluish-white light whose intensity seems to be under the voluntary control of the animal. All of the light organs of the krill are complicated and efficient, especially the large ones on the head. In front of the cells that produce the illumination is a convex lens, and behind them is a reflector, while black pigment shields the eye from the light. The use to which the krill puts this intricate organ is not known, but it may be that it serves as a searchlight to be focused on its prey. This suggestion is fortified by the fact that the strong lights on the head are placed so that when the eyes rotate the light follows them and illuminates whatever the krill is looking at. And when the abdomen of the krill is flexed, all the light organs focus on a spot ahead of the animal.

In the short antarctic summer the krill occur in vast shoals, tinting the water a dull reddish color. They crowd at the edge of the ice pack, feeding on explosive blooms of diatoms. Euphausids collect

FIGURE 7 – 5 . *Two deep-sea euphausid shrimps, both about an inch long. Euphausid means, roughly, "true shining light." The two strong lights on the shrimps' heads are probably used to focus on prey; when their eyes rotate, the light follows.*

N. B. Marshall, *Aspects of Deep Sea Biology*

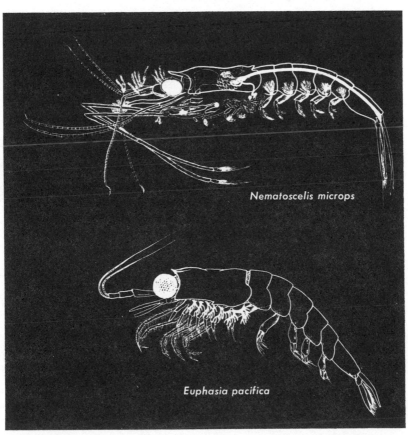

Nematoscelis microps

Euphasia pacifica

the floating plants by means of a fine filter basket formed from bristles and feathery hairs on their front limbs. Currents set in motion by their legs sweep the food into this net, from where it is passed to the mouth.

Just as the herring follow the copepods, the whales follow the euphausids. Among the great whales only the sperm whale has teeth and eats large creatures; the rest are largely plankton feeders, and the whalebone whales—like the blue, the fin, and the humpback whales—feed almost exclusively on krill. The blue whale is a bouncing 24-foot baby at birth, weighing 4,400 pounds. Two years later, almost entirely on a diet of krill, it has grown to a weight of 177,000 pounds; eventually it may reach 100 feet in length and weigh over 100 tons. The stomach of one blue whale taken in the Antarctic was found to contain 1,200 quarts of euphausids.

It is surprising that the herring can get enough copepods to fill its belly by picking them one by one out of the water, and it is too much to expect the whales to feed that way. Instead, they sweep majestically through the water, often at the surface, skimming up krill-rich waters. Accordion pleats in the throat allow the already gigantic mouth to expand greatly. Then the pleats are contracted, the mouth is closed, and the tongue forces the water out sideways through the great curtains of whalebone that hang like a "vast internal mustache" from inside the upper jaw. The whalebone sieve strains krill from the outrushing water, to be wiped off by the tongue and swallowed. In addition to feeding the whales of the Southern (Antarctic) Ocean, krill form the chief food of several species of petrels, penguins, seals, and many kinds of fishes. They have even been eaten by humans, and are reported to taste like shrimps—which is not too surprising considering their close affinity to the true shrimps.

Two groups of swimming crustaceans of less importance than the copepods or the euphausids are the ostracods and the amphipods (Figure 5-5). Ostracods are small, usually a quarter of an inch or less. The biggest of all ostracods, measuring nearly an inch in length, is found in the deep sea (Figure 13-2). The most remarkable feature of these little animals is a bivalved shell that makes them look superficially like clams rather than crustaceans. Ostracods are surrounded by these portable houses, often with only their antennae sticking out to serve a variety of functions: as sense organs, oars, legs, and food gatherers. Many ostracods reproduce parthenogenetically, that is, without the services of a male. In aquarium experiments generations of females have marched along in unending parade without the company of males for 8 years.

Amphipods are most familiar as the laterally flattened sand fleas of the seashore. There are swimming species in mid-water too, and the *Galathea* caught several kinds of bathypelagic amphipods, some with extra-large eyes and others with long, pointed heads. *Cystosoma* is a beautiful deep-sea form found down to 13,000 feet; its body is crystal clear.

Amphipods swarm in shallow water in some parts of the sea, especially in high latitudes. While they are not usually eaten by humans, they once saved the lives of seven explorers in the Arctic. The ill-fated Greely Expedition of 1882 was stranded for 3 years in the arctic wastes with food supplies designed for half that time. Two relief expeditions failed to reach them, and eighteen of the twenty-five men starved to death. The remaining seven stayed alive because the men learned to catch amphipods, using iron barrel hoops over which pieces of sacking were stretched. At first, bones and meat scraps were used as bait for the "shrimps," as the explorers called the amphipods. But later, when even the bones became treasured food items, little bits of sealskin were sewn into the hoops to attract the amphipods, which swarmed in countless numbers to every scrap of offal tossed into the water; 20 to 30 pounds of amphipods were caught in a day.

SCARLET PRAWNS

If they succeed in catching anything at all, scientists dragging deep nets in mid-water can usually rely on capturing brilliant scarlet shrimps and prawns. These spectacular animals are the abyssal representatives of a numerous and economically important tribe of crustaceans, most of whose members are inhabitants of shallower seas. Like many common names, "shrimp" and "prawn" are used loosely and inconsistently. In Britain and many other countries the name "shrimp" is used for the smaller individuals and "prawn" for the larger. In the United States nearly all kinds of these popular seafood creatures are called shrimps, irrespective of size. The shrimp industry is the most valuable fishery in the United States, having replaced salmon and tuna in 1952.

Shrimps range in size from small individuals the size of mosquitoes to *Hymenopenaeus,* huskies from deep water almost a foot

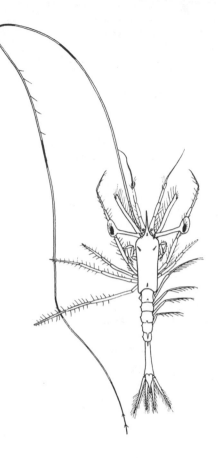

FIGURE 7–6. *Angler shrimp* Sergestes. *Its long antenna is like a fishing rod and line, stiff at the base and flexible at the end. Hooks are attached to the line, and it is possible that* Sergestes *actually fishes. This species is famous for its luminescence.*

N. B. Marshall, *Aspects of Deep Sea Biology*

long from the tips of their horns to the spikes on their tails. In most respects deep-sea shrimps found down to 16,000 feet differ very little from those in shallow water, unlike many fishes and other creatures whose deep-sea representatives seem queer-looking in comparison with their shallow-water relatives. The most characteristic attribute of many deep-sea shrimps is their red color. This is often brilliant, a clear scarlet or a blood red. Some species are partly red and partly transparent. The former have the color in the cuticle of the exoskeleton, while the particolored ones have the pigment in spots under the skin. In common with many other deep-sea crustaceans, the shrimps living in the depths are usually luminescent. Some reveal themselves by a pale green light. In many kinds the animal is capable of jetting a cloud of luminous material into the water when it is startled. This is produced in a gland under the eye and is emitted through a small pore.

[140]

Sergestes is a scarlet deep-sea shrimp. It is unique in the possession of long antennae, the outer part of which is flexible and whiplike while the base is stiff and stout, for all the world like a fishing line attached to a rod. This comparison is made more valid by the occurrence on the "line" of a large number of curved hooks. Sir Alister Hardy thinks the shrimp may actually use its antennae as fishing lines, catching prey on these hooks and drawing it toward the legs, which bear cruel clawlike spines.

A fixed position in the ocean does not seem to satisfy most of the deep-sea shrimps, and they move up and down in the water with a daily rhythm. The magnitude of these vertical migrations is surprising, with some shrimps mustering the power and swimming ability to alternate between a noontime position at 2,000 feet in the black depths and a midnight level of 650 feet. This means a twice-daily journey of 1,350 feet, one made against the pull of gravity.

8 Squids and Octopuses

of the Deep Sea

The squids and octopuses are the stuff of which legends are made. For reasons partly rational and partly mythical and irrational, these creatures are the objects of emotional reactions ranging from curiosity through fascination to repugnance and horror. The unlovely appearance of the octopus repels nearly everyone who sees it. The common attitude was expressed by Frank Bullen, a turn-of-the-century English naturalist: "In truth, even when very small, there is something ghastly about the appearance of an Octopus. The sombre brown of its body, the pustular skin, the eyes in which a whole inferno of hatred of everything living seems to be concentrated, the palpitating orifice at the top of the head which is the entrance to its body, opening now and then to show the parrot-like beak common to all the race, these are grisly features, but the eight arms, writhing, curling, clinging like a Medusa's hair, are the features of the Octopus which hold the imagination captive. . . . Now I would gladly, if I could, say a good word for the Octopus on my root-principle of justice for all. But I admit that it is very difficult. I do not see, I cannot see, why the Octopus is, except for the purpose of providing abundant succulent food for the shapely fish prowling along the shallow seabed . . . the Octopus is like the rest of his appalling family, a fellow of whom no one can conscientiously say much that is good; but as with the alligator, the mosquito and the louse, since the Lord has seen fit to create him and place him in his present position, it does not become short-sighted man to question that Supreme Wisdom."

Even William Beebe, a professional marine biologist whose breadth of experience with animals probably cannot be matched by many men throughout history, reacted in a strangely irrational manner to an encounter with an octopus. In his book *Galapagos, World's End* he relates how he came across a big octopus in a tide pool in which he was collecting: "We had been working about an hour, when I straightened up to ease an aching back. Almost at my side I saw what will be ever to me the most remarkable sight in the animal world. Frightened by our long continued splashing and tramping, a big octopus had crept quietly out of a crevice just behind me and was making his way as rapidly as possible over the seaweed shelf down to deep water. Nothing animate is comparable to this sight. The bulging mass of the head or body or both, the round staring eyes, as perfect and expressive as those of a mammal, and the horrible absence of all other bodily parts which such an eyed creature should have—nothing more but eight horrid, cup-covered, snaky tentacles, reaching out in front, splaying sideways, and pushing behind, while one or more always waved in the air in the direction of suspected danger, as if in some sort of infernal adieu. I have seen them before, but I have always a struggle before I can make my hands do their duty and seize a tentacle, at the same time protecting them from being drawn into the parrot-beaked mouth."

THE UGLY FAMILY

The repugnance generated by the appearance of the octopus is felt also toward other cephalopods, as the squids, the octopuses, and their allies are called. Cephalopoda means "head-foot," in reference to the circle of arms or tentacles surrounding the head, which is well defined and bears the two prominent eyes. The head is hidden behind the bases of the arms, and the mouth is enclosed by these arms.

The name of the octopus comes from its possession of eight arms. A mantle of thick skin and muscle surrounds a squat, dome-shaped or sometimes roughly egg-shaped body. At the "neck" an opening in the mantle leads into a cavity containing the gills. When the mantle dilates, water is drawn into the mantle cavity, oxygenating the gills; then the mantle closes and muscular contraction sends the water out through a siphon or funnel.

The squids are called decapods ("ten feet"), since in addition to the eight arms of the octopus they have two long tentacles, often with club-shaped ends. The body is cigar-shaped and has fins. Squids have the efficient eyes and the funnel of the octopus; but where the latter are crawling animals, the squids are active swimmers. The

FIGURE 8 – 1 . *An octopus. This picture, which clearly shows the prominent eye, funnel, and sucker-bearing arms, reveals why these creatures are so terrifying. The truth is disillusioning: octopuses are normally timid and harmless.*

Marineland of Florida

cuttlefish, of which *Sepia* is the commonest example, is a decapod and, while strictly not a squid, will be grouped here with them for simplicity. Cuttlebone is familiar to anyone who has kept a canary; this is the internal shell of the cuttlefish, lending it rigidity. One of its principal functions is to secrete gas that lends buoyancy to the animal.

SLIMY LORE

Even today we have much to learn about the cephalopods, and in ancient times they were so little known that mystery was added to the common repugnance toward them. Erroneous accounts as well as deliberate falsehood and exaggeration compounded the misconceptions. Pliny, writing in the century before the Christian Era, described a squid that was supposed to have attacked the fishponds of Carteia in Spain. Dogs were set upon the squid and these were "scourged with the ends of its tentacles" and "struck with its longer arms, which it used as clubs!"

A Norwegian cleric, Pontoppidan, Bishop of Bergen, has become notorious for his fantastic stories, one of which concerned an "island" that turned into a squid. Probably this was the blown-up tale of an encounter by fishermen with a giant squid at the surface. In Pontoppidan's tale a fleet of ships landed on an island where no island showed on the charts. To the horror of the sailors, around the island there arose a forest of snakelike arms, as tall as the masts, that clamped onto men and ships alike with deadly grip. Then arms, men, and ships and the island itself vanished into the sea forever.

In *Moby Dick,* Melville's classic tale of American whalemen, the crew of the *Pequod* sighted "A vast pulpy mass, furlongs in length and breadth, of a glancing cream color—innumerable long arms radiating from its center, and curling and twisting like a nest of anacondas, as if blindly to clutch at any object within reach." It was, said Starbuck, "the great live Squid, which, they say, few whaleships ever beheld and returned to their ports to tell it."

A French naturalist, Pierre Denys de Montfort, sent out from his office in the Museum of Natural History in Paris tales that were believed because of the reputation of that institution, but which were such a mixture of fact and fiction that the public have been misled to this day. One of his tales involved a *poulpe colossal,* which was said to have wrapped its tentacles around the masts of a vessel, the crew saving the ship and themselves only with the greatest difficulty. Another of his stories recounted that a gang of huge squids attacked and sank ten men-of-war in one night. Jules Verne embellished the facts in his famous account of the battle between the submarine *Nautilus* and a giant squid, but this may be excused as literary license; and so, of course, must be the romantic version of this incident in Walt Disney's movie, in which a 2-ton squid was manipulated by twenty-four men with the aid of electronics, compressed air, hydraulics, and remote controls.

As science has accumulated more information about the cephalopods, the truth turns out to be just as remarkable as the highly colored fiction of these and other accounts.

The squid and the octopus are related to oysters and clams, a connection that is easier to accept when we examine the ancient ancestors of the cephalopods: the great nautiloids, ammonites, and belemnites. In the Paleozoic and Mesozoic eras, between 500 and 100 million years ago, a great number of these creatures existed. Many of them had not yet doffed their shells, and rocks formed in ancient sea beds contain the remains of many species of cephalopods with curious shells that clearly proclaim their kinship with familiar molluscs of today.

When all the modern cephalopods except the nautilus discarded their external shells, they lost something in protection. But in exchange the octopuses and the squids—the latter especially—gained freedom of movement. Survival demanded that they achieve quickness and other weapons to compensate for the loss of their armor, and the consequence has been that the cephalopods have become the fastest and most intelligent of invertebrates, and among the most successful.

SWIFT SQUIDS

These superlatives probably refer more accurately to the squid than they do to the octopus; certainly the reference to speed does. The squid's cigar shape allows it to slip through the water smoothly. It swims backward or forward with equal ease, using either fins on the side of the body or, more spectacularly and swiftly, jet propulsion. The squid draws water into a spacious mantle cavity that opens to the outside around its "neck." The opening to the cavity is then closed, and powerful contractions of the muscular mantle discharge water with great force through the nozzle (or siphon) on the ventral side of the body, and the squid is propelled through the sea. The nozzle may be pointed forward to impel the animal backward or toward the rear to shoot it ahead; or it can be turned at any other angle to spin the squid around and give it great freedom of movement.

Sir Alister Hardy says that in short bursts some squids are the fastest of all aquatic animals. They move fast enough to have earned the name "sea arrows," and some species develop sufficient speed to shoot out of the water and sail through the air for all the world like flying fish. In 1948 the *Pequena* Expedition from Capetown to Tristan da Cunha encountered flying squids some 900 miles from Capetown. During a heavy rain and squall, "What was thought to be a flying fish was seen to take off out of a trough of a deep swell 40 to 50 yards away abeam on the port side, and came in at an angle at great speed, overtaking the ship with its wing-like pectoral fin all aquiver. It landed on the fore well-deck. When collected, everybody aboard, including the marine biologists, were amazed to find that it was a squid about 8 in. long and pale green in color. No-one had ever heard of a flying squid before." (From Frank Lane's *Kingdom of the Octopus.*) They had been seen previously, however; for example, the intrepid crew members of the raft *Kon Tiki* encountered large numbers of them during their 4,300-mile ride across the Pacific from Peru to Polynesia. Thor Heyerdahl, the Norwegian ethnologist who led the expedition, reports in his book *Kon-Tiki* of the flying squid: "They pump seawater through themselves till they get up a terrific speed, and then they steer up at an angle from the surface by unfolding the pieces of skin like wings. Like the flying fish they make a glider flight over the waves for as far as their speed can carry them. . . . We often saw them sailing along for 50 to 60 yards, singly and in twos and threes."

The Pacific squids observed by Heyerdahl flew chiefly at night but occasionally in broad daylight. They rose as much as 6 feet above the surface. It turns out that this is by no means the record for height, since there is a report of squid sailing 15 feet above the surface to land on a ship 300 miles off the coast of Brazil. W. J. Rees, an English expert on the squids, tells of a 7-inch individual that landed on the bridge of William K. Vanderbilt's yacht *Ara* as it sailed between Madeira and Casablanca, flying 20 feet above the surface to do so.

Douglas Wilson, the English biologist and photographer, watched a pair of squids in an aquarium. They moved backward and forward, never turning around but reversing direction as they reached the end of the tank. The pair sometimes moved side by side, one tail-forward and the other head-forward, "indifferent as to which end." Sir Alister Hardy put a 6-inch squid in an aquarium 2½ feet long. "The aquarium was of plate glass held together by a welded angle iron so that round the top edge of each side there was an overhanging flange of iron on the inside. In a matter of half a minute or less the squid had made repeated attempts to strike out in different directions, hitting all the walls in turn; then it shot upwards in each direction, up the side so that it came up against the angle iron edge and fell back; finally going to the middle of the tank, it suddenly shot upwards like a rocket, cleared the edge of the tank and hit the far wall of the deck laboratory." (From *The Open Sea: The World of Plankton.*)

The great speed and agility of the active species of squids is made possible partly by their possession of extraordinarily large nerve fibers. Speed of reaction depends on speed of conduction of nerve impulses, and these in turn depend approximately on the diameter of the mantle nerve fibers. The common squid has fibers about one twentieth of an inch in diameter, about fifty times the size of the fibers in most animals. As a result, reactions to disturbances are twice as fast in the squids. Researchers have made good use of these giant nerve fibers in neurophysiological investigations, and squids (especially the giant squids of the Peru Current) are sought for their exceedingly long and thick nerves. This work will contribute to the advance of human physiology and medicine.

STICKY EMBRACE

The arms of both the octopuses and the squids bear suckers. These are used for grasping prey, and in the octopuses also for climbing rocks and walking over the bottom. Lost arms may be regenerated, like the appendages of the crab and other crustaceans. The suckers of the octopus, located on the inner surface of the arms, are circular in shape. In a common species there are normally 240 suckers on each arm, a total of 1,920. The largest suckers, which in a Pacific species are about 2½ inches across, are near the middle of the arm. The suckers are muscular and are raised and strengthened at the rim. When the sucker is clamped onto the surface of a rock or an intended victim, the center is withdrawn by muscular contraction, forming a partial vacuum. This device holds with surprising strength; yet the octopus can release it instantaneously or glide it over the surface to grip a new area.

The suckers of the squid differ from those of the octopus in being raised on a short stalk. They vary greatly in size, from exceedingly small (1/250th of an inch) on a squid caught near Singapore to 2¼ inches in diameter on the giant squid *Architeuthis*. The suckers of the squid have horny rims, and some have teeth on their rims to grip the slippery surfaces of victims. These become fearsome weapons in certain species, being transformed into claws that, like those of a cat, can be sheathed or extended at will. Sperm whales have been caught with great circular scars on their skin, mute testimony to the power of the giant squid.

It is this curious ability of the octopus and the squid to grip with leechlike tenacity that engenders much of the fear and revulsion felt by humans encountering these animals. There is something peculiarly agitating about being gripped by a snakelike arm whose surface is glued so tightly that it often cannot be disengaged or, if it is, is replaced by another arm that clings just as tightly. Even small octopuses and squids produce something akin to panic in many people. The suckers feel like "hundreds of tiny, wet,

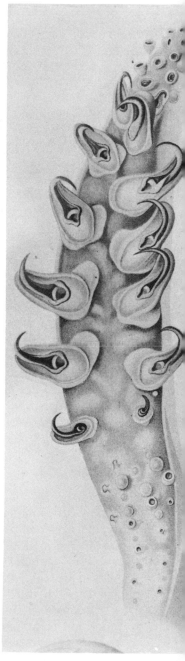

FIGURE 8 – 2. *Cat-claw suckers on a tentacle of the squid* Galiteuthis. *In some species the claws can be sheathed and unsheathed at will.*

Valdivia Expedition

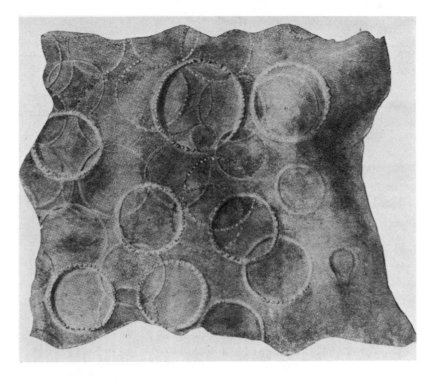

FIGURE 8–3. *Battle scars on a piece of skin from a sperm whale's jaw, telling of an encounter with a squid. When the whale was caught, a big piece of tentacle was found in its mouth, and many squid beaks in the stomach. In its death flurry it disgorged a squid arm more than 18 feet long.*

J. Murray and J. Hjort, *The Depths of the Ocean*

clammy hands pulling at the skin," according to Gilbert Klingel, an American naturalist. Guy Gilpatric, an experienced skin diver, says that if the arm of an octopus is stripped off a human arm "the sound as the suction breaks is that of tearing apart two sheets of flypaper. Also, flypaperlike, the arm which you peel from your leg will stick to your hand, so you'll have to begin all over again."

Despite the strangeness of the feel of an octopus clamped to the leg or arm of a human swimmer, ordinarily nothing is to be feared from an encounter with this creature, since the truth is that even big individuals are normally timid and harmless. Except in extraordinary circumstances, cephalopods are not dangerous to man. Frank Lane, in his book *Kingdom of the Octopus*, quotes the MacGinities (biologist husband and wife) as saying, "Although we have handled several thousand octopuses we have never been bitten by one, in spite of having tried to provoke them to bite." Max Gene Nohl, president of the American Diving Equipment Company, says: "In my opinion the chance of a diver being attacked by an octopus is as remote as the possibility of a hunter in the woods being attacked by a rabbit." But a man, out of his element, may encounter difficulty in a fight under water, and Lane recounts instances where men have had to struggle hard to avoid being drowned when gripped by a big octopus. It is clear, too, that even though the MacGinities escaped unharmed during many encounters with octopuses, these creatures can give a nasty bite. There is even one report that such a bite was fatal. The man involved was Kirke Dyson-Holland, a twenty-one-year-old seaman and skin diver. Spearfishing near Darwin, Australia, he caught a 6-inch, blue-colored octopus and allowed it to crawl over his back and shoulders. Although Dyson-Holland did not complain of being bitten, a companion noticed a small wound on his back, oozing with a trickle of blood. After leaving the water the swimmer complained that his mouth was very dry and that he found it hard to swallow. He began to vomit, collapsed when he tried to walk, and was carried in a weak condition to the hospital. Adrenaline and an iron lung failed to help him, and two hours after being bitten he was dead.

Squids are more dangerous than octopuses. Frank Lane tells "one of the most macabre cephalopod stories I know: When the troopship *Brittania* was in the Central Atlantic, 1,400 miles west of Freetown, she was attacked and sunk on March 25, 1941, by the German raider *Santa Cruz*. Lieutenant R. E. G. Cox, as he was then, and eleven other men, clung to a raft 'no bigger than a hearth rug.' The men took it in turns to sit on the raft, while the others clung to it with only their head and shoulders above water. One night a large squid threw its tentacles around a sailor, broke his hold on the raft and pulled him under. He was not seen again. Shortly afterwards Cox was attacked. He tells me: 'A tentacle quickly twisted around my leg and caused terrible pain. It removed itself almost immediately afterwards but left me in agony. The pain I think was caused by the suckers in the tentacle for later the next day I noticed that where it had gripped me large ulcers had formed, and after I was rescued I remember the constant treatment which was given me for these ulcers which seemed to eat right into the flesh and were red and raw. I still have these marks to this day.' "

The giant squids of the Humboldt Current off Peru are perhaps the most ferocious and dangerous of these animals. Big-game anglers like John Manning, Lou Marron, and Michael Lerner have had their prized marlins chopped to pieces by these squids, which grow to lengths of 12 feet and weights of 350 pounds. In revenge, and because the sport was exciting, the fishermen took to angling for the squid themselves. This produced some curious problems seldom encountered by sport fishermen, including the danger of being deluged with water and jet-black ink. Manning describes the water shot from the funnel of the squid as "about like the blast from a fire hose"; anglers and vessel alike were blackened by the ink. In defense the anglers took to wearing pillowcases over their heads, with holes cut for their eyes. They must have looked like eccentric ghosts, clinging to the rail of a pitching boat in the blackness of the Pacific night, flailing the sea with their fishing lines, while the giant squids shot inky water and hurled themselves with explosive speed through the sea, which exploded into brilliant phosphorescence as the animals blazed through it. The great squids bit through heavy leader wire with their powerful beaks, and one gaffed squid took chunks out of the handle of a tough wooden gaff hook. It is little wonder that the natives of that part of the coast fear the squids, and regard them as demons rather than animals. Lane insists that a man who fell overboard in such waters would not last half a minute.

This Humboldt Current squid, *Dosidicus gigas,* is called a giant squid with good reason: any animal 10 to 12 feet long and weighing up to 350 pounds is in truth a giant. There are still bigger squids, however, although it may well be that there are no more ferocious ones. The biggest of the squids is *Architeuthis,* which attains a length of 50 to 55 feet and perhaps longer. Most of this length is in the two slim tentacles. On a giant squid found in New Zealand the tentacles measured 49 feet 3 inches in a 57-foot animal. This disparity seems to have been extreme, however, and for it this creature was named *Architeuthis longimanus.* Dr. Hardy lists specimens of *Architeuthis princeps* 55 and 52 feet long whose bodies were 20 and 15 feet long respectively, and a specimen of *Architeuthis harveyi* of 52 feet with a 10-foot body.

[152]

FIGURE 8-4. *Giant squid* Architeuthis princeps *and a frogman, drawn to the same scale. The biggest specimens of* Architeuthis *seen have been about 55 feet long. Squids can jet forward or backward with equal ease, and some go fast enough to shoot out of the water like flying fish.*

Sir Alister Hardy, *The Open Sea: Its Natural History. Part 1. The World of Plankton*

COLOR BEARER

The chameleon is the standard in the animal kingdom for dramatic and rapid changes of color, but this little lizard is a sluggish beginner at the game of camouflage as compared with the cephalopods. Many of the octopuses and some of the squids have remarkable ability to match the color of their background, even being capable of causing a wave of color to flow across their bodies as they swim from one background to another. Often overriding this sort of color change is one produced by emotion. Amusingly like humans, octopuses will blanch ashy gray when frightened and flush red or purple when angered. In *Inagua,* Gilbert Klingel describes an octopus he observed off Great Inagua Island in the Bahamas: "It always seemed irritated at my presence. Its nervousness may have been caused by fear, for it certainly made no pretense of belligerency, and it constantly underwent a series of pigment changes that were little short of marvellous. Blushing was its specialty. No schoolgirl with her first love was ever subjected to a more rapid or recurring course of excited flushes than this particular octopus. The most common colors were creamy white, mottled vandyke brown, maroon, bluish grey, and finally light ultramarine nearly the color of the water. When most agitated it turned livid white, which is, I believe, the reaction of fear. During some of the changes it became streaked, at times in wide bands of maroon and cream, and once or twice in wavy lines of lavender and deep rose. Even red spots and irregular purplish polka dots were included in its repertoire, though these gaudy variations seldom lasted for long."

These amazing color changes in the octopus, and in all cephalopods, are produced by tiny spots of pigment in the skin. The pigment cells, called chromatophores, are very small and transparent and are filled with a fluid in which are suspended granules of pigment. Attached to the very elastic walls of the chromatophores are radially arranged muscle fibers. When these fibers are at rest, the pigment cells are contracted to a very small size, almost disappearing in the skin of the animal. The muscles can pull out the chromatophores to a much greater size, spreading the color over a wide area of the skin. In combination, many of the pigment cells lend their color to a large or a small area of the skin. Douglas Wilson says: "It is a fine sight to watch a shoal of these elongated creatures and see their skins shimmering as the tiny spots of color rapidly expand and contract."

The cuttlefish *Sepia* is probably the most skillful of the whole clever group at producing color changes. *Sepia* has chromatophores of three colors arranged in three layers beneath its transparent skin. The top layer contains bright yellow cells, the next layer orange-red cells, and the third layer dark brown, nearly black, cells. Beneath these layers of contractile cells is a basal layer of brilliant iridocytes whose beautiful iridescence forms the background for the other colors. *Sepia* can make itself white by contracting all the cells, or yellow by contracting all but the top layer of cells, or it can darken its skin by exhibiting only the third layer. It can also expand some of each kind of chromatophore, completely or partially, in infinite variation and thus produce a rainbow kaleidoscope of color to suit all occasions. Guided by the eye, the central nervous system controls the slim muscle fibers that work the chromatophores, and the color of the animal is therefore under its full control. The speed with which color changes can be made greatly exceeds anything elsewhere in the animal kingdom, and a color cell can change from a condition of full contraction to full expansion—as much as sixty times its original diameter—in two thirds of a second. The muscle fibers seem to be immune to fatigue. Scientists found that the muscles seemed to work as efficiently as ever after being stimulated for 30 minutes with electric shocks at the rate of thirty per second.

At rest the cuttlefish sends continuous waves of color over its skin, the shifting of shades and the generally striped pattern undoubtedly making it more difficult for a predator to keep the little animal in focus. While swimming, the cuttlefish can alter its color to suit the background. In *Kingdom of the Octopus,* Frank Lane reports observations by Joyce Allan, an Australian biologist, who "watched some small Australian cuttlefish swimming slowly about an aquarium containing variously colored objects. As the cuttles passed over dark, reddish brown rock, they matched its color perfectly. A few inches away there was some dead coral, and as the cuttles passed over it they turned a light grey. Then, in turn, light brown for sand, green-brown for weeds, and back again to the reddish-brown of the rock."

Dr. William Holmes, a British biologist, published results of interesting experiments on the color changes of *Sepia officinalis.* In a large aquarium tank at the Plymouth Laboratory the back of the squid was dark brown, nearly a solid color down the middle but with irregular stripes down the sides that broke up the outline of the animal against its background. When it lay on the sand the animal assumed a light mottled appearance like the sand itself; in addition, it threw up sand with its fins so that particles fell on the edges of its body, obliterating its outline. In a black tank the squid assumed a very dark color. If now a white piece of porcelain was put in the tank near it, the animal produced an astonishing square patch of white on its back against the dark skin to mimic the porcelain. In a well-lighted aquarium having a strongly contrasted background of light and dark stones on the bottom, a broad, very white stripe appeared laterally across the back of the animal. When disturbed, the squid produced two black spots on its back, always in the same position. If the annoyance persisted, the animal accentuated the black spots by lightening the rest of its skin so that the spots stood out vividly against the iridescent white body. Often at this point the squid would contract its body suddenly and shoot away with an abrupt squirt of its siphon, at the same time flashing a dark color over its body. This performance left the observer staring at the place where the white animal with the black spots had been an instant before, while the squid was safely elsewhere.

REPRODUCTION

The male cephalopod is often smaller than his mate. The disparity is greatest in the argonaut (*Argonauta*), where the female may be fifteen times as long as the male. The latter commonly has one and occasionally two of its arms modified as a device for passing sperm to the female. In some cephalopods this modification may be slight, consisting largely of the loss of some or all of the suckers, or it may change the arm greatly, making it into a long and whiplike affair or one with a spoon-shaped end.

The modified arm is called a hectocotylus, the name coming from an odd mistake made by the eminent French biologist of the nineteenth century, Baron de Cuvier. One of the mysteries of natural science in the early 1800's was how the argonaut reproduced itself. Then Cuvier found small bodies attached to the female. He thought they were parasitic worms, and because they had about a hundred suckers he called them *Hectocotylus* ("hundred cups"). Later it was discovered that these "worms" contained the sperm of the argonaut, and that they were really the ends of the long arms of the male. In transferring the sperm to the female the male sacrificed the end of its arm, which contained the sperm in a tubular spermatophore. The name applied by Cuvier to the false parasitic worm is perpetuated in the name of the specialized arm of all cephalopods.

Fertilization usually takes place in the mantle cavity of the female, and the eggs are then extruded through the funnel. In the common squid, *Loligo,* the egg-laying takes place when the female stands on her head, her arms grasping some firm surface to which the eggs are attached in strings by a soft, sticky substance which resembles jelly.

The octopus exhibits maternal care of the eggs, behavior that is rare in the invertebrates. In an aquarium a female of the common species *Octopus vulgaris* was seen to fasten her eggs to the side of the tank in strings, like bunches of tiny grapes, and then cover them with her body. She guarded the eggs fiercely, killing another octopus that came too close to suit her, and never left the eggs—even to eat—until they were hatched. She aerated the eggs by moving her mantle to pass a current over them, and she kept them clean by squirting water over them with her siphon. When the eggs began to hatch she became very excited and dashed at a hand inserted into the tank.

The young of some cephalopods do not go through larval stages, but hatch in the form of the adult. For a while after hatching, the babies swim in the plankton; later the octopuses go to the bottom, while the squids assume a pelagic existence.

THE INK TRICK

One of the more spectacular tricks of the cephalopods is shooting from its siphon a cloud of ink to confuse its enemies. *Sepia* is one of the cephalopods that uses this protective mechanism. When attempts to avoid detection through protective coloration fail, the cuttlefish shoots a cloud of blue-black ink into the water and hides beneath it. This ink has enormous diffusive powers, and one cuttlefish can spoil the color of the water in a 1,000-gallon aquarium tank; aquarium curators are understandably hostile to excitable cuttlefish in their care.

Octopuses and squids apparently use the ink not so much as a smoke screen as to produce a diversionary substitute shape for themselves in the water. Jacques Cousteau has seen many octopuses in his skin-diving excursions, and he says that the ink is "not a smoke-screen to hide the creature from pursuers. The pigment did not dissipate; it hung in the water in a fairly firm blob with a tail, too small to conceal the octopus. . . . The size and shape of the puff roughly corresponds to that of the swimming octopus which discharged it." Squids perform the same trick, simultaneously changing color and producing a silhouette in the water toward which a predator may lunge as the squid shoots away. The squid *Ommastrephes pteropus* produces not only a cigar-shaped cloud of ink of nearly the same length and diameter as itself but one that coagulates in the water and keeps its shape for several minutes. Another species, *Alloteuthis subulata,* has been seen to produce a dummy that maintained its shape for 10 minutes or more if it was undisturbed.

In addition to creating a smoke screen or a ghostly squid to confound enemies, the ink may deaden the olfactory organs of predators. Moray eels, mortal enemies of cephalopods, seem to lose their ability to smell their prey if they are deluged with ink. On the other hand, dilute concentrations of ink alert the eel to the presence of a meal, and they become wildly excited, dashing about in search of the squid or octopus. The odor of cephalopod ink has been described as a "fishy musk."

The ink itself is a thick, dark fluid. It is brown in the squids, blue-black or brown in the cuttlefish, and black in the octopuses. Diluted, the black ink of the octopus appears as dark purple, with tints of red and other colors. Sepia, the ink of some species of cuttlefishes, is a rich brown color famous as an artist's pigment. It has been used in pen and brush drawings by artists for centuries. It makes paintings and drawings look old when they are fresh, and older still later on, for sepia fades with age; some masterpieces done with this strange pigment have paled to oblivion. Paradoxically, the substance itself is amazingly stable chemically, and fossilized sepia from a squid dead millions of years has been used to make drawings. The artist who used this ancient pigment judged it to be "as rich in color as that manufactured from the ink of recent cuttlefishes."

THE WONDERFUL EYE

When biologists describe the eyes of the cephalopods they can hardly contain their surprise and enthusiasm, using adjectives and superlatives uncommon in ordinarily cool scientific language. Sir Alister Hardy calls them "wonderful eyes" and Dr. N. B. Marshall says they are "astonishingly like those of the vertebrate in structure." To the biologist it is a constant source of wonder that living creatures should have evolved the enormously intricate, efficient, and durable organs they possess, and the eye is the most marvelous of all of these. What the scientist finds hardest to understand in considering the squid eye and the human eye is that two entirely independent lines of evolution should have converged at the same point. Dr. N. J. Berrill, a Canadian biologist, puts it thus: "I think if you asked any zoologist to select the single most startling feature in the whole animal kingdom, the chances are he would say, not the human eye, which by any account is an organ amazing beyond belief, nor the squid-octopus eye, but the fact that these two eyes, man's and squid's, are alike in almost every detail."

Actually, in principle, the eye of the squid is superior to that of man. While both have a transparent cornea, an iris diaphragm, a clear lens, and a chamber filled with liquid and muscles to control the lens and the movement of the eyeball, the arrangement of the retinal cells in the squid eye seems more logical than that in man's. The light-sensitive cells of the retina in our eyes are covered by the ganglion (nerve) cells, which intervene between them and the lens. In the squid this is not the case, and it may therefore have a wider field of clear vision. The squid's eye, however, is inferior to the vertebrate eye in power of accommodation (the ability to adjust its focus from near to far objects and the reverse).

While the primitive shelled nautilus has a small eye, the eyes of most of the other cephalopods are large in relation to the size of the head. That of the giant squid *Architeuthis* may be 15 inches in diameter, as big as a sizable beach ball, and over twice the size of the eye of the 100-ton blue whale. One of the most inexplicable results of evolution is the development of eyes of greatly different sizes on the two sides of some squids. In *Histioteuthis* and *Calliteuthis* the left eye has about four times the area of the right eye. In *Calliteuthis reversa*, a squid from as deep as 3,300 feet, the right eye not only is smaller but is sunken and surrounded by a closely spaced ring of photophores (light organs), which are absent from the large left eye. Dr. Gilbert Voss of the University of Miami Institute of Marine Science suggests that the large eye functions in the dim light of the deep sea and the small eye in surface waters.

THE HUNTERS

The octopus relishes crabs above all other food, and it is perhaps the most efficient crab trap in existence. Patience is one weapon employed by the octopus: it lies in wait among rocks until a crab happens by. If no convenient hole is available at a suitable spot, the octopus will build itself a fortress of stones in which to lurk, enlarging the crevice by blowing out the sand or fine gravel with its siphon. From this ambush a long, sinuous arm flicks out and almost casually clamps onto a passing crab, which is drawn within range of the powerful jaws and toothed, ribbonlike tongue. Joseph Sinel, an English naturalist who observed octopuses for many years in the Channel Islands, describes the tentacle as starting coiled up, but unwinding itself gently toward the crab, eventually to clamp onto it with the suckers on the end. "I have seen a moderate-sized octopus," says Mr. Sinel, "thus catch 17 crabs in succession, just storing them in the custody of its manifold suckers, to await their turn." Its ability to change color to blend with the background rocks, and the extra purchase lent by its grip on the bottom with some of its arms, give the octopus advantages over its quarry. Sometimes an octopus will stalk its prey rather than ambush it. When it encounters a potential victim it will rise above it and descend like a parachute, the arms and the webs that join them spread to form a deadly blanket.

The cuttlefish, too, is an active pursuer of prey. Its favorite foods are shrimps, small fish, and small crabs. When these are sighted, the cuttle blushes and fades in excitement, circles the prey until it is in position, and then shoots out the two long tentacles. These act like a pair of tongs, nipping the victim between their ends, where the strong suckers hold it fast. If the intended meal is a crab, the cuttlefish maneuvers carefully to get behind it so that the claws cannot operate. If no prey is in sight the cuttlefish may hunt blind, squirting little jets of water into the sand at random in hope of uncovering a cowering shrimp. If the shrimp is wise enough not to move, it may still avoid capture though stripped of its sand cover, since it is efficiently camouflaged. But it usually panics and attempts to swim away, in which case it is lost, for the cuttlefish pounces on it and whips it to its mouth with the long arms.

Squids are fierce and merciless hunters. Their devastating attacks on fish caught by anglers off Peru has already been mentioned. Great triangular chunks are bitten out of anything hooked, including other squids, for these creatures are cannibals. Dr. Paul Bartsch of the United States National Museum describes watching squids at night in Philippine waters, in the harbor of Jolo, savagely catching small sardines and apparently killing for the fierce joy of it as well as for food. A light was shined over the side of the ship into the water, and "we soon found millions of creatures drawn into the vortex about our light . . . small fish of various kinds, a school of sardines dashing madly after the small crustaceans and worms. . . . It was a mad dance that this whirling, circling host of creatures performed. Then a new element entered: living arrows, a school of loligopsoids shooting across our lighted field, apparently attracted not so much by the light as by the feast before them. They were wonderful creatures, these squids, unlike anything else there; they shot forward and back like shuttles with lightning rapidity. Not only that, but they were able to divert their course to any new direction with equal speed. As each of them shot forward, his tentacles would seize a small fish, and instantly he would come to a full stop, only to dart backward like a flash at the least sign of danger. Kill, kill, kill; they were bloodthirsty pirates. A bite in the neck and the fish was done for; but the sport continued, and, likely as not, one fish would be dropped and another seized and dispatched." (From an article in *Smithsonian Scientific Series,* Vol. 10.)

Once caught by a squid or an octopus, a victim gets short shrift. Held helpless by the powerful suckers, the prey is killed by a chitinous black beak, curiously like a parrot's except that the underjaw is the longer instead of the other way about. It is this weapon that tears the triangular pieces from the body of the victim. A fish or other soft-bodied prey is quickly consumed; a crab takes longer—but not much. The beak crunches through the shell, and the legs are torn off. Many octopuses also use poison to kill their prey. This works rapidly on crabs, which lose control of their limbs, develop convulsive movements, and die within a few minutes. The octopus may also inject a digestive fluid into the crab that liquefies the tissues of the prey inside its shell, after which, as Douglas Wilson says, "The resulting soup is sucked out and the empty shell discarded, few solid pieces having been swallowed."

The feeding operation is assisted by the radula, a hornlike ribbon with a central row of fair-sized teeth, on each side of which are minute curved teeth. This serves to jam down the esophagus chunks of food bitten off by the beak.

The squids and octopuses account for a great many crabs, fish, and other victims, and at times become so numerous that they constitute a real menace to crab and lobster fishermen. A famous pestilence of common octopus occurred in 1899 off the coast of France. The beaches were littered with their dead bodies, some more than 6 feet across. Every stone on the beach harbored an octopus, and tons of these animals were carted off by the farmers for fertilizer. The octopuses practically wiped out stocks of crabs and damaged lobster and oyster populations too. In 1900 the plague reached the English coast, and there, too, the crab and lobster fishermen suffered. Since then several other outbursts of octopus abundance have taken place in Europe—moderately serious ones in 1950 and 1951. Scotland and Japan have also been visited by swarms of these animals at one time or another.

THE HUNTED

But if octopuses and squids are often active predators, they are also the prey of many animals—including man. They are the favorite food of many fishes; they are eaten by elephant seals and penguins, jellyfish and sea lions; they form the principal diet of the sperm whale; they are eaten enthusiastically by others of their own species. This passion of so many of the sea's creatures for the cephalopods "is the answer," according to Bullen, who held them in low regard, "to the question of how do they enjoy their lives, being only apparently born to be chased and devoured."

And there are surely very large numbers available to be devoured. This is especially true of the squids, although they are so fast that fishing gear and biological sampling devices probably fail to reflect their true abundance. Sir Alister Hardy says in his book *The Open Sea: The World of Plankton* that "they probably avoid most of our nets, or if once inside, may easily get out again. On the *Discovery* expedition to the Antarctic in 1925–27, when we spent much time studying the general biology of the whaling grounds round the island of South Georgia, we used almost every kind of net and trawl but caught only very few squid; yet the stomach of every elephant-seal we examined was full of their remains. The elephant-seals occur there in huge numbers; to keep that stock nourished the squids must be enormously abundant. . . ." The sperm whale, or cachalot, feeds principally (but not exclusively) on squids. Ambergris, the curious material produced by the digestive processes of the whale and sought by perfume manufacturers to "fix" the odor of their product, is regarded as having its basis in the partly digested remains of squid. Ambergris is cast out by the whale, and if it escapes destruction by the waves, may be found floating at the surface or on a tropical beach.

Included in the diet of the sperm whales are occasional giant squids, *Architeuthis,* but the principal species consumed are smaller kinds, mostly *Histioteuthis bonelliana* and *Cucioteuthis unguiculatus,* averaging just over 3 feet and ranging up to 8 feet in length. In 1895, when the *Princesse Alice I,* one of the famous ships of Prince Albert of Monaco, was working near the Azores, local whalers mortally wounded a whale, which died under the yacht. In its death throes it spewed up parts of a great squid new to science. Excited about the possibilities of finding still more specimens, the Prince had his yacht fitted out as a whale catcher. His investigations on the food of whales showed that some species feed mainly on squids, including some with tentacles as thick as a man's arm and studded with great suckers bearing powerful catlike claws.

Humans, knowing the eagerness with which many fishes regard squids, use them as bait. The squid's toughness on the hook adds to its value as bait, and tons of squids are caught and used all over the world by commercial fishermen. Not enough squid is caught to satisfy the demand. Half the cod taken on the Grand Banks is caught on hooks baited with squid, and when it is in short supply the fishery is seriously handicapped.

Fish may relish squids as food, but most North Americans regard cephalopods as inedible. This is a narrow point of view, as millions of people in Asia and southern Europe will testify. The ancients of the Mediterranean region considered cephalopods choice morsels. Of the octopus Dr. Paul Bartsch says: "The Greeks and Romans considered it the finest food furnished by the sea. Pliny tells us that the gourmands of Rome ate every variety of octopus known in the Mediterranean. The cooks baked the creature in a sort of big pie, cutting off the arms, and filling the body with spices; and they were so careful in their preparation of the animal for cooking that they used pieces of bamboo for drawing the body, instead of iron knives, which were supposed to communicate an ill flavor to the delicious morsel. How highly the cuttle was esteemed by the Greeks is evident from a story told of Philoxenus of Syracuse, who, desiring a delicious dinner, caused an octopus of three feet spread to be prepared for the principal dish. He alone ate it, all but the head, and was taken so sick in consequence of his surfeit that a physician had to be called. On being bluntly told that his case was desperate and that he had but a few hours to live, Philoxenus called for the head also. When he had eaten the last bit of it he resigned himself to his fate, saying that he left nothing on earth which seemed to him worthy of regret." (From an article in *Smithsonian Scientific Series,* Vol. 10.)

In Italy, street-corner stalls sell hot octopus on a fork. In Spain, dried octopus is sold looking like old tarred rope. In Cuba and Mexico and other Latin American countries, octopus and squid form important parts of many local dishes, while octopus in its own ink is canned in Portugal and sold in many parts of the world.

Apart from their appearance, which keeps squids and octopuses from being eaten by many people, the other principal deterrent is their toughness, unless they are properly prepared. Bartsch tells how the tentacle of an octopus captured by native fishermen in the Philippines defeated his efforts to chew it: "I chewed a single tentacle during the greater part of the following forenoon and relinquished it only, and that with regret, when my jaws, aching from overexertion, refused to operate any longer." To be edible, cephalopods have to be vigorously beaten, or tenderized in a pressure cooker.

In Japan, squids and octopuses are important seafoods, especially in the northern islands. The Japanese are reported to have

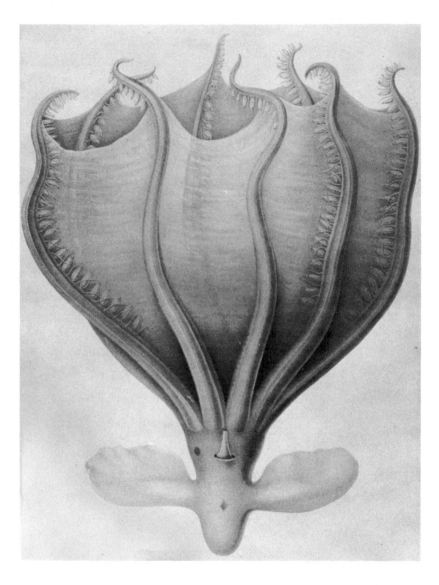

FIGURE 8 – 5 . *A swimming octopus of the abyss,* Cirrothauma murrayi, *about 6 inches long. This animal is so fragile and gelatinous that it might be mistaken for a jellyfish. It is the only blind octopus known.*

Valdivia Expedition

used squids in another and very curious way. The story is told by Dr. H. M. Smith, to whom it was first told by a professor at Imperial University: "More than a century ago a vessel laden with a very valuable cargo of porcelains from Korea destined for the imperial household was wrecked in the Inland Sea; the captain and other officers did what seems to have been a favorite amusement of the olden days; namely they committed suicide just before the vessel sank in deep water. Recently the fishermen have been recovering pieces of this pottery, which now has an appreciated value, by tying strings to octopuses and lowering them in the vicinity of the wreck. The animals enter the vessels and retain their hold of them while being drawn to the surface. Several pieces of this porcelain which I saw were gems, seeming but little the worse for their prolonged submergence." (Quoted in Paul Bartsch's article in *Smithsonian Scientific Series*.)

The deep sea probably shelters large numbers of squids, although relatively few appear in trawls and other fishing gear, as in the case of shallow-water representatives of the group. We might expect, however, that deep-sea squids would be less powerful and slower than those in the shallows, because nearly all of them are weakly muscled. Gelatinous tissue invades the muscles, and in addition there is often a layer of gelatinous material under the skin. So marked is this condition that many species of abyssal cephalopods are as soft and fragile as a jellyfish. Sometimes the comparison to a jellyfish goes another step, in that the cephalopod may be translucent. Internal organs shine through the body wall—especially the liver, which may be brightly colored.

This replacement of dense muscle tissue by lighter gelatinous material serves to reduce the specific gravity of the animal, and a reduction in heavy body parts accomplishes the same end. The radula may be weakly developed, and the ink sac small or absent.

Some cephalopods live in very deep water. Squids have been caught down to at least 11,500 feet. In the Weddell Sea off Antarctica an octopus, *Grimpoteuthis*, lives on the bottom at a depth of about 9,000 feet. In the upper levels of the deep sea (between 325 feet and 1,600 feet) there are luminous squids swimming with the myriad lantern fish, hatchet fish, *Cyclothone*, *Stomias*, and red prawns. Some deep-sea cephalopods are red too; others are black; still others are deep purple or translucent.

Of all the brilliantly lighted creatures of the deep sea—even

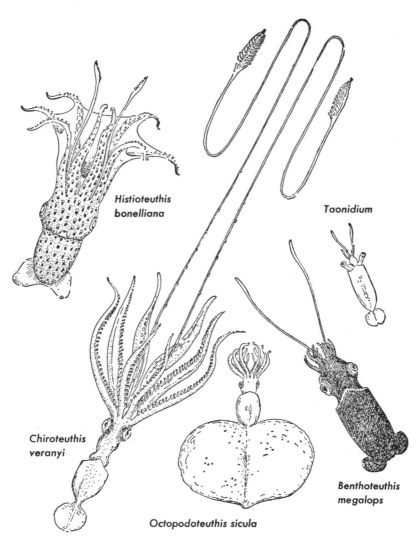

Histioteuthis
bonelliana

Taonidium

Chiroteuthis
veranyi

Benthoteuthis
megalops

Octopodoteuthis sicula

FIGURE 8–6. *Deep-sea squids have weaker muscles than shallow-water species, and many are soft and translucent like jellyfish.* Histioteuthis bonelliana *flashes brilliant blue and yellow lights; its left eye is much bigger than the right.* Chiroteuthis veranyi's *tentacles carry lights and hooks at their ends.* Octopodoteuthis sicula *has very large tail fins;* Taonidium *has eyes on long stalks.*

Sir Alister Hardy, *The Open Sea: Its Natural History.*
Part 1. The World of Plankton

including the spectacular lantern fish and stomiatoids—the squids make a strong bid for first place. They possess numerous multicolored photophores, flashing and gleaming these organs with virtuoso skill. More will be said of this in the chapter on luminescence.

The food of deep-sea squids is necessarily different from that of the shallow-water species, since the same prey animals are not available in the abyss. Small species of deep-sea cephalopods probably feed on the plankton. Specimens of *Lycoteuthis diadema* had in their stomachs the remains of pteropods and small crustaceans. Bigger squids undoubtedly eat hatchet fish and other small fishes of the deep

FIGURE 8 – 7. *Octopus* Opisthoteuthis extensa *from the deep-sea floor. It is quite formless and has been described as resembling a very dirty floor-mop.*

Valdivia Expedition

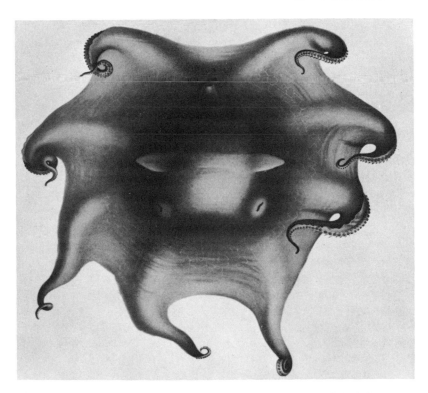

sea, as well as crustaceans, including prawns. Like so many of the inhabitants of the abyss, the squids migrate toward the surface at night, sometimes from considerable depths, to consume animals that are numerous in the shallower waters, descending into the deep sea again as light approaches.

Some of the cephalopods already described are deep-sea animals even though they appear at the surface at certain times. Sir Alister Hardy has sketched several curious examples of deep-sea squids; *Chiroteuthis veranyi* has two enormously long and very slender tentacles, several times the length of the body, and very much longer than the rest of the arms; *Octopodoteuthis sicula* has ludicrously broad tail fins; *Bathothauma (Taonidium)* (like many others from the abyss) has eyes on long stalks. Perhaps the oddest of an odd lot are the deep-sea octopuses, *Opisthoteuthis*, which are so shapeless and nondescript that when preserved "they resemble in about equal degree a soggy pancake or a very dirty floor-mop."

"Probably no group in the sea has undergone such weird modifications as the deep-sea squid," according to Dr. Gilbert L. Voss, squid expert at the Marine Laboratory of the University of Miami. "In some the head projects in front of the body on a long stalk. In others the eyes are on the ends of long stalks armed with searchlights, which can be flashed around in a curious manner. Some deep-sea squids have completely transparent bodies in which even nerves and blood vessels can easily be seen." The long tentacles of the deep-sea squid *Chiroteuthis veranyi* are armed at their ends with light organs and sharp hooks. Dr. Voss believes that this squid, a weak swimmer, dangles its tentacles with the luminous organs lighted, and thus attracts planktonic animals like moths around a candle. Then the hooks are employed like "jigs" to snag the little prey and save the squid the bother of pursuing its dinner.

9 Fishes of the Deep-Sea Floor

The uttermost depth of the ocean was penetrated in 1960 when Jacques Piccard and Lieutenant Donald Walsh of the United States Navy dived to the bottom of the Marianas Trench in the bathyscaph *Trieste*. One of the momentous events of this dive was the sighting of a living fish on the floor of the trench, at the prodigious depth of 35,800 feet, resting on the cream-colored ooze of the bottom. In the beam of the searchlight, "Slowly, very slowly, this fish—apparently of the sole family, about a foot long and half as wide—moved away from us, swimming half in the bottom ooze, and disappeared into the black night, the eternal night which is its domain."

The fish that Piccard and Walsh saw was a flatfish, one of the major groups that live at the very bottom of the sea, usually right on the oozy floor. Actually, it is risky to make an artificial division of the fishes of the deep sea into mid-water and bottom forms, since some groups have representatives in each of these major habitats and some species are not exclusive to one or the other. Besides, it is hard to be sure whether certain fishes were captured by a trawl when it was on the bottom or in mid-water. Nonetheless, there are certain fishes whose destiny is closely tied to the bottom of the ocean, and this habitat is so different from that of mid-water that it is worthy of separate treatment.

RAT-TAILS

The cod has been called the beef of the sea. Next to the swarming herrings, more cods and their allies (collectively, the family Gadidae) are caught in the world oceans than any other fishes—

some 11 billion pounds each year. It was cod that drew the early adventurers to the northeast shores of America, not gold or silver or gems. And these adventurers came early, before Cabot's voyage of 1497. Portuguese fishermen (who still bring fleets of sailing vessels to the Grand Banks of Newfoundland for cod) have a tradition that their countrymen started the Campanha Bachaloeira—the "Codfish Campaign"—in 1493. It may have been still earlier, since Cabot wrote shortly after his epic voyage, "There are plenty of fish and those very great, as seals and those we commonly call salmons: there are soles also a yard long, but especially there is a great abundance of that kind which the savages call Baccalaos." If Cabot was reporting accurately, Europeans must have been on American shores long before he landed, and before Columbus, since *baccalao* is the Basque name for cod. In any case, the cod fishery is a very ancient one in America, and still more ancient in Europe.

The common cod, *Gadus morrhua*, on occasion wanders into what we have defined as the deep sea, that is, waters deeper than 650 feet. Close relatives of the cods, the rat-tails (family Macrouridae) are more numerous still; they are, in fact, the commonest of all the deep-sea bottom-living fishes. These odd fishes, of which there are many kinds, have a big, heavy-looking head from which the body tapers rapidly to a long tail. This tail, which gives the group its common name, is compressed and often ends in a fine filament (Figure 13-8).

The head is heavily armored and is pitted with prominent canals containing the curious long-distance "lateral-line" sense organs. The snout is occasionally blunt, but more often it is extended forward into various shapes—slopes and overhanging projections or sharp, spikelike tips. A fleshy barbel ("little beard") projects from the chin, proclaiming that these are bottom fishes, for it is used to sense food organisms in the mud, where the rat-tails usually feed. The mouth is called "inferior" by ichthyologists. This is not any reflection on its quality, but indicates that it is on the underside of the head. The mouth is usually armored and may project as a wide tube.

The dorsal fin (the large fin on top of the body) is long and low in most of the rat-tails, appearing to be continuous with the long anal fin. The lateral line is prominent, and so are the scales. The latter are "keeled" or ornamented in various ways and are easily shed. Because of the latter characteristic, rat-tails commonly come out of the depths with most of their scales knocked off in the trawl, and their grotesque

appearance is made all the more unattractive. Most rat-tails are between 1 and 2 feet in length, with the adults of the various species
ranging from 8 inches to about 3 feet.

The eyes, which are strikingly large in nearly all species, are a
dominant feature of the rat-tails. Individuals from 10 to 30 inches
long have eyes with diameters ranging from ½ to 1½ inches. But
the eyes are also among the most sensitive in the whole animal
kingdom. One of the measures of visual acuity is the number of rod
cells present, since these are the retinal elements active in night
vision. In 1908 the German biologist August Brauer published results
of a careful examination of the eyes of deep-sea fishes caught by the
Valdivia Expedition. He estimated that one of the macrourids,
Lionurus (*Macrourus*) *pumiliceps*, had the prodigious number of 20
million long, slender rods in a square millimeter (less than one six-
teenth of an inch square). This number, the greatest for any fish
examined, seems to be incredibly large, and it may be that Dr.
Brauer's equipment was inadequate to count such small structures.
But even if his estimate was much too high, the fish's eye clearly
seems to possess far more rods than the human eye, in which there
are 20 million rods (plus 750,000 cones, the elements responsible for
color vision, missing in the deep-sea macrourids) in an area of 1½
square centimeters, 225 times the area in which Brauer counted
20 million rods in *Lionurus*.

The figure for humans is given by Dr. M. H. Pirenne of Oxford
University, who points out that with this complement of rods the
"dark-adapted" human eye (that is, one conditioned to the gloom)
is sensitive to amazingly low intensities of light. Says Dr. Pirenne,
"If we wake up in the middle of a very dark night and look at the
window of our bedroom, we can distinguish features of the window
against the lighter patch formed by the sky. . . . The luminance of
the sky at full moon is roughly 20,000 times that corresponding to
the absolute threshold of the eye [the lower limit of ability of the
eye to perceive light]. . . . The minuteness of the quantity of energy
involved . . . is illustrated by the fact that the mechanical energy
required to lift a pea one inch would, if converted into light, be
sufficient to show a faint light signal to all the men who ever lived."
(From an article in *Scientific American.*) For all this fantastic sensitivity, the human eye probably is outperformed by that of the rat-tail
in perception of low intensities of light. The macrourid eye, with its
exceedingly large number of rods, its large lens, and its wide pupil,

is designed to detect faint glimmerings in the deep sea that would be quite indistinguishable to the human.

It is a matter of great curiosity that the rat-tails should possess these magnificently sensitive sense organs of light when they usually live in an environment void of any solar light. Some species live at the lower edge of the continental slope at 1,000 to 3,000 feet. In such depths some faint blue rays of sunlight may still exist in clear tropical waters, and perhaps the rat-tails living there make use of it. But other species live much deeper, down to 13,000 feet, where no sunlight exists. Why, then, do the eyes of various species of rat-tails grow proportionately larger the older they get and the deeper they swim, instead of the opposite, as is the case with other deep-sea fishes, including the anglers?

One possible explanation, of course, is that they perceive the living light of deep-sea luminescent animals, since many of the creatures that serve as their food are luminescent. Some species of rat-tails have light organs on their bodies too, although not in the numbers or the spectacular colors of the lantern fish or many other mid-water deep-sea species. From his bathysphere Beebe saw macrourid fishes in waters off Bermuda with six good-sized lights on their bodies and others with single lights under each eye.

Not all the creatures of the abyss are lighted, so even the best eyes need help if the rat-tails are to be successful in locating food. That they are eminently successful may be in large measure due to their possession of the complex lateral-line organs comprising a sixth sense not possessed by ourselves. These sense organs are located in canals along the sides of the animal and on its head. Little cushions of sensory cells pick up the energy produced by water currents and transmit the message of their presence to the brain. More is said of this remarkable sense organ in a later chapter.

Stomachs of rat-tails have been found to contain crabs, shrimps, and many other kinds of crustaceans, sea squirts, sponges, and foraminiferans. The rat-tail probably roots in the ooze, plowing through it with its pointed nose to uncover prey that it then locates with the help of its sensitive barbel. Some years ago a photograph of the bottom of the sea taken in 2,300 feet south of Cape Cod showed curious furrows in the bottom that mystified scientists. Dr. N. B. Marshall, an English authority on the deep sea, believes these may very well be trails left by rat-tails as they pushed their way through the ooze in search of food. The fact that a macrouridlike fish appears in the photograph tends to confirm this possibility.

While the rat-tails usually feed on the bottom, they may often swim in mid-water. Beebe saw them in deep water off Bermuda far from the bottom, and deep-sea research gear has caught them in mid-water. Their food is known to include pelagic creatures—euphausids, prawns, copepods, lantern fish, and even their own kind.

Rat-tails live in shallower water when they are young than they do later in life. They seek the great depths as they approach adult-hood, but seem to retain the ability to swim upward at will, as is the case with many other deep-sea fishes. To assist them in these vertical migrations they have well-developed swim bladders that serve to buoy them up in the water.

The swim bladders of rat-tails have another function—to produce noise. In a number of species, muscles attached in pairs to the forward end of the swim bladder vibrate the bladder to produce grunts and booms and other sounds. Few rat-tails have had their sounds identified positively, owing to the difficulty of making observations in the great depths where they live. But as long ago as the 1830's an Italian biologist named Bonaparte reported that the Mediterranean rat-tail *Coelorhynchus* made a grunting noise on the rare occasions when fishermen brought it to the surface. The structure of the air-bladder muscles suggests that the rat-tails have full control over the production of sound. Clearly, then, it must serve some useful purpose. Marshall thinks the purpose may be recognition—that is, keeping schools together in the blackness of the deep sea in the same way that the drumming noise of catfish in murky rivers substitutes for vision in maintaining contact with other members of the school. Sound production implies the ability to hear sound, and anatomical evidence indicates that the rat-tails have great acuity of hearing. No one has been able to devise direct experiments to measure the sensitivity of their ears, but in many species the acoustic centers of the brain are very highly developed, and the ears are large.

BROTULIDS

Members of the next family to be considered, the Brotulidae, have big heads tapering to long, compressed and pointed tails. The dorsal and anal fins are continuous with the tail fin. Many species have large, prominent eyes, and barbels under their chins. They are ordinarily from a few inches to a foot or so long, with the maximum length 3 feet. Does this description sound familiar? It should, for it would do perfectly well for the rat-tails, the group just considered.

It is of considerable interest that these two families of fishes should be so similar that in some species it is difficult for anyone but an expert to tell them apart. Yet some scientists think they are not closely related; whereas the rat-tails are relatives of the cods, the brotulids are nearer to the live-bearing blennies of shallow shore lines. Clearly, the deep-sea environment has taken two different kinds of material and pressed it from the same mold into remarkably similar creatures—a phenomenon that students of evolution call "convergence." In the harsh environment of the deep sea some distinct advantage must be derived from the odd shape and the other attributes common to the rat-tails and brotulids and some other groups of fishes.

The brotulids are not like the rat-tails in all respects. The eyes in almost all the rat-tails are well developed, whereas in the brotulids the size of the eyes appears to vary inversely with depth. Brotulids that live on the bottom in shallow depths, where some sunlight still penetrates, have large eyes with big lenses. Some brotulids caught below the limit of sunlight also have efficient eyes, but they are apparently in the habit of swimming to higher levels, to judge by their possession of big air bladders. Species living all or most of their lives in waters much over 3,000 feet have eyes of smaller size and lower efficiency; the eyes are apparently in the process of disappearing through evolution, and may now be used only to detect luminescence in the depths. When the depth increases considerably, to 6,500 feet, species like *Tauredophidium hextii* and *Leucicorus lusciosus* have degenerate blind eyes or very small eyes. Deep-sea brotulids are forced to depend on a lateral-line system that is very sensitive (although not as efficient as that of the rat-tails) to detect their prey. They can detect even faint water currents produced by animals that are breathing, feeding, or swimming nearby. The brotulids apparently do less rooting in the mud than the rat-tails, since their snouts are usually blunt.

While some brotulid species lay eggs like most fishes, others bear their young alive. In these species the developing embryo is protected in the mother's ovary by a special tough capsule. Unlike many other deep-sea animals, the young seem to be produced at

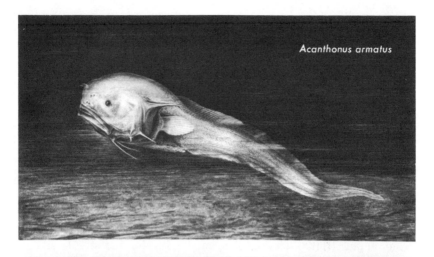

FIGURE 9–1. *Brotulids. These deep-sea fishes have big heads that taper to long, compressed, pointed tails. Their length varies from a few inches to 3 feet.* Acanthonus armatus *is semitransparent and has large spines on its gill covers; it was caught off West Africa in the Gulf of Guinea, 8,365 feet down.* Typhlonus nasus *is completely blind; it was caught in the Celebes Sea, 16,300 feet down.*

Poul H. Winther, *Galathea* Expedition

about the depth where the adults usually live, instead of higher in the water column. Off the Azores the English research ship *Discovery II* caught a female of the species *Parabrotula plagiophthalmus* with the tails of the young fish protruding from her body. Apparently the act of birth was under way when the fish was caught. This was at 2,300 feet, about the depth where the species is usually encountered.

If conquest of a variety of habitats can be used as a criterion, the brotulids are among the most successful of all fishes. Most species are dwellers of the deep sea, but a considerable number inhabit shallow waters right up into the intertidal zone, and there are even fresh-water cave-dwelling species in the West Indies. These last are among the blind members of the family; they are also colorless, a common attribute of animals living in lightless caves. In one haul from about 10,000 feet in the middle of the Indian Ocean, the *Galathea* brought up a little transparent brotulid, *Typhlonus,* that resembled these cave dwellers more closely than do most deep-sea fishes. It was quite blind, its eyes being reduced to two small and functionless black specks, barely visible, deep under the skin. Dr. Anton Bruun remarked on the little fish's resemblance to the species

FIGURE 9 – 2 . *Abyssal brotulid* Grimaldichthys profundissimus. *In 1901 Prince Albert of Monaco caught this fish in 19,800 feet of water south of the Cape Verde Islands. For half a century it held the record as the deepest-dwelling fish caught. But in 1952 Danish scientists caught a brotulid in 23,400 feet in the Sunda Trench, and in 1960 Piccard and Walsh saw a live fish from the bathyscaph in 35,800 feet, the deepest part of the ocean.*

Monaco Reports

in the subterranean caves of Cuba, and described its appearance as "nature's reply to the life conditions in the perpetual darkness, where other senses than sight are important, and where in the struggle for existence colours are quite useless."

The *Galathea* caught a brotulid in the Sunda Trench in 1952, in water about 23,400 feet deep, setting a depth record for the occurrence of a live fish that lasted until 1960, when it was broken by Piccard and Walsh. The *Galathea*'s little brotulid, about 6½ inches long, broke a record that had stood for half a century: Prince Albert of Monaco had caught a fish, *Grimaldichthys profundissimus* (from the Prince's family name, Grimaldi), at a depth of 19,800 feet in the early part of the century. This fish, too, was a brotulid.

After the rat-tails, the brotulids are the most numerous of deep-sea bottom fishes. Those living at the edge of the deep sea on the continental slope are caught occasionally in the trawls of commercial fishermen, and as fishing gear improves and the trawls probe deeper, the brotulids may become of commercial importance.

RAY FINS

The ray fins, members of the family Sudidae, were of special interest to early deep-sea explorers because of their extraordinarily elongated "feelers." The *Challenger* caught numbers of these strange fishes, and the biologists on that great expedition speculated as to the manner in which the fish could use such extravagant appendages. It was supposed that they were organs of touch, but it was hard to visualize why they should be so long.

Like many other fishes of the deep-sea floor, the ray fins are elongated, with very small eyes. They are different from the rat-tails and the brotulids, however, in having a large tail instead of a long, pointed filament. There is a small, fleshy fin on top of the tail like that of the salmon and trout.

The most curious feature by far is the long rays that project from the fins—the feelers that have fascinated biologists from the beginning. In nearly all species the feelers are composed of the greatly elongated first fin rays of the pectoral fins, the paired fins on each side behind the head. In some species certain rays of the pelvic fins (paired fins placed lower on the body and behind the pectorals) and of the caudal or tail fin are also enormously drawn out. These long rays are stiff feelers capable of being moved independently.

They may have the function of probing the deep-sea oozes for food, and they seem to be used as stilts. The contents of stomachs of some of the ray fins caught from considerable depths suggest that the fish hunts by moving over the bottom of the sea, "fingering" the mud and rooting out the little animals that live there. Annelid worms, copepods, amphipods, and other kinds of crustaceans constitute the bulk of this food.

The great development of these tactile organs seems to compensate for poor eyes. In some species there are very small eyes, so ineffective that they are incapable of projecting an actual image, and apparently serve only as indicators of the glow of luminescent light. In others the eyes have lost even this limited function and show only as spots on the head, reminders that ancestral fish were once able to see. *Bathymicrops regis* has its vestigial eyes covered with scales; a related genus, *Ipnops*, is famous for its complete lack of eyes. At the front of the flattened head in *Ipnops* are two large, thin, transparent bones, and under these are symmetrical white organs that some investigators have thought might be luminescent organs: a number of short, glistening white columns standing on a basal membrane. These organs are so different from the luminescent structures in other deep-sea fishes, however, that it is unlikely that they are light organs. In any case, this fish gets along without the slightest ability to detect light.

The ray fins are found mostly in the deep sea, although a few species are occasionally caught on the continental slopes. They have been encountered at 23,000 feet, over 4 miles deep. In the Gulf of Panama, *Galathea* scientists caught forty-two in a single haul at a depth of 3,002 feet. In the Mozambique Channel they made another noteworthy catch: a blind ray fin of the genus *Benthosaurus*, which had previously been caught only in the Atlantic Ocean and of which only five specimens had been taken. The *Galathea*'s was perhaps the best-preserved specimen ever caught, since it came up in the trawl protected with a covering of mud, so that the enormously elongated fin rays were intact. Another ray fin, *Bathysaurus*, was caught at the great depth of 11,155 feet. It is remarkable that in the same haul there were quantities of decaying vegetation from the land—sunken pieces of branches and roots from rivers and mangrove swamps. Whenever the *Galathea* trawl came up with such debris, the catch of animals was likely to be good, suggesting that the deep-sea creatures congregated where this material collected.

Sydney Hickson, the Cambridge biologist, quotes Sir John Murray on the appearance of *Bathypterois* in the catches of the *Challenger*: "When taken from the trawl they were always dead, and the long pectoral rays were erected like an arch over the head, requiring considerable pressure to make them lie along the side of the body." Hickson thought the fish "probably carries these organs directed forward, seeking with them in the mud for any worms or other animals upon which it preys, or receiving through them warning of the approach of an enemy from whom it is necessary to make an immediate escape." But he decided, unwisely, that "We shall never know . . . with any degree of certainty, how *Bathypterois* uses its long feeler-like pectoral fins, nor the meaning of the fierce armature of *Lithodes ferox* [a spiny deep-sea crab]; why the deep-sea crustacea are so uniformly coloured red, or the intensity of the phosphorescent light

F I G U R E 9 – 3 . *The ray fin, or tripod fish, Benthosaurus, apparently uses its enormously elongated fin rays as a kind of landing gear, and springs across the bottom like a cricket. The other fish in the picture is* Haloporphyrus, *casting a shadow on the bottom. This photograph was taken from the French bathyscaph F.N.R.S. No. 3 at a depth of 7,380 feet off Toulon.*

Station Marine d'Endoume et
Centre d'Océanographie, Marseilles

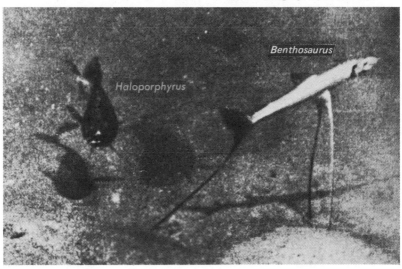

emitted by the Alcyonaria and Echinodermata. These and many others are and must remain among the mysteries of the deep-sea."

Hickson did not give enough credit to the curiosity of scientists or to their ingenuity. Some 60 years after he wrote these words we know a good deal about the colors of deep-sea crustaceans (including the fact that they are not universally red) and about the intensity of the luminescent light of the sea fans and starfish. We have, furthermore, remarkable pictures of *Bathypterois*, taken on the bottom of the deep sea from a bathyscaph, showing it standing on its great fin rays as though on a tripod. Contrary to Hickson's opinion, we are certain of finding out clearly how this fish uses its fins and of solving all the other mysteries he despaired of.

SHARKS

Sharks, with their relatives the rays, are mostly dwellers of the shallow seas, with relatively few representatives from the deeper parts of the ocean. Whether from the deep sea or the shallow, the sharks possess only a skeleton of cartilage and have no true bones. Even their teeth, which are among the hardest structures manufactured by animals, are not made of true bone. Sharks differ from the bony fishes in other ways, including the lack of gill covers and absence of scales of the familiar type. If scales are present they are in the form of hard denticles with a flat base and an enamel-covered spine curving up out of the skin and pointing backward. The projections on the scales make the skin of a shark rough to the touch —so much so that shagreen, as sharkskin is called when it is prepared, was formerly used as sandpaper. The reinforcement of the scales makes sharkskin very durable, and excellent shoes and belts and other products are made from it.

On the floor of the deep sea the sharks and rays are outnumbered by the bony fishes. The *Michael Sars* Expedition, fishing on the Atlantic slope in depths between 2,600 feet and 8,500 feet, caught mostly rat-tails, deep-sea cods, and fishes of the family Alepocephalidae, the latter being odd fishes related to the salmons and trouts. Only 4 per cent of the catch was sharks and rays (and chimaeras, which used to be classified with the sharks). Like many of the other species described in this chapter, not all species of sharks are bottom dwellers, and even those that are predominantly so may range into mid-water at times. This is deduced from the fact that individuals are commonly caught by mid-water gear, and from the objects found in the bellies of some specimens. The food of the deep-sea sharks is largely fish, cephalopods, and crustaceans. Occasionally, unexpected items turn up, as in the case of a shark of the deep-sea genus *Centroscyllium* that had eaten a large jellyfish, *Atolla.*

Feeding habits of some of the deep-sea rays are even more unusual compared with those of shallow-water species. The rays are ordinarily eaters of such creatures as clams and other hard-shelled molluscs; for this purpose they have broad, flat, pavementlike teeth for crushing their prey. In sharp contrast to this, a deep-sea ray, *Raja hyperborea*, has peculiar teeth that are long and slender, irregular and widely set. They are adapted for seizing and holding active prey, showing that their owner has abandoned the practice of its shallow-water relatives of subsisting on sedentary shellfish. Without this change in feeding habits the ray probably could not have made the successful transition to an abyssal existence. *Raja hyperborea* is also remarkable for its tolerance of cold water. It lives in depths of about 550 to 1,000 or more feet, in waters whose temperature sometimes is below that of the freezing point of fresh water.

The Greenland shark, *Somniosus microcephalus*, is another of the cold-water group. It has been caught in depths of about 1,000 feet in the Norwegian Sea. It is also of interest in that it lays unprotected eggs, unlike nearly all the members of its group. The frilled shark, *Chlamydoselachus anguineus*, is a deep-water shark that deserves mention. It is a curious-looking fish with a long, slender body, low fins, and frilly gills behind the head. It lives off the coasts of South Africa and northern Europe, in water several hundreds of feet deep. It is of special interest because it resembles very closely forms extinct for millions of years.

CHIMAERAS

Fishes that have changed very little from ancient times are the chimaeras. Whereas ichthyologists used to group these strange fishes with the sharks and rays, the chimaeras are now dignified by being placed in a separate group, although they are still called ghost sharks in some parts of the world, including South Africa. The more common name—chimaeras—is derived from their appearance, which is grotesque enough to have reminded someone of the strange mythical beings of ancient Greece. The chimaera is described in the

Iliad as a female monster with a lion's forepart, a goat's middle, and a dragon's hinder end. To add to its strangeness, it breathed fire. The fish to which the name has been attached does not breathe fire, but its odd appearance warrants some sort of special appellation. Chimaeras are robust fishes with outsized heads, from which the body tapers rapidly back to a very narrow tail, sometimes drawn out to a remarkably long and slim tip. The tail in some species is shaped like that of the sharks, and this is one of the reasons for having grouped the two together at one time. Other sharklike characteristics are a cartilaginous skeleton and the male's possession of "claspers" with which to fertilize the female. The fins are large, and the first dorsal fin usually has a stout, erectile spine in front of it, covered with sharp serrations—apparently for defense.

The eyes of the chimaeras are large, as is common with fishes from moderate parts of the deep sea. The snout in most species is rounded, but in some it is pulled out into a long proboscis. The head is covered with lateral-line canals that are often very prominent, being lighter in color than the rest of the head. The mouth of the chimaera is underslung and bears teeth that have become united to form bony, chisellike plates. The male has a curious device called a tenaculum attached to his head. This is a cartilaginous hook, armed with prickles; it points forward at times, but it can be folded into a pit in the head. Its function is not known, but it may be sexual.

A number of the chimaeras live in the deep sea. One of the most interesting of these is *Harriotta raleighana*, which lives in the North Atlantic at depths below 3,300 feet. It was first caught on the American side of the ocean, and was named by the famous ichthyologists George Brown Goode and Tarlton H. Bean in honor of Thomas Harriott, "the most eminent philosopher and naturalist of his day." He was sent by Sir Walter Raleigh (who is honored in the second name of the fish) as part of the group colonizing Roanoke, Virginia, in 1585. Harriott's book on the natural history of Virginia was the first published account of the living creatures of what is now the United States.

Harriott might not have been flattered to know that this species of chimaera was named for him. Among a grotesque group, *Harriotta* is one of the strangest. The snout is greatly elongated and ends in a point. It is so long that it is supported by cartilaginous stiffening, and it spreads out at the end in a leaflike lateral expansion. *Harriotta* grows to a length of about 4 feet.

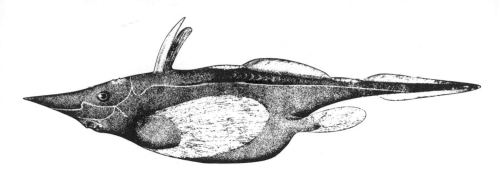

FIGURE 9 – 4 . *Chimaeras, or ghost sharks, look the way they did in ancient times. Shown here is* Harriotta raleighana, *a deep-sea chimaera, about 4 feet long.*

J. Murray and J. Hjort, *The Depths of the Ocean*

For all their repulsive appearance, the chimaeras are timid and harmless creatures. Perhaps because of their lack of aggressiveness, they seem to be moving slowly toward extinction in some parts of the ocean. It may be that the species that have adopted the deep sea as their home will manage to survive longest.

EELS

The common eel spends most of its adulthood in fresh water, but it is to some extent a deep-sea fish, since it spawns at great depths in the Sargasso Sea. There are many other kinds of eels besides the familiar one, and some of them are more truly of the deep sea, spending all or most of their lives there. Some inhabit very

FIGURE 9 – 5 . *Spiny eel* Notacanthus bonapartii, *from about 6,500 feet. In spite of its name, this fish is not an eel at all—the big head, underslung jaw, and tapering tail are typical of many kinds of deep-sea fishes.*

G. B. Goode and T. H. Bean, *Oceanic Ichthyology*

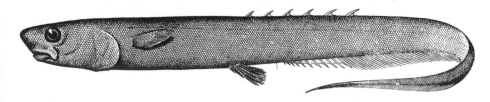

great depths; an example is *Histiobranchus infernalis,* which has been caught between 5,900 and 10,000 feet deep. Most deep-sea eels have the long, slender, small-finned shape familiar in the common eel of Europe and America, but some representatives of the group are very different. The *Galathea* caught five specimens of an eel, *Coloconger,* so unusual that it was hard to believe it was really an eel. It was short and fat and reminded the ship's scientists more of a turbot, one of the flatfish.

There is a group of deep-sea fishes called the spiny eels, but the common name is misleading since they are not really eels at all. Their family name is Notacanthidae, and they are eellike only in the possession of relatively slender, elongated bodies. The dorsal fin—the unpaired fin along the top of the body—consists of a series of single spines, unconnected by skin or membranes as they are in most fishes. As with so many other deep-sea fishes of the ocean floor, the spiny eels have very long, tapering tails, many-rayed fins, and jaws slung under a projecting snout. There is a sharp spine concealed at each corner of the mouth. The character of the mouth suggests that the spiny eels feed in the mud of the bottom, and the food found in their stomachs confirms this. The food seems to consist largely of sponges, hydroids, and starfish (a prickly fare), as well as copepods and small squids. The spiny eels have been caught at depths of 6,500 feet and more.

Rarely seen relatives of the spiny eels are more fishes without common names, whose family is called Halosauridae. These again are long-bodied fishes with elongated, tapering tails, big eyes, and the inferior mouth of bottom feeders. The biggest of these fishes seem to be about 20 inches long; they have been caught at depths of 5,200 feet by the *Galathea.*

Most of the fishes described so far are among the spiny-rayed fishes, generally regarded as the most advanced group. There are some deep-sea fishes, however, that belong to the soft-rayed group, of which the salmons and the herrings are the most familiar shallow-water representatives. Fishes of the family Alepocephalidae are among these. They are small, a few inches to a little over a foot in length, with bodies not unlike those of herrings, although the head is commonly proportionately bigger than a herring's and is remarkable for being always darker than the body, often deep violet or bluish-black. The body is either naked or covered with thin, deciduous scales. Some species have light organs on the body. These fishes are rarely seen, partly because they occur in deep water—as a rule well below 2,000 feet.

THE FLATFISH

The flatfish comprise a large group whose shallow-water representatives enter the commercial catches of fishermen of many countries. The halibut and the flounder are the most familiar of the flatfishes in America, while the plaice and the sole are commonest on European restaurant menus. (Several fishes are called "sole" in the United States and Canada, but there are no true soles among the commercial fish of the New World.) In addition to these more familiar members of the flatfish family, there are others with interesting names: witch, dab, turbot, brill, megrim, topknot.

The flatfish exhibit beautiful adaptations to life on the bottom of the sea. "Flat as a flounder" has become a standard phrase, for flatness is the characteristic that sets these fishes apart from all others. Rays are flattened also, but in a different manner; it is as though the rays had had a weight pressed on them from above, so that their sides spread laterally. The flounders, on the other hand, are flattened from side to side—"compressed" instead of "depressed."

With this habit have come some other major modifications of structure. An ordinary fish lying on its side would have one eye buried in the sand, quite useless. The flatfish have both eyes on the upper side, one eye higher and farther forward than the other. The upper side of the body is colored to blend with the sand bottom, while the underside is colorless. The colored side is usually somewhat convex, and the underside is flat. In *A History of Fishes,* Dr. J. R. Norman, the distinguished English ichthyologist, recounts two legends associated with the difference in coloration of the upper and lower sides of the flatfish. An Arab legend has it that "Moses was once engaged in cooking a flatfish, and that when this had been broiled until it was brown on one side the oil gave out; this so annoyed him that he threw the fish into the sea, when, although half cooked, it promptly came to life again, and its descendants have preserved this curious arrangement of color ever since." The other legend is Russian. The Virgin Mary was eating a brill [*Rhombus*] when she heard the tidings of the Resurrection. "Incredulous and as one of little faith she flung the uneaten half of the *Rhombus* into the water, bidding it, if the message be true, come back to life whole! And lo! this it instantly did!"

The color of the skin of the flatfish results from the presence of chromatophores. Since these are usually missing where the least light strikes the skin, the underside is generally white. But occasionally there are spots of color on the underside, or even part of the pattern that appears on the upper side of the fish. In some cases only the underside of the head is left white, and even this can be colored on rare occasions. Halibut fishermen sometimes refer to fishes with a partial colored pattern on the underside as "circus halibut." Albinos occasionally occur among the flatfish too; these have the upper as well as the lower side white.

In some of the flatfishes, including the halibut, the plaice, and the lemon dab, the eyes and the colored surface are on the right side, while in others, including the turbot and the brill, it is the left side that is on top. In rare instances (about once in 5,000 times in the case of the halibut) the colored side is the opposite of what is normal for the species.

The influence of light in coloring the upper side of a flatfish and leaving the underside white has been shown by experiments in Plymouth, England. Professor J. T. Cunningham rigged a series of mirrors in an aquarium so that light would be thrown onto the underside of flounders. Slowly the white side became colored like the upper side, and after periods of one to three years the fish was nearly the same on both sides!

The normal colors of the upper side of the flatfish are various shades of brown or gray, spotted and marbled, blending with the tints of the sandy bottom. Some species, such as the plaice, have reddish-orange spots; others have shades of yellow, in addition to the more subdued colors. Like those of the squids, the chromatophores of the flatfish are under the animal's nervous and muscular control. As a consequence, the flatfish are skillful in altering their color to make it blend with the background. They are quite as adept at this as the chameleon, although they do not have the extreme virtuosity of the squids and octopuses.

Brill and turbot are especially good at changing their color and matching it to the bottom. They do this by first examining the bottom with their remarkably mobile eyes, which protrude from the head and can be pivoted in all directions. (Blinded fishes cannot change their color.) Then the flatfish vary the relative expansion and contraction of their various color cells. As the cells are expanded, the skin becomes dark; as they are contracted, the white underskin shines through and the fish lightens in color. These changes take only seconds, or minutes at most. In the Naples aquarium, Mediterranean flounders (*Bothus*) were placed in a series of glass dishes on the bottom of which were painted patterns of black and white squares and circles and other combinations of shapes. The fish showed remarkable ability to adapt to these unfamiliar backgrounds; with practice they cut the time required to change color from half an hour or so at first to a matter of a few minutes. In later experiments with the American summer flounder (*Paralichthys dentatus*), other colors than black and white were added: brown, blue, green, pink, and yellow. The fish succeeded remarkably well in imitating these colors, although they had more success with the yellows and browns than with the greens and blues, and the least success with the red shades.

Concealment through camouflage is enhanced by another flatfish ability—that of snuggling down into the sand as it lies on the bottom. It does this by fluttering its fins to stir up sand and gravel, which fall back on top of the fish, especially along the edges, and break its sharp outline. Sometimes only the eyes protrude from the sand, swiveling around like miniature gun turrets.

The fins of the flatfish, supported by numerous soft rays, extend like a fringe around the whole body. The ancestors of the flounders were spiny-rayed, but the sharp spines have become soft again, making possible the flexible, wavelike motion of the fringing fins. These are often used to grip the bottom. The smaller flatfish depend on their fins for swimming, using an undulating motion that passes from head to tail. The fish "skims forward like a billowing magic carpet," to use Sir Alister Hardy's phrase, and it comes to rest on the bottom after short swimming efforts followed by a graceful glide. The much bigger halibut does not depend so much on the fringing fins, but swims by vigorous movements of the body and powerful tail. It leaves the bottom frequently, pursuing active prey through the water and even following fish to the surface from considerable depths.

Among the flatfish tribe the halibut is one of the few active and voracious predators. It is large-mouthed and sharp-toothed, consuming crabs, lobsters, and molluscs, but especially fish—herring, flounders, cod, skates, ratfish, and many others. Its voracity is shown by some of the strange objects occasionally found in its stomach. These have included a block of wood a cubic foot in size and pieces of floe ice. Dr. G. B. Goode, formerly Commissioner of Fisheries of the United States, reports that fishermen have told him the halibut sometimes kills fish by repeated blows of its tail, but apparently no other observer has confirmed this.

In keeping with its feeding habits the halibut has a symmetrical mouth, with strong teeth on both jaws. In contrast, many of the other flatfish have twisted jaws adapted to their habit of feeding in the bottom sediments on shellfish and crustaceans. Lying partially buried in the bottom, such fishes as flounders and soles "can still eat comfortably, like so many Romans dining from their couches." The lemon dab *Microstomus kitt* feeds exclusively on annelid worms of the kind that live in tubes in the bottom of the sea, and the manner of its hunting is described by Dr. G. A. Steven of England. The worms poke their heads cautiously from the ends of their tubes when they think there is no danger near; if they are alarmed they whip themselves back with lightning speed. The dab must therefore be very discreet about its movements if it is to surprise the worms partly out of their burrows. It moves restlessly and constantly over the bottom, swimming for short distances and then pausing quietly while it searches the terrain. To increase its field of vision the fish arches its back so that its head is raised as high as possible, while its prominent and highly movable eyes scan the bottom. If it sees an incautious worm with its head protruding from a burrow, it pounces suddenly with a kind of forward leap, bringing its mouth down almost vertically by a strong arching of the forward part of the body, and swallows the worm unbroken.

The sole, a small flatfish of England, hunts in still another fashion, described by William Bateson: "In searching for food the sole creeps on the bottom by means of the fringe of fin-rays with which its body is edged, and thus slowly moving, it raises its head upwards and sideways, and gently pats the ground at intervals, feeling the objects in its path with the peculiar villiform papillae which cover the lower (left) side of its head and face. In this way it will examine the whole surface of the floor of the tank, stopping and going back to investigate pieces of stick, string and other objects which it feels below its cheek. The sole appears to be unable to find food that does not lie on the bottom, and will not succeed in finding food suspended in the water unless it be lowered so that the sole is able to cover part of it with the lower side of its head, when it seizes it at once."

The flatfish do not start life with one side uncolored, or with both eyes on the same side. Instead, the larvae are like those of most fishes: quite symmetrical, with one eye on each side of the head and swimming upright in mid-water. The remarkable change that transforms them into the odd, twisted creatures they finally become tells us a lot about the evolutionary history of the group.

The hazards faced by the tiny flatfish after they hatch from the egg are so tremendous that in order to maintain the population the female is obliged to produce very large numbers of eggs. In the plaice this can be as many as half a million; in the halibut it reaches the huge number of 2,750,000. In the main spawning grounds of the plaice in the North Sea, between the Thames Estuary and the Dutch coast, an estimated 60 million fish congregate each year to spawn. The eggs are cast into the water to be fertilized by the males and then drift in the sea, developing as they use up the accumulated yolk. In about 15 days they hatch, at a length of approximately a quarter of an inch, and feed actively on the tiny creatures of the plankton. It is about a month before the strange metamorphosis begins. The left eye begins a migration over the top of the head, eventually ending up on the right side. The whole skull becomes twisted in this process, and the jaw becomes asymmetrical too. After the eye has moved over the top of the head, the dorsal fin grows forward so that eventually there is a fringe of fin rays around the whole body. (In other species, where the migration of the eye is slower than the development of the fin, the eye may have to move right through the base of the fin.) In the meantime the little fish drops out of the plankton and begins its adult existence on the floor of the sea.

In the North Pacific the halibut spawn off Kodiak Island during the winter, casting the eggs at depths of about 900 feet. They drift with the great counterclockwise eddy that circles the Gulf of Alaska. At first they are transparent, and it is not until 4 or 5 months have elapsed that they begin to take on color. At this time the left eye, which was normally placed, begins its movement across the head of the little fish. As the transformation takes place, the young halibut are carried by currents to shallow-water "nursery" grounds where they spend the first years of their lives, near the beaches of the Queen Charlotte Islands and nearby areas. Later they move offshore to deeper water. The halibut is one of the longest-lived of fishes, some females living at least thirty-five years. These elderly fishes may reach a length of 10 feet and a weight of 490 pounds.

The halibut is the flatfish of greatest economic importance in North America. It was mentioned very early, Captain John Smith having this to say in his *History of Virginia*: "There is a large sized fish called Halibut, or Turbot; some are taken so bigg that two men have much a doe to hall them into the boate; but there is such a plenty, that the fisher men onley eat the heads & fins, and throw away the bodies; such in Paris would yeeld 5. or 6. crownes a peece: and this is no discommodity." For a time the Atlantic and the Pacific halibut fisheries flourished and then withered. That of the Atlantic is now small compared with the great haddock and ocean-perch fisheries, but the halibut stocks of the Pacific have been restored by the brilliant research and management of the International Pacific Halibut Commission, an agency of Canada and the United States.

The halibut can be regarded as a deep-sea fish, since it is found at times in water to at least 3,600 feet. Most are found much shallower than this, however, and, as we have seen, the young occupy very shallow nursery grounds. Some flatfishes even occupy estuaries, while the flounder *Pleuronectes flesus* is sometimes found in freshwater streams. As a group the flatfish have not adapted as successfully

FIGURE 9 – 6 . *Deep-sea sole* Aphoristia marginata, *with both eyes on one side of its head. Its fins are continuous almost all around the body. While the upper side is striped, the underside is colorless.*

G. B. Goode and T. H. Bean, *Oceanic Ichthyology*

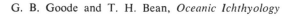
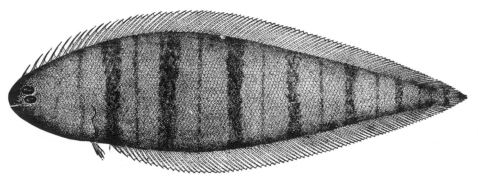

to a deep-sea existence as many other kinds of fishes. Nonetheless, there are deep-sea representatives, including the one seen by Piccard and Walsh in the Marianas Trench. Another deep-sea sole, *Aphoristia wood-masoni*, has been caught in the Indian Ocean at 2,500 feet. The *Galathea* caught a flatfish described by J. R. Pfaff, a biologist of the expedition, in *The Galathea Deep Sea Expedition,* as one of the most fascinating fishes captured during the whole voyage. This was the flounder *Azygopus pinnifasciatus*, taken in 2,000 feet in the Tasman Sea. "On the tail of this fish are some dark spots, two of which resemble eyes, and the theory has been advanced that their purpose is to make pursuers mistake the tail for the head. However, the fish lives at a depth that renders this camouflage illusionary, as it is totally dark. If this theory is correct, it provides us with one of the proofs that deep-sea fish must have originated on the continental shelf."

The *Galathea* caught another unusual flatfish off Natal, in a depth of about 1,600 feet. This was the deep-sea turbot *Chascanopsetta*. Fourteen specimens were captured of this creature, which was unlike other turbots in the possession of an immense gaping mouth with long, movable teeth. Its shallow-water cousins eat clams and other bottom creatures, but *Chascanopsetta* has adapted to the pursuit of active prey.

10 *Deep-Swimming Fishes*

If a group of surrealist artists had engaged in a contest to see which could create the most grotesque and improbable monstrosities, surely they could not have approached the astonishing shapes exhibited by the fishes of the deep ocean, or have invented details of form and behavior as improbable as the real creatures. What jocular imagination could have conjured up a jet-black fish whose body is a mere appendage to a huge mouth, or a hideous, lurking "fish trap" that dangles a lighted lure to bemuse its victims, or a fierce creature whose greed encourages it to stretch and engulf another fish considerably larger than itself? And then, having had the audacity to invent such creatures, which of our competing surrealists would have had the imaginativeness to reduce the monsters to diminutive size, making them objects not of awe but of absurdity?

BIG MOUTHS

The dark fishes that are "all mouth" are gulpers, or pelican eels. They live in bone-chilling cold and gloomy darkness as much as 9,000 feet and even more below the surface. At these depths the numbers of inhabitants have thinned out markedly, so that one animal meets another only very infrequently. This may suit the hermits among the sea's creatures, but it raises severe problems in keeping body and soul together. Plant food is almost completely absent, since there is no sunlight to support it; the few animals that exist are mostly of a relatively large size—as large as their pursuers. The result is that many of the creatures in this region of the sea are simultaneously pursuer and pursued. It is common for fishes of the deep to eat their own kind or even to eat fishes bigger than themselves.

The gulpers are marvelously well fashioned to compete success-fully in this pitiless watery jungle. All their body functions and struc-tures seem subordinated to the fierce necessity of seizing, holding, and swallowing whatever comes along in the way of food: the mouth is huge, the jaws are fantastically distensible, the belly can expand to several times its normal size, and there is some evidence that the long tail holds prey while the mouth goes into action.

Eurypharynx pelecanoides is one of the oddest members of this improbable family. A measure of its success in meeting and con-quering its difficult environment is its wide occurrence. It is found in the oceans of the entire globe, in tropical and subtropical waters, nearly always deeper than 6,500 feet. Among the small dwellers of the very deep sea, *Eurypharynx* is big, at about 2 feet long, but is smaller than closely related gulpers such as *Saccopharynx ampul-laceus*, which attains lengths of at least 6 feet, and Harrison's gulper, *Saccopharynx harrisoni*, which reaches 4½ feet. However, a large proportion of this comparatively grand length is taken up by the slender, whiplike tail, leaving a mere few inches for the rest of the body. And the body, in turn, is largely a truly magnificent mouth.

The gulper's mouth is a roomy bag (whence, of course, the name *"pelecanoides"*) beset with teeth. *Eurypharynx* is not an aggressive hunter, as its weak, soft muscles and feeble, thin bones attest. It subsists by stuffing into its mouth whatever comes along, holding the catch with long, slender, curved teeth, widely set to grip and hold but not to chew. It must bolt its prey quickly, before the victim can struggle free. Dr. Albert C. L. G. Günther, the celebrated German ichthyologist of the nineteenth century, said that these fishes do not use the muscles of their pharynx to swallow, but that "deglutition [what a marvelous word for swallowing!] is performed . . . by the independent and alternate action of the jaws, as in snakes. These fishes cannot be said to swallow their food, but rather draw them-selves over their victim. . . ."

The gulper's very flabbiness is an adaptation to an environment poor in nutrient. On the small amount of food available in the deep sea the fish could not support the same kind of body equipment as that of species living in surface waters. Lacking the lithe muscles and stout bones of the surface predator fishes, the gulper falls back on guile and on unique sense organs to attract and detect its victims. At

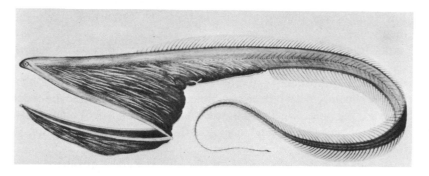

FIGURE 10-1. *Gulper, or pelican eel* (Eurypharynx pelecanoides). *This fish is all mouth and practically no brain: a 10-inch specimen had a brain case ¼ of an inch long. Gulpers live in tropical and subtropical oceans, usually below 6,500 feet.*

Dana Expeditions

the end of the whiplike tail there is a reddish light that is presumably used to attract the attention of other creatures and to tempt them within range of the gulper's jaws. No one has yet been lucky enough to observe the gulper in action, but perhaps this tail sweeps back and forth seductively in the black depths of the deep sea, bemusing other animals so that they begin to swim toward it. Unable to see its potential victim in the gloom, the gulper now depends on its lateral-line system, that series of sensitive organs strung out over the flanks and head and capable of detecting the swimming movements of prey. If the victim approaches from the rear, it may be seized by the tail of the gulper, which apparently winds tightly around the swimmer and holds it fast. Sir Alister Hardy caught *Saccopharynx ampullaceus* with its tail tightly coiled around the filaments of a small animal in the net.

In the latter part of the nineteenth century only a handful of gulpers had ever been seen, all of them individuals that had been picked up on the surface of the sea when their greed got the better of them. They had apparently been forced to the surface after ingesting fish bigger than themselves, unable to survive the struggles of their victims, or incapable of digesting such a gargantuan meal. Dr. Günther described them in 1880 as "deepsea congers" and stated

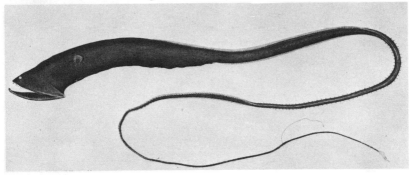

FIGURE 10-2. *Gulper* Saccopharynx schmidti. *A tiny eye about 1/10 of an inch wide is set over a gigantic mouth with many teeth. This fish grows to about 4½ feet, including its whiplike tail. It has a tail light, a reddish, glowing spot that attracts prey.*

Dana Expeditions

that only three specimens were known, all from the North Atlantic. Since then others have been caught, some in deep-sea trawls dragged by research ships. One individual of *Saccopharynx* managed to tuck into its belly a 9-inch codlike fish called *Harargyreus*; the gulper that accomplished this feat had a body length (not counting the tail) of only 6 inches.

The eyes of the gulper are relatively tiny. A 10-inch individual of the genus *Eurypharynx* has an eye less than one fifteenth of an inch in diameter; *Saccopharynx harrisoni,* 4½ feet long, has eyes of less than a quarter of an inch; and *Saccopharynx schmidti,* of the same length, has eyes even smaller—a mere tenth of an inch in diameter. But if the eyes are small in relation to the head, they are large compared to the size of the brain case. The one-fifteenth-inch eye of *Eurypharynx* is fully one quarter the size of the tiny brain—which says very little for the mental capacity of the gulpers! Like Pooh Bear in the Milne books, these are fishes "of very little brain."

The swallowers are deep-sea fishes that have developed the ability to engulf enormous prey to a degree surpassing even the gulpers. The black swallower *Chiasmodus niger* is only a few inches long (6 at most) yet it can swallow fishes 8 or 10 inches in length.

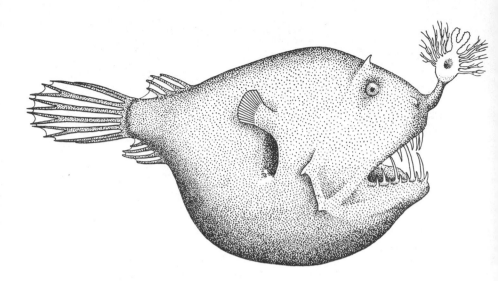

FIGURE 10-3. *Deep-sea angler* Borophryne apogon, *from 11,500 feet down in the Pacific and Atlantic oceans. While most fish are compressed from side to side, deep-sea anglers look squashed down from above; they are as wide as they are high. They move slowly through the water, propelled mainly by their tail fins.*

Dana Expeditions

FISHES THAT FISH

Human beings always display a special kind of startled amusement when animals behave in any way like men. We pay very little attention to a horse that is handsome or strong: this is what a horse should be. But let the horse appear to smoke a pipe or count to ten and it becomes a celebrity. The bravest or silkiest-coated dog is still a dog, and hardly unusual—but let it wear a hat and walk on its hind legs like a man and it becomes a circus star. Monkeys are surely the most popular animals in the zoo because they behave most like their visitors. Turning to the sea, we expect fish to hunt and eat other fish—but we are startled and amused to find that some of them appear to use human techniques, using a rod and line, with a worm-like bait on the end!

Fishing with a baited line is only one aspect of an efficient adaptation to life in the great ocean depths on the part of the angler fish, which are perhaps the most successfully specialized of any creature that lives in the deep sea. The female (for it is only she who has a "fishing license") has been shaped and honed into a proficient machine for catching prey.

Deep-sea anglers first came to the attention of science in the 1830's when Lieutenant Commander C. Holboell of the Royal Danish Navy found a bizarre little fish washed up on the shores of Greenland. In the next few years he found two more, different but clearly related to the first. Jet black, with an enormous head and mouth, and bearing a long, flexible appendage dangling from the head, these were obviously fishes of the abyss. It was not until the voyage of the *Challenger* in 1873 that anglers were caught in their natural habitat, and even today their numbers in the specimen bottles of the museums of the world are not large.

We have enough specimens, however, to have made some good guesses as to the kind of life they lead in the numbing and lightless waters of the very deep ocean. We are aided in this study of the deep-sea anglers by their relationship to the well-known frogfish, the angler of the shallow seas. This ugly, blotched creature, commonly a yard long and sometimes as big as 5 feet, conceals itself among the rocks and seaweed of the bottom of the Atlantic. It waves a lure resembling wriggling worms on the end of a rod. It gulps down any curious creature that approaches its bait, and has been recorded as ingesting remarkably large fish—as well as full-grown ducks. The deep-sea anglers are descendants of the frogfish, having moved in the course of millennia into deeper water, and in most cases having changed from a bottom-dwelling existence to life in mid-water. The frogfish is an inhabitant of the northern seas, but the deep-sea anglers are mostly found in tropical and subtropical oceans.

If the frogfish is ugly, the deep-sea anglers are grotesque. While most fishes are somewhat compressed from side to side, the anglers are "depressed"—squashed down so that they are as wide as they are high. Even more curious, they are only as long as they are wide, so that the whole effect is ludicrous: a short, pear-shaped lump with a ridiculous air of ungainliness. At the same time, paradoxically, these absurd fishes manage to appear formidable—no small trick when they are rarely as large as a man's fist. Several examples of angler fishes are shown in Figures 10-3, 10-4, 10-5, and 14-4.

Much of the ferocious appearance of the angler is derived from the enormous mouth, whose wicked-looking jaws occupy much of the head. These jaws slant downward in most species so that, when its mouth is closed, the fish assumes a cruel expression. With the mouth open, the savage mien of the angler is multiplied, for now the teeth are revealed. Long, curved, and sharp, these are formidable spikes whose obvious function is to stab the angler's victims. (From his bathysphere Beebe saw one angler with faintly luminous teeth— which would seem to be a doubtful asset to the fish.)

FIGURE 10-4. *Two deep-sea anglers.* Photocorynus spiniceps, *from the Gulf of Panama, has a clublike fishing apparatus.* Linophryne macrodon, *about 2⅛ inches long, has a luminous bait and barbel (appendage on the under jaw). The light attracts prey.*

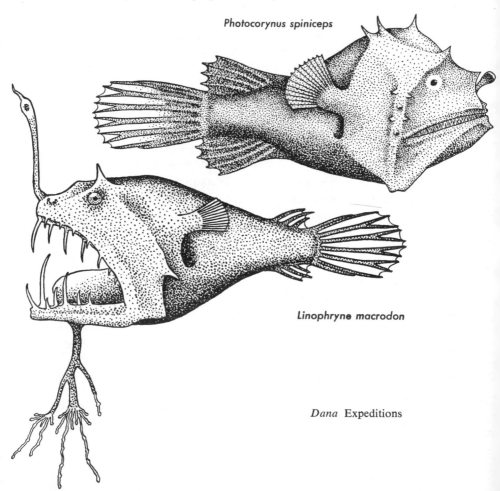

Photocorynus spiniceps

Linophryne macrodon

Dana Expeditions

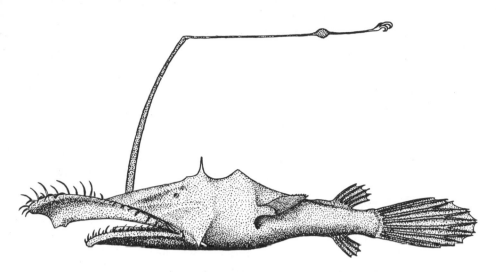

FIGURE 10-5. Lasiognathus saccostoma, *the "compleat angler" of the abyss. This 3-inch fish has a movable fishing rod, a line, a float, a lighted bait, and 3 hooks. The hooks are not, however, used to catch prey.*

Dana Expeditions

The tail in most species is roughly rectangular (or occasionally oval) and is greatly overshadowed in size by the massive head. The swimming movements of the angler are produced principally by the tail, and these movements are leisurely, if not labored; the tail simply cannot push the roly-poly mass through the water with any speed. The paired pectoral fins behind the gill covers are used for steering rather than propulsion, and the pelvic fins, placed near the under-side of many fishes, are missing in the anglers. It is certain, then, that the angler does not mount a hot pursuit of its prey but rather coasts slowly through the water, waiting for a victim to come to it. Probably the anglers are partly jet-propelled—not with the whoosh-ing speed that this term connotes, but slowly and sedately, as a small stream of water taken into the mouth is forced through little gill slits in the process of respiration.

Like other very deep-sea fishes, most anglers are either jet black or dark brown in color (although *Chaunax*, caught in 1,290 feet off the Fiji Islands, is a surprising pink). Their skin is usually scaleless and is covered with rough folds, warts, furrows, and protuberances of many kinds. In some species there is a barbel suspended from

[199]

the underjaw, and in one genus, *Linophryne* (Figure 10-4), this takes the form of an extravagantly branched appendage. The barbel is faintly luminous in some species and may serve as an accessory lure, although in our present imperfect state of knowledge there is uncertainty about its true function. It may serve instead as a sensory organ, picking up the faint energy generated by the swimming movements of deep-sea dwellers, possible prey or predators of the angler.

The eyes of young anglers are of normal size in relation to the bulk of the body because the larval stages are passed at the surface, where abundant light makes the presence of eyes useful. At metamorphosis, when the larva changes into the adult form and the young fish sinks rapidly into the abyss, the eyes do not grow. The rest of the body increases in size so that eventually the eyes are relatively so small that they are sometimes hard to find. This is not unexpected, since there would seem to be little point in the possession of functional eyes in a lightless environment. The fact that in many species the eyes, tiny as they are, retain some ability to perceive light argues that anglers spend part of their time in depths where there is at least a glimmering. This suggestion is reinforced by the position of the eyes in some species: near the top of the head, pointing upward as though to catch the last faint gleams of sunlight struggling down through the enormous bulk of water above. This upward glance lends these species a particularly amusing look of wistfulness—an appearance quite contrary to their true character as competent and merciless predators.

None of these curious aspects of the structure of the anglers is as remarkable as their fishing apparatus, which varies widely from one species to another. It is stubby in some, and several times the length of the fish's body in others; in some species it is straight, in others it is bent in a right angle.

The fishing line, or illicium, is the modified first ray of the dorsal fin. In soft-rayed fishes like the salmon and the herring, the dorsal fin on the back of the fish is supported by pliant rays, while in fishes like the yellow perch and the sculpin the first rays are spiny, and in some species these are capable of inflicting a painful stab. The fin rays pivot on a tiny ball and socket, rotated by the pull of muscles attached to them. In the angler fish this first spiny ray has left the rest of the dorsal fin behind and has migrated forward from its usual position to form the front of the fish's head.

On the end of the illicium hangs the bait, called the "esca." Sometimes this is only a small swelling on the end of the line, but usually it is more complex: a flashing, frilly bait with filaments and tempting devices designed to hoodwink prey into thinking it sees something edible. Sometimes the fishing line is actually equipped with hooks (Figure 10-5)! (The hooks are, alas, not actually for catching prey. Their shapes are apparently purely fortuitous.) Since the inky blackness of the abyss would prevent potential victims from seeing the bait of the angler fish otherwise, the latter has thoughtfully provided a light organ on the end of the illicium. The light shines through a transparent window, where the skin has lost its pigment. Pale shades of yellow or yellow-green, blue-green, and orange tinged with purple are emitted in short gleams followed by longer periods of darkness. Deep-sea creatures must find these colored lights irresistible as they flicker and flash faintly in the dark waters.

The ultimate in efficient placement of the lighted bait was encountered by scientists in 1952. In that year the *Galathea* caught a new angler fish off tropical America in nearly 12,000 feet of water. Instead of dangling the lure from a fishing line on the head, *Galatheathauma*, as the new fish is called, suspends a forked light organ from the roof of its mouth (Figure 14-4). Any prey curious enough to investigate this lure would be in a convenient spot to be snapped up by the angler.

The light in the lures of angler fish seems to be supplied by luminous bacteria. These enslaved bacteria are stimulated to gleam briefly as the fish supplies the esca with oxygenated blood, which the bacteria require to luminesce; when the blood is cut off by the fish the bacteria subside again into darkness.

When the illicium was a more prosaic first ray of the dorsal fin in the ancestors of the anglers, it was attached to a vertical bone on which it rotated in a ball and socket. As the fin ray moved forward to become a fishing line, the basal bone tipped from its upright position to lie flat on the skull of the fish in a deep furrow. In some anglers the basal bone can be moved forward to push out the base of the fishing rod so that it dangles as far as possible into the gloom to attract its prey, or moved backward a little to tempt the quarry closer to the capacious mouth. Muscles attached to the fishing line can also swing it forward ahead of the fish, or back toward the tail to keep it out of the way when the mouth goes into action. In some species the basal bone can move very little, but in others it is capable of carrying the illicium back and forth considerable distances. This development is notable in *Ceratias holboelli*, in which the basal bone has grown so long that the groove on the head is no longer able to contain it, and it extends backward over the trunk of the fish to occupy a groove between the great muscle folds of the back, over the vertebral column. When this long rod of the basal bone moves forward, it pushes out past the profile of the fish, covered by a little sock of skin. At the same time the hind end of the rod pulls forward and creates a little dimple where the end of the rod is attached to the skin of the back. In *Lasiognathus* (Figure 10-5), this remarkable arrangement is carried to an extreme, and the basal rod is a full four fifths of the total length of the fish!

From his lifelong studies of the angler fish the eminent Danish biologist Dr. E. Bertelsen has reconstructed their probable method of hunting and devouring prey. "Hunting" is perhaps the wrong word to use in this case; it is like one form of tiger hunt in India: a patient wait while the prey is attracted to a tethered goat. A feeble swimmer, the angler cannot dash after its prey. Instead, it floats almost motionless in the darkness—3,000 to 12,000 feet down—perhaps barely moving its fins or tail to keep its position, or drifting slowly forward as it passes water through its mouth and out the narrow gill slits. Never in schools but always alone, the angler flutters and twitches its bait in ways that it hopes resemble the motions of a deep-sea copepod or other tempting abyssal morsel. Its colored lights gleam and flicker, enticing prey to have a closer look.

An animal attracted by these antics will probably not be seen by the angler, since its eyes are rudimentary. Rather, the approach of the prey will be sensed by the platelets on the papillae of the lateral line. If the quarry is interested in the bait, the angler may craftily draw the lure backward as the victim follows it closer to the jaws, the basal bone moving back in its slot to permit this. Now the vast mouth suddenly yawns open, and the suction created by this movement creates a current that sweeps the luckless victim to its death. The whole apparatus of the mouth cooperates to produce as strong a suction and current as possible. The floor of the mouth sinks, the jaws turn outward and upward, and the gill covers expand suddenly; all of these actions increase the size of the cavity. The sharp, curved teeth of the jaws, folded down when the mouth was closed, spring into position as the mouth gapes, and the victim is stabbed and held securely when the jaws clamp shut. The teeth and the jaws move the victim farther into the mouth, and other sets of teeth on the floor of the mouth and in the throat stoke it into the great belly. The latter is huge and capable of enormous expansion to take creatures as big as the angler itself.

Not all the food found in anglers' stomachs has been other fish. Shrimplike crustaceans of various sizes as well as arrowworms and squids have been found. But lantern fish, hatchet fish, and other angler fish are common items of food. One angler, *Melanocetus johnsoni*, 3½ inches long, was captured with a large lantern fish, *Lampanyctus crocodylis,* coiled neatly inside its belly, the meal being considerably larger than the diner. This greed has its perils, since it is apparently not unusual for an angler fish to seize booty too large and—unable to release the victim because of the great teeth, which point backward—to die along with its meal. Many of the finds of angler fish have been made when they have been forced to the surface of the sea after engulfing too large a prey.

This whole description of the structure and behavior of the angler fish has referred to the females; the males are so different that they hardly seem to belong to the same group. Indeed, for many years they were placed in a separate family, and it was remarked with surprise that only females of the deep-sea anglers were encountered. It is only recently that biologists have begun to sort out the angler fish and pair males with the females of the same species.

The striking variations between the males and the females of many anglers are due to their different modes of life. The male angler fish does not angle. It obtains its food by the surprising method of attaching itself to the body of a female, spending the rest of its life as a parasite on her! More is said about this astonishing behavior in a later chapter.

VIPERS AND RELATIVES

The next group of fishes to be described is a very large one, both in numbers of species and individuals. These are the stomiatoids. We are handicapped for lack of a common name here, since they are so unfamiliar that they have not received one. Some of the tribe are called viper fishes, and occasionally members of another group are called dragon fishes. But for the great majority we are obliged to use the rather formidable name stomiatoid, derived from the name of one of the principal genera, *Stomias*.

If the name is formidable, so are the little fishes themselves. Unlike the angler fish, the stomiatoids lack spiny rays in their fins. This links them with the soft-rayed salmons and herrings, and indeed they are sometimes referred to as deep-sea herrings. In many respects, however, they could hardly be more different from herring, which are small and mild-mannered plankton feeders that swarm in prodigious numbers in the cool surface seas of the temperate zones and are probably prey for more marine creatures than any other fishes. The stomiatoids, in sharp contrast, are not mild-mannered prey at all, but slashing predators with cruel fangs and huge mouths to gulp down unlucky victims that cross their paths.

There are six families and scores of species of the stomiatoids, and there are so many differences among the members of the group that it is not easy to paint a general picture of them. It is nonetheless convenient to treat the group as a unit, with the reservation that there are many exceptions to general statements about them.

Whereas the angler fish live in the very deep sea, the largest concentration being found at about 6,000 feet, the stomiatoids dwell in waters of the shallower twilight zone or, in some cases, in some-

FIGURE 10–6. *Deep-sea stomiatoids, unlike anglers, are without spiny rays in their fins and are sometimes called deep-sea herrings.* Aristostomias grimaldii, *about 5½ inches long, is from the Atlantic Ocean and Caribbean Sea.*

Dana Expeditions

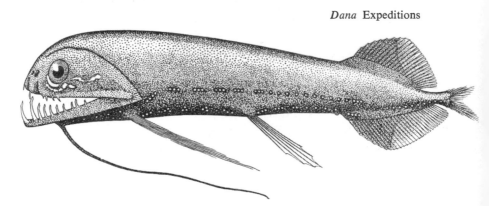

what deeper waters, at the edge of the area where perpetual night exists, at depths of about 500 to 2,000 feet. These slender little fishes, described as "beautiful" by some biologists and "fearsome" and "appalling" by others, usually taper from a large head to what appears by ordinary fishy standards to be a disproportionately undersized tail. Wisps of paired pectoral fins, sometimes only a single ray, spring from behind the gill covers; while far back on the body the other paired fins slant down like the stabilizers on a submarine. Still farther back, crowded almost against the tail, are the dorsal and anal fins. Thus these fishes somewhat resemble the muskellunge, a familiar carnivore of northern fresh-water lakes, and the predatory barracuda of tropic seas.

But no barracuda or muskie, both famous for driving strength and rapaciousness, exhibits the ferocity of the stomiatoids, if we can judge from the appearance of the latter. This formidable aspect is derived chiefly from the huge jaws armed with fangs, and a chin barbel that is often long and marvelously formed. The mouth is deeply slashed into the big head, carrying the jaws far back and permitting the same kind of prodigious gape that we saw in the gulpers and the angler fish. Matching this capacity to open very wide, the jaws are equipped with strong muscles, giving them a vicious snap and bite, and the ability to hold tightly to any animal seized. The teeth are unequal in length, many of them are fearsome and the greatest of them are cruel-looking barbed fangs. These canine teeth are so large that they cannot find room in the closed mouth of the fish; in all probability the mouth is not often closed, but wide agape, to give frightened prey a chilling view of impending doom.

All but a few species of the stomiatoids sport a barbel, a long filament streaming from the chin or throat. Occasionally this is only a stub, as in the codfish, but usually it is much more than this, and in a few species it is really showy. In *Ultimostomias mirabilis* the barbel is ten times the length of the body of the fish, in *Grammatostomias flagellibarba,* nearly five times.

The great variety of the barbels is especially evident in their terminal ends. Some end only in a slight swelling, but in others filaments are added, then branched filaments, and then rebranched filaments, followed by a second and a third series of knobs. Bunches of grapes, strings of beads, bouquets of flowers, tapering whips, limbs of spring-blooming trees—all these are suggested by the intricate combinations.

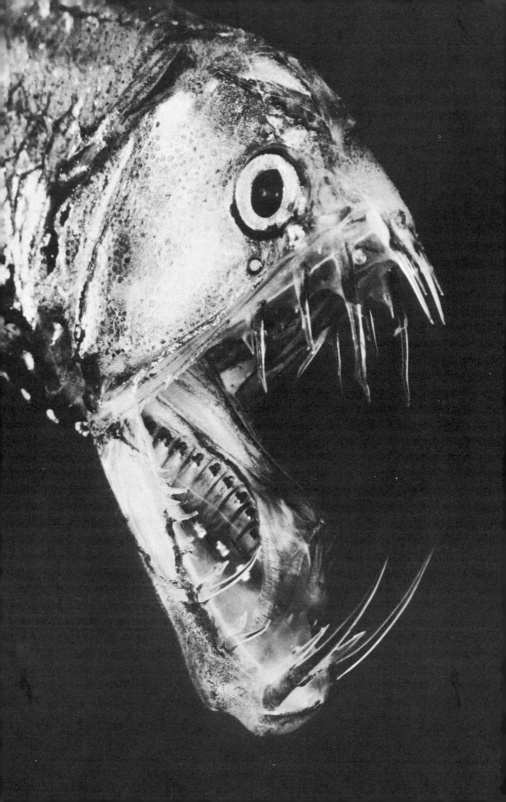

Biologists will have to see the stomiatoids in action far more than has been possible so far before they can be certain of the function of these fantastically varied chin decorations. It seems likely, however, that they serve the same purpose as the illicium does for the angler: to lure prey. The swellings that are the nearly universal component of all barbels, regardless of what else may be present, are usually light organs, glowing or flashing with multicolored illumination. Sir Alister Hardy says that the swollen tip on the end of the barbel of *Stomias boa* (perhaps the best known of all the stomiatoids) "looks very much like a small red copepod and presumably serves to lure other smaller fishes to destruction." In addition, some of the streamers of the barbels of various species are luminous.

The seemingly overelaborate variation among the barbels of different species may also serve to provide distinctive recognition signals, allowing individuals to link up with members of their own tribe or sheer away from unfriendly members of a gang on the prowl. A third possible function of the chin whiskers is sensory. The cods use their short barbels to investigate the bottom and to detect any of its inhabitants that might serve as food. But most of the stomiatoids are not bottom grubbers and could not use their barbels as feelers. Indeed, it is hard to believe that an enormous trailing whisker like that of *Ultimostomias mirabilis*, ten times the length of the little 1½-inch fish, could be anything but a serious hindrance rather than an aid. But the barbel of the stomiatoids may well act as an antenna, like the lateral-line system of the gulpers and angler fish. In some species this organ does seem to be exceedingly sensitive. William Beebe described one of the rare experiences that man has had with a live stomiatoid: "Even a slight stirring of the water near the barbel would arouse the fish to the utmost so that it threshed about and snapped, striving to reach and bite the source of irritation. Again and again we proved the astonishing sensitiveness of this organ."

FIGURE 10–7. *Saber-toothed viper fish. Although small (2 or 3 inches to a foot), this fish is as terrible a predator as its fearsome aspect suggests. Luminous patches inside the mouth undoubtedly serve as a lure. The teeth bend under pressure; nevertheless, they are quite capable of piercing the soft bodies of deep-sea victims.*

Photo by Paul A. Zahl, National Geographic Society

[207]

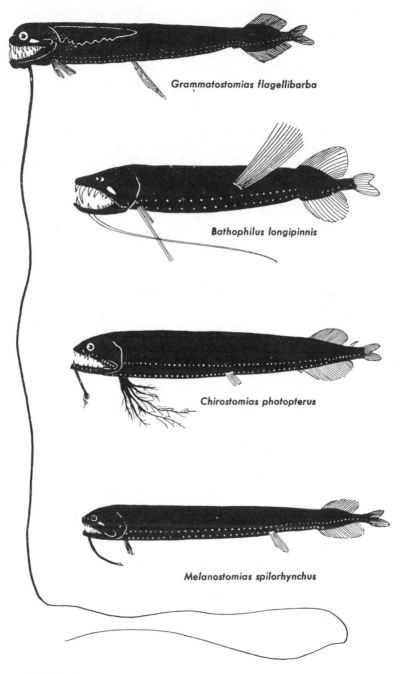

Grammatostomias flagellibarba

Bathophilus longipinnis

Chirostomias photopterus

Melanostomias spilorhynchus

Some of the stomiatoids, since they are dwellers of the very deep sea, have the velvety black and brown colors that characterize the anglers and the other inhabitants of this region. More of them are lighter, however, taking on the tints produced by waters with at least a glimmering of light: gray, silvery, light brown. Many have a shifting iridescence, like *Stomias boa*, whose basic color is silver. *Chauliodus sloani* has no scales, and opalescent pigment is arranged on the skin of its back in hexagonal patterns that resemble scales; on the sides the body is gray.

It is not primarily the skin that lends these fishes color, however, but the brightly colored light organs—hundreds or even thousands of them on a single individual. Most of the stomiatoids have a double row of luminescent buttons on each side, along the underside of the body from just behind the gill cover to the tail (Figure 10-8). On the tail itself there are one or two luminescent organs on each side. In most species the small organs are almost evenly spaced, but in others there are groups of three or four, separated by gaps; the pattern is constant for a given species. In addition to these easily noticeable light organs, many species have other smaller ones in various positions on the body—on the head, the fins, the barbel, even in the mouth and on the eyeballs. There may be many thousands of these, and the combined effect makes the fish resemble a miniature lighted ocean liner or a decorated Christmas tree. The smallest of the luminous organs are mere pinpoints of light, while those on the lower flanks are complex, with a lens and a dark-pigmented reflector. A few species have very large light organs behind the eye, the male often sporting one considerably larger than that of the female.

Light is produced in the luminescent organs of the stomiatoids

FIGURE 10 – 8. *Most stomiatoids have a barbel streaming from the under jaw. This ranges from a short stub to a whip that stretches 10 times the length of the fish's body. There is a double row of lights on each side of the fish, and luminous patches behind the eyes of a few species. Shown here are* Grammatostomias flagellibarba *(about 8½ inches long),* Bathophilus longipinnis *(about 2 inches long),* Chirostomias photopterus *(about 8 inches long), and* Melanostomias spilorhynchus *(about 9 inches long).*

N. B. Marshall, *Aspects of Deep Sea Biology*

FIGURE 10-9. *Bristlemouth or roundmouth* (Cyclo-thone). *The bristlelike teeth show clearly in the fish on the right. These rapacious little fishes (1 to 3 inches long) open their trapdoor mouths to nearly 180 degrees to snap up prey in the deep sea. Light organs shine brightly on their under-sides, and less so on their heads and jaws.*

Photo by Paul A. Zahl, National Geographic Society

by gland cells, and not by luminous bacteria as in the angler fish. The large light organ behind the eye in some species is located in a pocket covered by transparent skin. The fish has the ability to rotate this organ so that it can be turned down and off.

With such equipment for attracting prey, along with the great jaws and extensible stomachs, the stomiatoids are efficient exploiters of their section of the deep sea. Their food is small planktonic crustaceans, such as copepods, and large numbers of fish—often as big as or bigger than themselves. Lantern fish are favorite prey of many species of stomiatoids.

Not the least remarkable thing about this group is its migratory behavior. The movements are vertical, from deep water to the surface and back again. *Idiacanthus* (Figure 14-3), which is distributed through all three of the major oceans of the world, lives as an adult in deep water, from about 2,800 to 5,600 feet. At night it rises from these great depths to within 600 feet of the surface. The daily journeys are made easier by the lack of a swim bladder, so that there is no necessity to spend energy in equalizing the pressure of such an organ.

The curious young of *Idiacanthus* also make it a worthy object of attention. Spawning is a summer activity. Very large numbers of small, delicate, almost transparent larvae have been caught in late summer off Bermuda in about 550 feet of water. The larvae were for many years not recognized as the young of *Idiacanthus*, but were first described as a completely different genus and named *Stylophthalmus paradoxus*. Biologists may be readily excused for this error, for the larvae are very different from their parents. It was William Beebe who finally untangled the puzzle and linked them to *Idiacanthus*.

The most curious feature of the larval *Idiacanthus* is the exceedingly long eyestalks. The eye is placed on the end of a cartilaginous rod protruding from the head of the little fish. It is a fully functional eye, being put to very good use in the lighted upper areas of the sea where the larvae live. It seems to be capable of being turned, judging by the long muscle fibers that run the length of the stalk. The stalk itself can be turned and twisted too, and the whole curious apparatus gives the strong impression that eyes must be very important to the fish at this stage of its life. At a length of about 1½ inches the fish begins to descend to greater depths, and the cartilaginous rods begin to draw in. They finally twist and turn into loops that are knotted into little capsules that lie in front of the eyes—as though there were no really good place to get rid of the curious structures.

The stomiatoids are small, mostly only 6 or 7 inches long and generally no more than a foot. Beebe did see a 6-foot stomiatoid from the window in his bathysphere, but no such giant has ever been caught. Maybe it is just as well that they are rare, considering the rapacious character of this group and the equipment they have for indulging their inclinations.

FIGURE 10–10. *Larva of the stomiatoid fish* Idiacanthus. *This peculiar larva has eyes sticking out on long stalks. As the fish grows, the eyestalks twist into loops and are eventually knotted into small capsules in front of the eyes.*

N. B. Marshall, *Aspects of Deep Sea Biology*

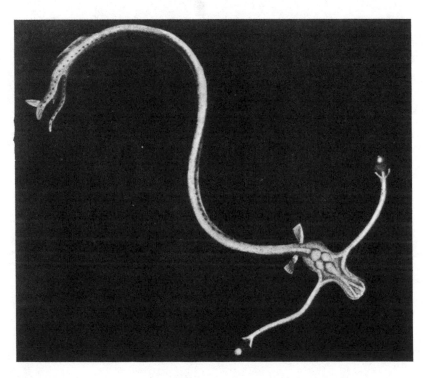

LANTERN FISH

When we come to the lantern fish we deal with a group that is more acceptably fishlike by normal standards. Perhaps their more ordinary appearance is related to the fact that they are only "partly" of the deep sea, since at night they are frequently encountered on the very surface of the ocean. They are, nevertheless, deep-sea creatures, since they are swimmers in the dark waters during the daytime; some species of the large group are found as deep as 3,500 feet. In common with most deep-sea fishes, the lantern fish are small. When fully grown, these "minnows" are often less than the length of a man's little finger, and a big one is only 6 inches long. They are compact, with fins of ordinary size placed in normal positions. Their eyes are considerably bigger than those of most fishes of their size (Figures 10-11 and 13-7).

The part-time residence of the lantern fish in the lighted upper layers of water is revealed in their colors. Most species are light shades of brown or gray, or have a silvery-white sheen, especially on the underside. Some are a beautiful iridescent blue, characteristic of surface-layer fishes, and the silver color of the flanks is sometimes tinged with pink or green like mother-of-pearl. Beebe described *Myctophum affine*, a species caught as deep as 2,500 feet, as having a deep purple color. Although the lantern fish resemble familiar shallow-water fishes, they are clearly set apart by their batteries of lights. There are two sets of these, one consisting of small, round organs arranged in two rows on the underside of the body and on the head. These photophores look like black-edged pearl buttons, and they are arranged in various patterns, which are constant within a particular species but differ sufficiently among species so that an expert can tell by its lights which of the 170-odd species he is looking at (Figure 14-5). The other series of light organs consists of a small number of platelike photophores on the tail, just in front of the caudal fin.

The lights on the flanks glow dimly, but when the lantern fish flash their caudal lights it is obvious that they come honestly by their name. Beebe says of the caudal light: "The power of this glandular light is very much greater than that given off by all the other organs together. The entire dark room of an aquarium was momentarily illuminated when they flared out." On another occasion he says, "a single fish would have illuminated the whole dark room, and enabled one easily to read fine print."

At least one lantern fish, *Neoscopelus,* has strong lights on the underside of the body that may illuminate the dark waters where the fish swims, and may also attract prey. This little fish (about 5 inches long) has pushed the business of luring victims another curious step forward, developing a series of small but bright photophores on its tongue. And *Diaphus* has large "headlights" that shine brilliantly (Figure 14-5).

Everyone who has been lucky enough to see lantern fish alive has expressed delight over the beauty of these little creatures, ablaze with jewellike lights. Green seems to be a common color of the light organs, but many other colors have been reported. In his book *Half Mile Down,* an account of his historic dives in the bathysphere in the 1930's off Bermuda, Beebe tells how "a school of brilliantly illuminated lantern fish with pale green lights swam past within three feet of my window, their lights being exceedingly bright . . . the lateral lights were sometimes dim, sometimes bright, but without exception seemed to glow steadily. Their tint was usually pale yellow, but several times a reddish tinge was observed." At the surface at night, schools of lantern fish sometimes flash red when they are caught in the lights of ships. Sir Alister Hardy points out that this is not the light from a photophore, however, but a reflection from the fish's eye as from a cat's eye in the beam of an automobile headlight.

Lantern fish can be attracted to a beam of light on the water at night, although they ordinarily shun light. Their vertical migrations from the depths to the surface take place only when darkness descends, and the little fish dive for the protection of the deep sea when daylight approaches. So sensitive are they to the strength of illumination that catches made at the surface are much smaller on moonlit nights than on dark nights: Beebe captured forty individuals on a cloudy night but only two in the same length of haul the next night, when the moon was almost full. Yet a powerful searchlight beam on the surface of the sea will sometimes attract great

FIGURE 10-11. *Lantern fish spend the day in the deep sea; when darkness comes, they rise to the surface. Rows of lights on their flanks glow dimly, but the tail lights are very strong. One species of lantern fish has lights on its tongue.*

G. B. Goode and T. H. Bean, *Oceanic Ichthyology*

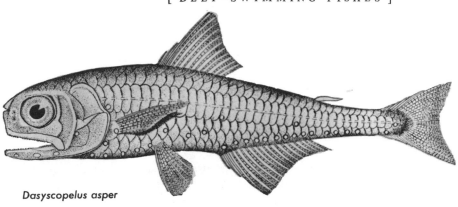

Dasyscopelus asper

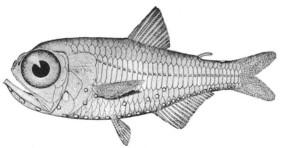

Electrona rissoi

Rhinoscopelus coccoi

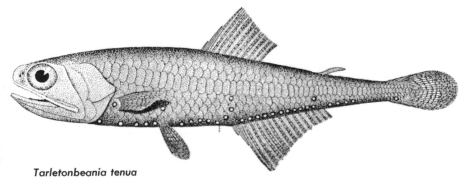

Tarletonbeania tenua

crowds of lantern fish, lured like moths. Actually this reversal of reaction to light is not uncommon among many kinds of animals; apparently after a certain threshold of light intensity is exceeded, the reaction to light changes from repulsion to attraction.

The hordes of lantern fish sometimes encountered at the surface are apparently there to feed, taking advantage of the much greater abundance of food at the surface as compared to that in the deep sea. The lantern fish are among the most numerous of deep-sea fishes, probably coming second only to *Cyclothone* (one of the stomiatoids) in numbers. The most dramatic account of these numbers is given by Marshall, who describes how the British weather ship *Weather Observer,* returning to Glasgow from a period of duty at sea, steamed through lantern fishes, *Myctophum punctatum,* for 5 hours one night. In the Antarctic, lantern fish rise to feed on the krill. The whalebone whales also relish krill, and unlucky lantern fish are sometimes found in the bellies of the whales, engulfed as the big animals scooped up the krill. Dr. Giles Mead of Harvard University discovered that sea lions of the Pribilof Islands of Alaska eat lantern fish as a main item of their diet, undoubtedly capturing them during the fishes' nighttime excursions to the surface.

Dr. Rolf Bolin of Stanford University's Hopkins Marine Station noted that lantern fish taken from the stomachs of deep-sea predators (an especially rich source of scientific specimens) were nearly always males. By contrast, the catches of deep-fishing trawls were more often females. Dr. Bolin suggests that the male, which possesses a brighter set of lights than the female, deliberately flashes its photophores to attract the predator away from its mate, sacrificing itself for the sake of the race. The trawl, however, is impervious to the flash of the brave male, and catches both sexes.

"Imagine," says Beebe in an article in *Zoologica,* describing *Myctophum coccoi,* one of the very common lantern fishes, "a minnow which is iridescent copper above and silvery white below, not over two inches in length, with large eyes and moderate fins. A full-grown fish weighs a gram, which means that it would take about 450 to make a pound. It feeds on copepods, amphipods, floating snails and other minute plankton fry, and from this food it generates sufficient energy to swim, to make daily migrations up and down, to illuminate one hundred lights and to deposit upwards of two to three thousand eggs."

[216]

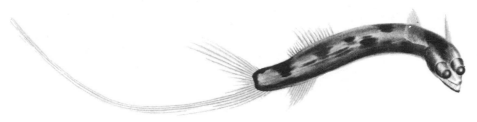

FIGURE 10–12. *Giant-tail* Gigantura. *The distinguishing feature of this fish is its very long tail fin. Tubular eyes, which stick out from both sides of the head, enable the giant-tail to see with both eyes at once—an ability that most fishes lack.*

N. B. Marshall, *Aspects of Deep Sea Biology*

GIANT-TAILS

The giant-tail *Gigantura* has enough strange characteristics to be set apart by some ichthyologists in a separate order. A large tail fin with the lower lobe greatly elongated gives the group its name. In gluttony the giant-tails are commonly compared to the deep-sea anglers, since they are nearly equal to the latter in their capacity to engulf huge meals as big as or bigger than themselves. Unlike nearly all other deep-sea fishes, the giant-tails are scaleless. They also lack the common dark colors of fishes from the depths. Instead, they have bright, metallic colors. Another peculiarity is their possession of long, telescopic eyes protruding from the sides of the head; these make it possible for the fish to see the same object with both eyes at once—a skill we take for granted but one that most fishes lack.

HATCHET FISH

The last of the deep-sea fishes we will discuss are the hatchet fish. These are greatly compressed so that they are sometimes as thin as coins, and some species, especially *Sternoptyx*, are remarkably like hatchets in appearance: the tail represents the handle, and the straight-edged belly is the blade. As a group these fishes are even smaller than some of the other Lilliputian fishes of the deep sea,

FIGURE 10-14. *Head-on view of swimming hatchet fish* (Argyropelecus). *Their skin is covered with a silvery iridescent pigment, and light organs stud their bodies, usually giving off blue rays. This remarkable photograph was taken in the Straits of Messina in the Mediterranean Sea.*

Photo by Paul A. Zahl, National Geographic Society

including the lantern fish and the stomiatoids. Many are 1 to 2 inches in length when full grown, and the biggest species reaches only 4 inches. When the fish is seen head-on, as in the stunning photograph by Paul Zahl, two features seize our attention—the huge, almost round, gaping mouth with the corners drawn down sharply in a dour expression, and the bulging, dark eyes close together on top of the head, gazing upward in the water. The eyes are often tubular, like telescopes. These chunky, high-backed little fishes lack scales over most of their bodies, the skin being covered with a silvery pigment that iridesces. The lower edge of the tail is peculiarly transparent,

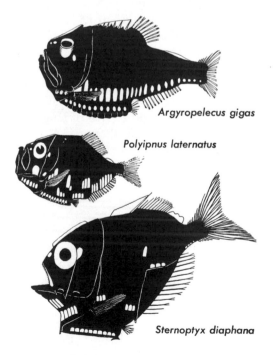

Argyropelecus gigas

Polyipnus laternatus

Sternoptyx diaphana

FIGURE 10-13. *The shape of* Sternoptyx diaphana *shows why these are called hatchet fish. In addition, they are so greatly compressed that they are sometimes as thin as coins. The eyes usually point upward and are often tubular, as in* Argyropelecus gigas. Argyropelecus *is about 3½ inches long,* Polyipnus *about 1½ inches, and* Sternoptyx *about 2½ inches.*

N. B. Marshall, *Aspects of Deep Sea Biology*

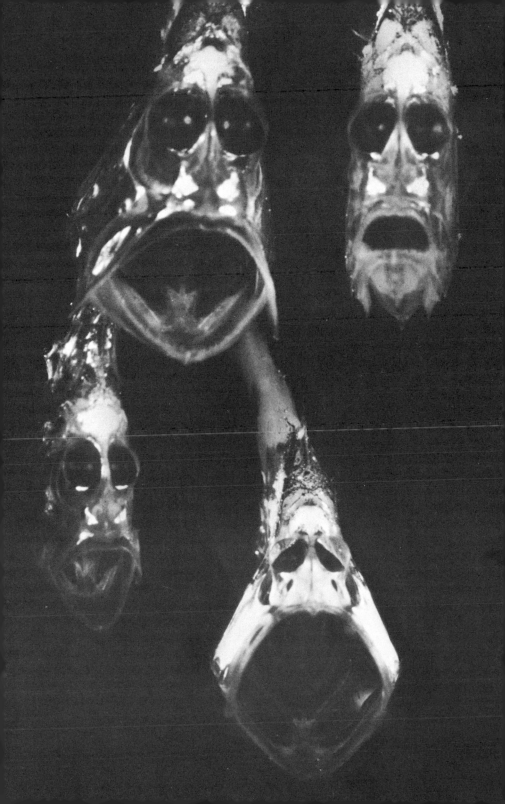

so that the fin rays shine through the muscle. Hatchet fish live in moderate depths, from about 880 to 1,600 feet, similar to those occupied by the lantern fish. Like the lanterns, the hatchet fish have a colorful collection of light organs. Although the fish's eyes point upward, the light organs shine their beams from the lower edge of the body, slanting downward into the dark water. The photophores do not appear as pearly spots, like those of the lantern fish, but as elongated ovals upright on the flanks of the fish, adjacent photophores almost touching each other in many species. The light given off by the photophores of the hatchet fish is blue in most cases, although bright ruby-red organs have also been seen.

The curious behavior and form of the deep-sea fishes illustrates the plasticity of living tissue—its ability to be molded by environmental conditions into patterns of action and structure best calculated for survival. The extremes of conditions in the deep sea have induced extremes in its inhabitants, and we regard these with the interest and awe of inhabitants of one world receiving a glimpse of the dwellers of another. Most of these fishes, especially the gulpers, the swallowers, and the angler fish, are monsters of the deep sea. This is not because of their size: as we have seen, few of them exceed the size of a pear and the biggest are something over a yard or two long. But they are monstrous in the sense of deviating widely from the normal.

But if these are not sea monsters in the usual sense, are there such things? This is the question to be considered next.

11 *Sea Monsters*

While many of the animals of the deep sea are monstrous and bizarre in form, they are almost universally of small size. What, then, ever happened to the sea serpents? Are the stories all myths, and must we dismiss any romantic hope of discovering huge monsters in the ocean?

Not quite. Many scientists would probably reject the idea that authentic sea monsters of huge dimensions are ever likely to be discovered in the deep sea, yet Anton Bruun mounted the *Galathea* Expedition of 1950–52 with the serious purpose of testing his theory that such creatures do in fact exist. Of course more prosaic purposes involving series of oceanographic and biological observations were also included, but the expedition was given its first push when a Danish author, Hakon Mielche, became interested in Bruun's lecture on the possibility of the existence of sea serpents. Dr. Bruun's argument is that the evidence against the existence of sea serpents is all negative—we have never caught any. But, he says, this does not prove that they do not exist, and with a rapidly increasing number of expeditions exploring the deep sea with more efficient kinds of traps, who knows but what someday not one but a whole galaxy of strange sea serpents may be found?

THE ONES THAT GOT AWAY?

Certainly it would not be surprising if large, agile animals were able to avoid the clumsy gear with which oceanographers have so far sampled the deep sea. We know for sure that specimens of some creatures of the depths have not yet been caught, for no one has hauled to the surface the 6-foot fishes of which Beebe got such tantalizing glimpses from the bathysphere off Bermuda. And of course Beebe was by no means in the true deep sea; we have no way of knowing whether not only 6-foot fish but perhaps 60-foot fish may live there, someday to be snared by an astonished and triumphant marine scientist.

GIANT SQUIDS

Some biologists believe that if sea serpents are ever found it is most likely that they will turn out to be giant squids. We have good evidence, from huge animals that have turned up from time to time, that truly enormous and powerful squids, or kraken, exist in deep water. Although it is hard to separate fact from fiction in some of the old stories, it is known with certainty that there are giant squids, since individuals measuring 57 feet have been examined. These are the biggest of the invertebrate animals, and are as long as that largest of fishes, the whale shark. One encounter with the giant squid *Architeuthis* took place in 1861 between Madeira and Tenerife, in the Canary Islands. A French corvette, the *Alecton,* came upon the monster lying on the surface of the sea. The warship's commander, Lieutenant Frédéric-Marie Bouyer, wrote: "Finding myself in the presence of one of those strange creatures which the Ocean brings forth at times from its depths as if to challenge science, I resolved to study it more closely and try to catch it." A heavy swell and evasive action by the giant squid made effective pursuit difficult, and it was some time before the ship could be brought close enough to hit the squid with a harpoon and to throw a noose around it. A violent movement by the animal broke the harpoon loose and tore off part of the tail fin. The maimed squid disappeared. The piece torn from it was estimated to be about a tenth the weight of the whole animal; the squid's length was estimated at 15 to 18 feet. The monster was described as "frightful, its color brick-red, . . . this colossal and slimy embryo presented a repulsive and terrible figure."

We can only guess why the giant squid was at the surface on this occasion. There are enough accounts of sightings of these animals after this early report to suggest that they sometimes appear on the surface as part of their normal behavior. It may be, on the other hand, that this particular squid was injured or sick and was lying on the surface as a consequence.

FIGURE 11–1. *A sea monster come to life. This is the immense artificial squid used in Walt Disney's version of Jules Verne's* Twenty Thousand Leagues Under the Sea. *It weighed 2 tons and required 24 men to operate it with hydraulics, compressed air, electronics, and remote controls.*

Walt Disney Productions

Another actual encounter with a giant squid took place in 1873 in the waters off Newfoundland. Two cod fishermen, Daniel Squires and Theophilus Piccot, and the latter's son, twelve-year-old Tom Piccot, saw a large object in the water. Rowing over to it in their small dory, they saw that it was alive. They struck it with a boat hook and the squid retaliated by attacking them. Rushing toward the boat, it struck the gunwale with its parrotlike peak and wrapped a long, thin tentacle and a shorter, stouter one around the boat. The squid seemed about to draw the boat beneath the water when Tom Piccot, rising to the occasion more bravely than his elders, cut off the long tentacle and part of the other one. On shore, the short arm was devoured by dogs, but the long tentacle was preserved. Its length was 19 feet, and its largest diameter 3½ inches; it was exceedingly strong and tough.

The tentacle was taken to St. Johns and shown to the Reverend Moses Harvey, a clergyman well known for his interest in natural history. This was a fortunate move, for it quickened Harvey's already lively interest in giant squid and prepared the way for the next episode in the story of the kraken off Newfoundland. When Harvey saw the tentacle he said, "I knew that I held in my hand the key of a great mystery, and that a new chapter would now be added to Natural History."

That same year, in Logy Bay near St. Johns, close to the place where the three cod fishermen had the encounter with the giant squid, four astonished herring fishermen entangled a kraken in their net. The violent thrashings of the squid, which shot long tentacles through the meshes of the net toward the boat, nearly caused the fishermen to cut it loose. Instead, one of the more intrepid of the crew chopped off its head behind the eyes.

Moses Harvey was quick to hear about this newest event and he offered the handsome sum of "10 dols." for it. He was feverishly eager to get "the dead giant, and 'rolling as a sweet morsel under my tongue' the thought of how I would astonish the savants, and confound the naturalists, and startle the world at large. They evidently thought me cracked to be paying away so much money for a nasty brute that nearly cost them their lives before they could get it into their herring net. To allay their curiosity and make them more careful, I hinted that I wanted it for *a present for the Queen!* Next day, to my great satisfaction, a cart arrived at my door almost filled with the hideous corpse-like creature, which I speedily stowed away in an out-building, in a huge vat filled with the strongest brine."

Commendably determined to serve the interests of science, Harvey took the squid from its pickle vat and draped it over his sponge bath to photograph and measure it. Its tentacles were 24 feet long and the over-all length was 32 feet.

Other great squids were caught in surprisingly large numbers off Newfoundland in the next few years, fifty or sixty being reported by the fishing boats. Some change in conditions in the ocean appears to have driven these giants from their usual habitation in the deep sea, a circumstance that does not seem to have been repeated since to the same degree.

It is certain, then, that very large squids have been seen, and we are led to believe that there are still others, perhaps larger and more powerful, that have never been captured or even seen dead. *Architeuthis* is big but not strong, but in Great Britain a museum has the arm of another kind of squid which must have been approximately 24 feet long, and both strong and swift.

SEA SERPENTS?

Another distinguished marine scientist besides Dr. Bruun firmly believed that huge "sea serpents" exist in the deep sea: Professor J. L. B. Smith of South Africa, famous as an ichthyologist and well known for his descriptions of the coelacanth, a fish that was considered extinct until it turned up dramatically off the coast of South Africa in 1938. If a representative of a group of animals supposedly vanished from the earth for 70 million years has actually been swimming in fairly shallow water near shore, what might the vast expanses of the deep sea hold?

Professor Smith did not limit sea monsters to the giant squids, but confidently expected that research will eventually turn up other strange animals, including ancient kinds of elongated whales that paleontologists have thought were represented only by fossil remains millions of years old. He cited an impressive number of cases in which trustworthy witnesses have reported sighting "monsters" and supplied convincing details.

Two Scottish fishermen recently reported being severely frightened by the appearance of "a horrible monster out of prehistoric times," which swam past their small boat in the sea near Aberdeen. With a reptilian head 2½ feet long, large protruding eyes, and a sharply sloping back ("like an overturned boat"), it resembled descriptions of the Loch Ness monster, the pride and one of the chief claims to fame of Inverness-shire in the Scottish Highlands and undoubtedly the most famous of all sea serpents. Even some dim but apparently authentic photographs have been exhibited of the Loch Ness monster, whose head has been described as resembling that of dragons of old, and whose enormously long body undulates vertically in the water. Professor Smith accepts the existence of the Loch Ness sea monster as certain. He suggests that it may be the same as the Aberdeen serpent, since the two areas are close and are connected by the Caledonian Canal.

One apparently unimpeachable witness who soberly reported sighting a sea serpent was Hans Egede, a Norwegian missionary who became a bishop and whose reputation among his contemporaries was notably high. On a voyage to Greenland in 1734 he reported seeing "a terrible sea animal" whose head reached above the maintop. It had a long, sharp snout, broad flippers, and a tail as long as the ship. It blew like a whale, and Professor Smith is struck by

FIGURE 11-2. *An old engraving shows a sea serpent attacking a ship. There have been tales of great sea monsters through the ages, and while many scientists reject the idea that they exist, others are not so sure. One theory is that the monsters are giant squids.*

Sea Frontiers

the similarity between Egede's monster and a primitive type of whale, *Basilosaurus,* that inhabited the sea millions of years ago. Fossils of this whale show that it had a long, snakelike body, a pair of flippers near the head, and a blowhole toward the front of the head, which would permit it to blow in the manner described by the missionary.

The captain of a British man-of-war had several cowitnesses to the sighting of a monster on August 6, 1848. Captain M'Quhae was on the bridge of his ship H.M.S. *Daedalus* with several of his officers when they and members of the crew clearly saw an enormous eel-shaped animal nearby. It was estimated to be at least 60 feet long, with head and shoulders measuring about 4 feet and held above the water as it swam past. "It passed rapidly, but so close that had it been a man of my acquaintance I should easily have recognized his features with the naked eye," the captain wrote. Smith thinks it "outrageous even to suggest" that the captain of a British warship could concoct such a story and persuade his officers and men to be a party to such an elaborate fabrication.

In 1893 a number of people saw a sea monster off the west coast of Africa. On December 4 the S.S. *Umfali,* a small South African ship, was heading south for the Cape at about 5:30 P.M. when the master, Captain R. J. Cringle, his officers, several members of the crew, and some passengers saw a very large animal traveling north at a rapid pace. It was about 500 yards away and was estimated to be 80 feet long. It was described in one account as having the shape of an enormous conger eel with huge jaws 7 feet long, armed with large teeth. But Captain Cringle says, "I saw full fifteen feet of his head and neck on three separate occasions. The base or body from which the neck sprang was much thicker than the neck itself, and I should not therefore call it a serpent." Nor, presumably, could it accurately be compared to a conger eel. The captain ordered the ship turned about and the creature was pursued until darkness halted the chase. The *Umfali* traveled at 10 knots but could not overtake the swift animal. Whether conger eel, serpent, or seallike mammal, the *Umfali* monster seems unquestionably to have existed.

Two biologists later sighted a sea serpent whose description tallied closely with that of the *Umfali* monster. Mr. E. G. B. Mead-Waldo and Mr. J. Nichol saw this animal from the deck of the yacht *Valhalla* while on a cruise as guests of the ship's owner, the Earl of Crawford. They were off the coast of Brazil in 1906 when the serpent appeared. It had a long neck springing from a thicker body, and the head, which moved from side to side in a peculiar manner, was described as resembling that of a turtle. A big oblong fin rose above the surface of the water.

South Africa seems to be a happy hunting ground for sea serpents and living fossils. Another monster was seen outside Durban harbor in 1884, by the captain and others aboard the ship S.S. *Churchill.* The serpent carried an enormous head 7 or 8 feet out of the water. This same animal was sighted from shore by military officers some time after it was seen from the *Churchill.*

Professor Smith recounts another encounter with a sea serpent, this one occurring during World War I. It is based on the account of a U-boat skipper, Baron von Forstner, who reported in a German newspaper how he torpedoed a British ship, the *Iberian.* She sank in the water stern first, and then blew up with a loud explosion. "The commander and five other officers in the conning tower watching this were astounded to see a huge creature thrown some twenty or thirty yards in the air together with a lot of debris. According to them it was a good sixty feet long, had four short legs and resembled a giant crocodile. The creature was struggling and contorting during its flight in the air, and when it hit the water again it struggled and splashed in the surface for some twenty or thirty seconds before disappearing for good." (Quoted in an article by Professor Smith in the magazine *Outlook.*)

MIRAGES

Most of the accounts of sightings of strange giant creatures do not seem to be as trustworthy as those quoted. In many cases they are stories by single individuals, and, without confirmation by another person, there is an inclination to ascribe them to careless observation, ignorance of what was seen, or deliberate attempts to mislead through desire for notoriety. Other accounts may be dismissed as resulting from inebriation or homesickness that blurred and misled the vision.

Among the honest mistakes made by many people reporting the sighting of sea serpents, surely many can be explained by the undoubted existence of giant squids. Then such rather prosaic explanations as a flight of birds near the surface of the water, or actually swimming on the surface, may account for some. Seaweed, especially the giant kelp of the Pacific, whose enormous shiny stems undulate

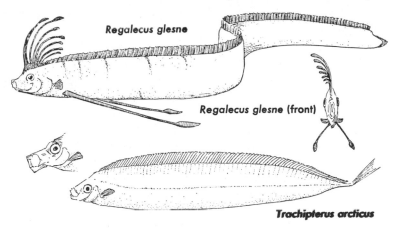

Regalecus glesne

Regalecus glesne (front)

Trachipterus arcticus

FIGURE 11 – 3. *Glimpses of the oarfish* Regalecus glesne *are thought to have led to many of the reports of sea monsters. It has been described as having the head of a horse and a flaming red mane. The oarfish drawn here was 13 feet long, but others may reach 50 feet. A relative is the dealfish* Trachipterus arcticus, *which has a curiously upturned tail and a mouth that can be thrust out.*

Sir Alister Hardy, *The Open Sea: Its Natural History.*
Part 2. *Fish and Fisheries*

and quiver on the surface under the influence of the currents, could fool some observers. Porpoises are familiar enough to most mariners, but their curious behavior of leaping from the water in graceful loops and following one another like children playing "follow-the-leader" might mislead some into thinking they were seeing the undulating coils of an enormously long serpent.

The oarfish or ribbonfish, *Regalecus,* is credited by Dr. J. R. Norman as accounting for many of the reports of sea serpents: "The monster described as having the head of a horse and a flaming red mane is the Oar-fish or the Ribbon-fish, a species which probably grows to more than fifty feet in length and may sometimes be seen swimming with undulating movements on the surface of the sea." Oarfishes are especially common in the Mediterranean, and sea serpents reported there may have been this animal in many cases.

The basking shark, *Cetorhinus maximus,* a very large species, is sometimes the cause of serpent reports. This shark gets its common name from its habit of lying on the surface of the sea as though basking in the sun; another name sometimes applied to it is "sunfish." It is an enormous creature, reaching a length of 40 feet, and is exceeded in size among the sharks only by the whale shark. Basking sharks have the curious habit of swimming one behind the other in pairs, showing two large dorsal fins 40 or 50 feet apart. Anyone could be forgiven if he took such a "creature" to be an enormous sea serpent, and without doubt many have. On shore, too, the basking shark has been called a sea serpent, since the carcass cast up on the beach looks unlike anything most people have encountered. Upon occasion, monsters on the beach have been described as having an eellike body, a head "like a camel with an upturned nose," and a body covered with coarse white hair. A basking shark looks especially monsterlike after it has lain on the beach for a while, since much of the flesh and familiar features are rotted away, leaving the frayed ends of the muscles to provide the "hair," the long backbone to serve as the "eel," and the odd-shaped cranium to baffle observers into thinking that it resembles a camel's head. A basking shark cast up on the beach at Stronsay in the Orkney Islands early in the nineteenth century was described as a new species with the name *Halsydrus pontoppidiani,* after Erik Pontoppidan, Bishop of Bergen, who had gained fame earlier with an account of a giant kraken.

But after we have made allowances for these errors of observation, and conceded that without doubt most of the alleged encounters with something unknown turn out to be explicable in terms of familiar creatures, there remain the accounts of the *Daedalus,* the *Umfali,* the *Iberian,* the *Churchill,* and a few others for which there is no satisfactory explanation. It is because these cases are so few and no *"corpus delicti"*—no body of a sea monster—has been found in all history that most scientists consider the case for sea serpents not proved.

It would seem foolish, nonetheless, to deny the possibility, for man has certainly not explored the deep sea so thoroughly that he can claim to know everything there. It does seem unlikely that any very large animal that frequently comes to the surface could fail to have been caught by this time. This would seem to eliminate mammals such as whales, or reptiles as sea serpents, for both of these must come to the surface to breathe at frequent intervals. Besides, there is no fossil of a mammal that could answer as a sea monster of today

—unless the ancient whale *Basilosaurus* does. There are fossil reptiles, it is true, the last of which disappeared some 80 million years ago. While this is not much older than the coelacanth fossils, it makes the existence of a live ancient reptile decidedly unlikely.

If a genuine sea monster does exist then, it is likely to be not a reptile or a mammal but the already familiar giant squid or an unknown fish. The creatures described in several cases sound like enormous eels, and some of the others reported could have been fish. The eel starts life as a curious leaflike creature with a small, toothed head. This larva, called a leptocephalus, is about one tenth or less the length of the full-grown fish; the 3½-foot conger eel has a leptocephalus of 3 or 4 inches. In recent years the deep sea has yielded a leptocephalus larva not inches in length but 6 feet long! This suggests the existence of an enormous eel of 60 or 70 feet, a length that would surely qualify it as a sea monster. It is proposed, then, that if sea monsters really exist they will probably turn out to be eels or other fish.

Contrary to the popular conception of the layman, science does not supply "proof," but deals in probabilities. As more evidence accumulates, the probability increases that an event will happen in a particular set of circumstances. In our present state of knowledge we cannot yet abandon the possibility that some as-yet-unknown monsters exist in the deep sea.

12   *Fossils of the Sea*

Creatures dragged from the remote depths of the ocean have the special fascination of inhabitants of a strange world, almost like visitors from outer space. One reason is the intriguing possibility that the deep sea might be a refuge for prehistoric animals—animals known only from their fossilized remains or perhaps completely unknown to man.

It is not easy to catch animals deep in the ocean. The scientist must fumble for them blind, letting down crude dredges and nets from the uncertain platform of a pitching vessel. He cannot even be sure that his gear has landed right side up or that it is operating properly. He fishes at random, unable to guide his gear toward animals he cannot see or otherwise detect. It is little wonder that the pace of under-sea exploration has been slow or that deep-sea collecting has often drawn a blank. Wyville Thomson describes the disappointment of losing a great catch after the labor of trawling in the deep sea at a depth of 7,710 feet: "The trawl was lowered, and on heaving in it came up apparently with a heavy weight, the accumulators being stretched to the utmost. It was a long and weary wind-in on account of the continuous strain; at length it came close to the surface, and we could see the distended net through the water; when, just as it was leaving the water, and so increasing its weight, the swivel between the dredge-rope and the chain gave way, and the trawl with its unknown burden sank quietly out of sight. It was a cruel disappointment, everyone was on the bridge, and curiosity was wound up to the highest pitch: some vowed that they saw resting on the beam of the vanishing trawl the white hand of the mermaiden for which we had watched so long in vain; but I think it is more likely that the trawl had got bagged with the large sea-slugs which occur in some of these deep dredgings in large quantity, and have more than once burst the trawl net."

But not all deep-sea hauls turn out like this, and when they are successful the rewards may be very great. Accounts of deep-sea trawling, including even some dry technical descriptions of the operation, commonly mention the excitement that accompanies the arrival of the net at the surface. Fishing is not by accident one of the most popular recreations in the world; there is always expectancy that this is the time of the big catch. Perhaps that is one reason why commercial fishing is a popular trade despite the hardships involved and the generally low income. Fishermen can never get over the excitement of wondering whether this will be the big haul, and whether something of more than usual interest will be in the net. Consider how much more exciting trawling miles deep in the ocean must be. If anything at all is captured, it will almost certainly be of great scientific interest. And there is always the chance that the net will contain a creature never before seen by humans.

The area deeper than about 20,000 feet is referred to as the "hadal" zone. This is appropriate, since in many characteristics— utter and perpetual darkness, icy-cold, crushing pressure—it qualifies as a type of Hades. It is only in very recent years that oceanographers have had boats and gear capable of sampling the creatures of this region.

FISHING HOLES

Waters of very great depths occur only in the trenches, which are situated close to great island arcs, for the most part in the Pacific (Figure 2-3). The trenches are elongated and narrow, and it is extremely difficult to drag them in order to sample their inhabitants. From his *Princesse Alice II* the Prince of Monaco first trawled successfully in the hadal zone in 1901, collecting animals from a depth of just under 20,000 feet. After his efforts there was a long gap until 1948, when the Swedish deep-sea expedition of the *Albatross* trawled on the east slope of the Puerto Rico Trench in depths up to 26,000 feet. That great expedition was quickly followed by the cruise of the Danish *Galathea* in 1951. The *Galathea* scientists were the first to prove that life exists even in the greatest depths when they brought animals to the surface from as deep as 32,565 feet in the Philippines

Trench. Later the *Galathea* investigated four more deep trenches. The Soviet research ship *Vitiaz* made a single haul in 1948 at 26,565 feet in the Kurile-Kamchatka Trench, and since then it has continued work in that trench and in the Marianas and Tonga trenches.

Trawling in the trenches is difficult, first of all, simply because of the immense depths involved. It requires tremendously long cables and powerful winches and engines to support these miles of cables and the nets attached to them. Even before the gear is put out, painstaking echo soundings must be taken to determine the exact configuration of the bottom and its precise depths, so that the right length of cable can be let out. The speed of the ship, the resistance of the cable, the angle of the cable as it leaves the ship, and the resistance of the water to the trawl are variables that must be carefully calculated before the trawling begins. The *Galathea* once put out nearly 7½ miles of cable, and her trawl touched bottom 1⅓ miles behind the ship. It required between 5 and 6 hours to pay out this wire. Once the trawl was lowered, there was a struggle to keep the gear on the bottom in the narrow trench against the influence of crosscurrents, which may have been going in different directions at different depths, and against winds that tended to push the ship off course. During the early trawling in the Philippines Trench the wind, the currents, and the waves all conspired to force the *Galathea* at right angles to the north-south direction of the trench. The ship had to be headed directly into the current, and it took considerable skill to keep the trawl on the narrow, smooth bottom of the trench. The trawl was of course much too far down to bump perceptibly on the bottom or otherwise signal to the winch man that it was at the proper depth; only continuous calculations could find the exact level that would avoid empty nets or even the loss of a trawl. Skill was not sufficient to beat the elements every time, and there were occasions when the long and tense work was rewarded with no catch—or no gear.

A composite list of animals caught in the hadal zone from successful deep trawling has been compiled by Torben Wolff, one of the scientists aboard the *Galathea*. He lists 310 species, including sponges, jellyfish, several kinds of worms, small crustaceans (but no large shrimps), sea spiders, snails, clams, sea urchins, starfish, sea cucumbers, and fish. So far only five species of fishes have been captured in the hadal zone; they may represent only a minor part of the deepest animal communities. The dominant groups in very deep water are the sea anemones, the polychaete worms, small crustaceans, snails and clams, and sea cucumbers. Some hadal creatures have been found only in the deepest trenches, and in many cases only in one trench. This may mean merely that the sparse sampling has not happened to catch them in more than one trench; but it might mean instead that some animals have become isolated in the enormously deep clefts of the trenches and have evolved unique forms, different from those occurring anywhere else.

"LIVING FOSSILS"

Early in the history of the exploration of the deep sea the exciting thought struck scientists that perhaps the abyss sheltered animals unchanged over millions of years; that the sea was a repository for "fossils"—the aquatic equivalents of the dinosaurs and mammoths of the land but, unlike these, still alive. As it was gradually revealed that even great depths harbored living creatures, this idea took a tighter grip on the imaginations of many biologists, including some of the most distinguished scientists. Louis Agassiz, the famous Swiss-American pioneer of oceanography and marine biology, hoped to find the deep sea populated with all manner of forms known then only as ancient fossils, and T. H. Huxley, one of Britain's most famous scientists of the nineteenth century, remarked on the large number of archaic forms in the sea.

PRIMORDIAL LIFE

Huxley, whose fertile mind leaped joyously at the ideas of evolution presented in Darwin's *Origin of Species,* became one of the militant champions of Darwin's theories and sought in paleontological and other studies for further evidence for the gradual genesis of modern animals. It seemed logical to him that the abyssal regions of the sea might harbor ancient forms that had escaped extinction because of isolation from modern, more successful competitors. His zeal trapped Huxley into one error, which he was later forced to admit publicly: he accepted the existence of "bathybius," supposedly an exceedingly primitive form of life from the deep sea. The British vessel H.M.S. *Cyclops* made a series of soundings across the Atlantic in 1857, collecting samples of the sediments of the sea floor. In these samples of ooze a strange gray gelatinous material was found, which Wyville Thomson and other distinguished scientists declared to be a primitive form of life.

Thomson wrote: "If the mud be shaken with weak spirit of wine, fine flakes separate like coagulated mucus; and if a little of the mud in which this viscid condition is most marked be placed in a drop of sea-water under a microscope, we can usually see, after a time, an irregular network of matter resembling white of egg, distinguished by its maintaining its outline and not mixing with the water. This network may be seen gradually altering in form, and entangled granules and foreign bodies change their relative positions. The gelatinous matter is therefore capable of movement, and there can be no doubt that it manifests the phenomena of a very simple form of life." Huxley was moved to give this "organism" a name, calling it *Bathybius haeckelii* after the German evolutionist Ernst Heinrich Haeckel, whom he greatly admired for his ardent support of Darwinism. Haeckel believed that bathybius was the "mother of protoplasm"; it had no organs but seemed to show reaction to stimulus and to be capable of assimilating food. Wyville Thomson was impressed by its apparent wide distribution: ". . . whether it be continuous in one vast sheet, or broken up into circumscribed individual particles, it appears to extend over a large part of the bed of the ocean." When the *Challenger* dredged up the same material, however, the "mother of protoplasm" came to an inglorious end. The expedition's chemist, Dr. J. Y. Buchanan, analyzed the gray material and could find no trace of organic material. Buchanan finally showed that bathybius was nothing more than a precipitate of calcium sulphate deposited when alcohol was added to the water—as it was when marine creatures from the nets were preserved.

Huxley admitted his mistake, saying "Bathybius has not fulfilled the promise of its youth," but he was unshaken in his hope that the deep sea might yet yield primitive forms of life. He was on the Committee of the Royal Society that planned the marvelously productive voyage of the *Challenger,* starting in 1872, and the scientists on that famous ship kept a constant lookout for living fossils. They were quite confident, for example, that they would drag from the abyss deep-sea representatives of the ammonites, great shelled relatives of the squid and octopus which flourished in the Mesozoic Era, 100 million years ago. They were even more alert to spot belemnites, Mesozoic fossil relatives of the squids and cuttlefish, whose principal characteristic was

FIGURE 12–1. *Sea lily* Rhizocrinus lofotensis *is not a plant but an animal related to the starfish. First caught in 1864, it astonished zoologists, who had previously seen it only as a fossil as old as 160 million years.*

J. Murray and J. Hjort, *The Depths of the Ocean*

a chambered internal shell. Dr. H. N. Moseley, one of the biologists on the ship, says that "even to the last every cuttlefish which came up in our deep-sea nets was squeezed to see if it had a Belemnite's bone in its back." No such "bones" were found. Or at least none seemed to be. One of the truly fabulous finds of the *Challenger* Expedition was a single specimen of an antique little squid, *Spirula.* Some scientists now regard *Spirula,* with its many-chambered internal shell, as the lone living representative of the belemnites. Wyville Thomson, Moseley, and the other eager scientists of the *Challenger,* as well as their adviser Huxley, to whom *Spirula* was entrusted for scientific study and description, seem to have let slip through their fingers the very thing they were excitedly seeking.

ANCIENT "FLOWER"

One of the first recognizedly antique forms from the deep sea came from Norwegian waters. In 1864, G. O. Sars dredged from 1,800 feet a sea lily unlike any previously known to be alive but the same as familiar forms from ancient rocks laid down in the Jurassic and Cretaceous periods, from

160 to 80 million years ago. This exciting creature, a legitimate "living fossil," was named *Rhizocrinus lofotensis,* after Norway's Lofoten Islands on whose fjords it was caught.

The sea lilies are echinoderms, which is to say that they are related to the starfish, sea anemones, and sea urchins. Their plant-like appearance and their common name are misleading, since they are animals. Another common name, stone lilies, is apt since it draws attention to their brittle, chalky shell, which permits only the slowest kind of stiff movement of the feathery crown. Modern representatives of the crinoids, as they are called by biologists, are nearly all without a stalk, but it is clear from abundant fossils that stalked sea lilies were very numerous in the Cretaceous Period, 100 million years ago. These antique creatures must have existed in dense forests, waving gently in slow currents and slowly twisting and opening the five arms around their heads as they brought food to their mouths. Louis Agassiz, writing to Professor B. Pierce, his mathematician friend at Harvard, describes a sea lily brought alive from the deep sea off Barbados in 1872 as "very much like the *Rhizocrinus lofotensis.* When disturbed the pinnules of the arms first contract, the arms straighten themselves out, and the whole gradually and slowly closes up. It was a very impressive sight for me to watch the movements of the creature, for it not only told of its own ways, but at the same time afforded a glimpse into the countless ages of the past, when these crinoids, so rare, and so rarely seen nowadays, formed a prominent feature of the animal kingdom."

The brittleness of the stone lilies cannot survive the force of even moderate currents, and their survival is possible only where the water is nearly motionless. This is doubtless one reason they are rarely to be found today except in the deep sea. After the epic find by Sars in 1864, many another ocean explorer has brought sea lilies to the surface. Thomson caught them from H.M.S. *Porcupine;* they were dredged up by the *Challenger* in 1872–76 and by the French *Talisman* in 1880–83.

Thomson caught another animal of ancient lineage during the cruise of the *Porcupine.* "This haul was not very rich," Thomson wrote of one drag in the North Atlantic, "but it yielded one specimen of extraordinary beauty and interest. As the dredge was coming in we got a glimpse from time to time of a large scarlet sea urchin in the bag . . . as it was blowing fresh and there was some little difficulty in getting the dredge capsized we gave little heed to what seemed to be an inevitable necessity—that it should be crushed to pieces. We were somewhat surprised, therefore, when it rolled out of the bag uninjured; and our surprise increased, and was certainly in my case mingled with a certain amount of nervousness when it settled down quietly in the form of a round red cake, and began to pant—a line of conduct, to say the least of it, very unusual in its rigid, unde-monstrative order. . . . I had to summon up some resolution before taking the weird little monster in my hand, and congratulating my-self on the most interesting addition to my favorite family which had been made for many a day." Thomson's delight and astonishment were because the other sea urchins of this kind had been known only from fossils chipped from the chalk deposits of the Cretaceous Period.

The theory of evolution supposes that all animals have had a series of more primitive ancestors. Life is commonly compared to a tree, with the trunk representing the original ancestral creature and the various branches the classifications into which animals and plants are divided. The main branches are the primary divisions, the phyla; the smaller branches represent the classes, orders, families, and gen-era. The twigs on the ends of the branches are the species, the ultimate divisions.

Many of the ancestral forms of living animals have disappeared from the earth, leaving, in some cases, fossils: petrified remains or impressions in ancient rocks. The genealogy of many groups of creatures is clouded because their early forms are missing from the fossil record, and scientists can only speculate as to what the primi-tive ancestors of certain creatures looked like. Agassiz drew atten-tion to "the antique character of the new genera discovered in deep water, and especially their resemblance to Cretaceous genera," and to "the probability of our finding in deeper water examples of earlier geological types." He thought this likely because "the depths of the ocean alone can place animals under a pressure corresponding to that caused by the heavy atmosphere of earlier times."

In the years that followed, despite the occasional capture of an archaic form, this expectation was only to a small extent fulfilled. The great majority of animals brought from the darkness of the abyss proved to be close relatives of forms common in shallow water. Con-servatism set in, and most scientists came to the conclusion that the hopes of Huxley and Agassiz and their contemporaries had been misplaced.

Since then, however, some exciting finds of ancient animals have been made in the sea. Among the most spectacular of these are a fish

FIGURE 12–2. *The most famous of living fossils, the* coelacanth fish Latimeria chalumnae. *Its fellows were abundant 220 million years ago, but the last one was thought to have died 70 million years ago. Yet in 1938 a 5-foot specimen, very much alive, was caught off East London, South Africa.*

Union of South Africa

supposed to be dead for millions of years and a primitive "presnail" thought to have been extinct for 350 million years.

THE COELACANTH

Sober-sided scientists were shaken in 1938 by the spectacular capture of a living coelacanth off the coast of South Africa. This species belongs to a group of fishes that is very prominent in the fossil record but that was thought to have become extinct 70 million years ago. The coelacanth was caught by a trawler near East London, South Africa. It was so strange in appearance that the fisherman took it to a nearby museum, and for this science should be grateful to him. His name is not recorded, but that of Miss Courtney-Latimer, the curator of the museum, is now well known since it is preserved as the scientific name of the fish, *Latimeria chalumnae*. A luminous blue in color, and 5 feet long, the fish is most remarkable for its long limblike fins, tasseled on the ends. These are undoubtedly used for crawling about on the bottom of the sea, and give the fish its common name, the "tassel fin." The tail is an odd shape, pointed with a small extra appendage at the extreme end.

The coelacanths have an ancient lineage indeed. They first put in an appearance on the earth some 350 million years ago, were successful, and spread over the oceans of nearly the whole world. About 220 million years ago the tassel fins were the most abundant fishes in the ocean, and they left a great many fossils. Then they apparently disappeared as completely as the dinosaurs. Scientists wrote *R.I.P.* over them and constructed theoretical models from the abundant fossil remains. Thus it was like having a stone statue come to life when *Latimeria* turned up, a statue cleverly constructed from bits and pieces and dim, scarred "photographs." Scientists were not only astounded to see this lively ghost, but were delighted to discover how close they had come in their reconstructions to the actual animal in the flesh.

The living-fossil coelacanth cannot be regarded as a deep-sea fish, since the first specimen was fished from a depth of only 240 feet, and the few others found since 1938 (all in the waters off South Africa) were recovered from similar depths. Its light blue color is characteristic of shallow-water fishes too, since deep-sea denizens are usually black or other dark colors. Nonetheless, it made scientists re-examine the possibility that the sea might be a repository for ancient forms that had disappeared from the shallow waters and remained unchanged over immense stretches of geologic time.

ANCIENT MOLLUSCS

Paleontologists and evolutionists delight in going through the exercise of reconstructing the appearance of the ancestors of modern animals. The reconstruction of such a "missing link" was attempted in 1952 by the biologist Brooks Knight in the case of the molluscs. These are the snails, the clams, the octopuses, and their various allies, the most primitive of which are the chitons, which cling to the rocks of the seashores of the world. Knight's ancestral mollusc resembled a limpet, with an oval shell, a fleshy foot, simple gills, and a radula, or toothed tongue. Imagine the amazement of biologists when out of the abyssal deep sea there came, later in 1952, a creature so like the theoretical mollusc that Knight could almost have had the actual animal in front of him when he created his model!

This epochal capture was made by the *Galathea* at the very end of her cruise, off the coast of Mexico in the Pacific. From a deep dredge, fishing on dark, muddy clay in 11,878 feet of water, she brought up ten live specimens (and three extra shells) of a limpetlike creature thought to have been extinct since the Devonian Period, 350 million years ago. This archaic mollusc, neither clam nor snail but more primitive than either, is a member of a group called the Monoplacophora. It was named *Neopilina galathea* in honor of the ship. The shell of this relic mollusc was spoon-shaped. The biggest individual was 1½ inches long, 1¼ inches wide, and ½ inch high. The oval shell, thin and fragile, semitransparent and pale yellowish-white, rose to a little peak, with the point tilted over. Inside, the shell had a very thin, lustrous layer of mother-of-pearl. The greater part of the body was taken up by a large, almost circular foot of a bluish color, with a diffuse pink central area. Five pairs of primitive gills surrounded the foot. In front of the foot was a small triangular area that has no counterpart in other living molluscs. Around the mouth were fleshy structures; these probably help to collect the bottom deposits that are the food of *Neopilina*. Radiolarians are their favorite

FIGURE 1 2 – 3 . *This primitive mollusc* Neopilina galathea, *which is neither clam nor snail, was thought to have been extinct for 350 million years. It was caught in 1952 in 11,878 feet of water.*

Galathea Expedition

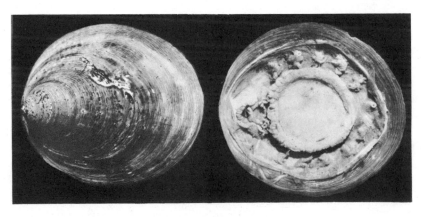

food, judging from the gut contents of the creatures captured off Mexico.

This discovery was so unexpected that Dr. Henning Lemche, the biologist entrusted with the description of *Neopilina,* waited some years before announcing that here indeed was a living fossil. Looking at its weakly muscled foot and other details of its anatomy, he decided that *Neopilina* must lie on its back waiting for food. This idea is disputed by Dr. C. M. Yonge, the eminent British marine biologist. He doubts that the limpetlike animal could feed in this manner, and, although he concedes that the foot is not equipped with enough musculature to carry it over the bottom, he believes that its broad base serves to keep the animal from sinking into the bottom oozes, while other structures may serve for creeping locomotion—the same structures, curiously, that act as gills.

Clearly, we do not have enough specimens to resolve questions like this. But it may not be too long before we do, for in 1958 another catch of *Neopilina* was made. This was by the research vessel *Vema* of the Lamont Geological Observatory of Columbia University, dragging in nearly 19,200 feet (about 7,300 feet deeper than the *Galathea*) off the coast of northern Peru. This was a second species, different in minor respects from the first.

To the biologist this discovery is exciting not merely because it brushes away the dark curtains of millions of years of time and presents to the researcher flesh-and-blood representatives of animals known before only in cold rock but also because it begins to answer nagging questions as to the origin of the whole mollusc group. There has been spirited and sometimes acrimonious debate as to which of the great branches of the tree of life the molluscs sprang from. One theory has held that they are derived from ancient segmented worms represented today by the familiar earthworm or the green fishing worms that are scraped from among the barnacles on a piling. It is of immense importance, then, to discover that in the gill-bearing region *Neopilina* has segments that are closely similar to the typical segments of the annelid worms, supporting the contention that the molluscs are indeed derived from the segmented worms.

This is looking backward in evolution. Looking ahead, the scientist finds in *Neopilina* characteristics that link it to the snails (a slight but unmistakable asymmetry in the shell) and to the cephalopods (a radula and other characteristics like those of the primitive *Nautilus*).

It is little wonder, then, that the finding of this relic has caused a real stir in scientific circles. Dr. Yonge says that "the discovery . . . is a zoological event of the first order . . . ; these dredge hauls off the west coast of Mexico . . . in themselves alone more than justify the voyage of the *Galathea* around the world."

A NEW PHYLUM!

Another startling zoological event was the discovery of the Pogonophora, which are not "living fossils" but something every bit as remarkable: representatives of a completely new phylum. New species of animals from the deep sea are of interest enough; new genera, families, orders, classes, or phyla are progressively more exciting, since they are higher-ranking divisions, embracing increasingly larger numbers of representatives. For example, all the many thousands of molluscs—the clams, the oysters, the snails, the squids, the octopuses, and many other related forms—constitute the phylum Mollusca. The phylum Arthropoda consists of a million or so species —the copepods, the barnacles, the shrimps, the crabs, the lobsters, the insects, and a host of other creatures. All the backboned animals —the fish, the amphibians, the reptiles, the birds, and the mammals, including man—constitute a single phylum, the Chordata. To find animals representing a hitherto-unknown phylum after so many centuries of collecting and examining the earth's creatures was an extraordinary event.

The elusiveness of the Pogonophora for thousands of years has been due to their residence in the deep sea. The first to be discovered was dredged up off Indonesia by the Dutch research ship *Siboga* during her cruise of 1899–1900. They were strange, elongated, worm-like animals that greatly puzzled their captors. They were sent to the French zoologist Maurice Caullery. It was 14 years before he was willing to publish a description of them, and even then he was so unsure of their identity that he refused to place them with certainty in any of the known animal groups.

The strangest aspect of the Pogonophora is their complete lack of a digestive tract. Caullery at first thought that part of the animal must be missing; he considered that the specimens might be broken individuals of a colonial creature, sharing a common digestive tract. But these are self-sufficient single animals. Each long and thin creature inhabits its own tube, which it forms for itself. A crown of tentacles surrounds one end of the animal and gives the group its name, which means "beard worms." Instead of taking food through a mouth like most animals, the pogonophorid forms a tubelike structure with its tentacles and with it simply surrounds food particles that come its way. These are digested inside the crown of tentacles but outside the animal. Absorption of the digested food takes place through small structures on the tentacles.

The tubes of these "worms" stand upright in the ooze of the deep sea. The animal never leaves the tube, which is longer than the worm, but moves freely up and down in its tube, thrusting its anterior end outside to intercept food. Some species are nearly 5 feet long but very thin, the tube being a tenth of an inch or less in diameter.

Pogonophora seem to live in dense groups on some parts of the ocean floor. In the Kurile-Kamchatka Trench, in 26,000 to 30,000 feet, one species, *Zenkevitchiana longissima,* is so abundant that trawls have come up with the white tubes wound around the frames of the gear.

Our knowledge of the Pogonophora comes to a considerable extent from the work of Russian scientists. The Soviet research ship *Vitiaz* has caught many of these animals, and forty-three different species had been described by 1960. Dr. A. V. Ivanov of the Soviet

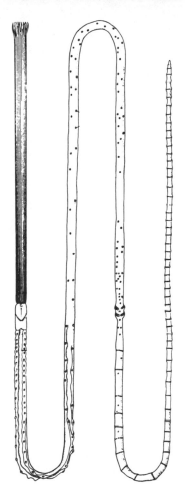

FIGURE 12 – 4. *This thin, wormlike animal is one of the Pogonophora, or beard worms. Pogonophora were not recognized until the twentieth century, largely because they live in the deep sea. They inhabit horny tubes that stand upright in the mud. They have no alimentary canal: food is surrounded by, and digested inside, the crown of tentacles, or "beard."*

Traité de Zoologie

[245]

Union has produced most of the recent information on them, and it was he who showed in 1955 that they constitute a distinct new phylum.

Dr. Libbie Hyman, the distinguished American biologist, says: "The finding of an entirely new phylum of animals in the twentieth century is certainly astounding, and ranks in zoological importance with the finding of the coelacanth fish and the archaic gastropod [*Neopilina*], both belonging to groups believed to be extinct for hundreds of millions of years."

SPIRULA

Few animals have a clearer fossil record than the cephalopods (the squids and the octopuses). This is because their ancestors had sturdy shells that have been preserved in the ancient rocks derived from primeval ocean sediments. Through fantastic overspecialization most of the ancient forms of the cephalopods have disappeared. The sole externally shelled representative, the survivor of thousands of species that swarmed the seas of Mesozoic times between 80 and 200 million years ago, is *Nautilus,* represented now by six species.

From an ancient *Neopilina*-like ancestor with a single, cap-shaped shell, the primitive cephalopod developed an external chambered shell. The animal built a new chamber as it outgrew its old one, moving into roomier quarters as needed. The old chambers were filled with gas, making the shell buoyant and permitting the animal to move much more briskly than when it had to drag its clumsy shell around unaided. The fossil cephalopod *Piloceras* probably represents this stage of the development of the group. From here two main lines of evolution took off.

One of these, leading to the belemnites, produced animals in which the shell became internal and remained straight (or nearly so). These creatures flowered into a particularly successful group during the Mesozoic Era. They toppled over, in the evolutionary sense, shortly after the beginning of the Cenozoic Era, and disappeared in the Eocene Epoch, about 50 million years ago. Their fossil shells are common in the rocks in certain parts of the world. They resemble stone bullets, and have been known as "thunderbolts," supposedly flung from the sky.

There is only one lively ghost of the belemnites. This is the little deep-sea "squid" *Spirula spirula,* one specimen of which was such a prized catch of the *Challenger* that it was entrusted to T. H. Huxley for description. Huxley, although on the lookout for modern belemnites, did not consider *Spirula* to be one. There is still uncertainty among scientists as to its relationship with the rest of the cephalopods, but Sir Alister Hardy declares that it is indeed the "lone surviving member of [this] once flourishing race."

Spirula's most obvious link with the belemnites is a chambered internal shell, and it is the only living cephalopod possessing this. The squids, cuttlefish, and octopuses are not direct-line descendants of the belemnites but offshoots of that ancient group. They show their relationship by their possession of the internal cuttlebone in the cuttlefish, the pen a (thin, horny internal skeleton) in the squids, and a few calcareous grains in the octopuses.

The chambered cell of *Spirula* is in the posterior end of the animal. There are from twenty-five to thirty-seven chambers, filled with gas that buoys the animal up in the water so that it floats with its head and tentacles dangling. When the animal dies, the little shell maintains its buoyancy and floats to the surface, frequently to be cast on the beaches of the tropical

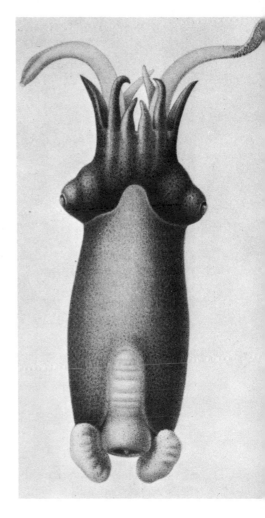

FIGURE 12–5. Spirula *has an internal shell divided into 25 to 37 gas-filled chambers, which buoy up the animal. When it dies, the shell floats to the surface. Some biologists think that* Spirula *is the only living representative of its fossil group.*

Valdivia Expedition

[247]

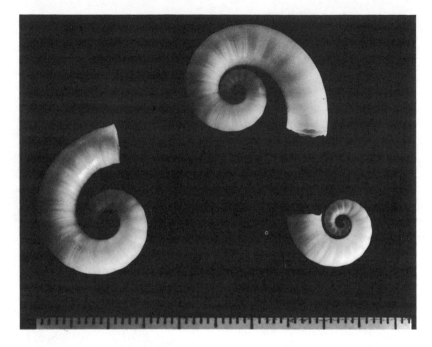

FIGURE 12 – 6. *Shells of the living fossil* Spirula. *When found in ancient rocks, these shells were sometimes regarded as thunderbolts from the sky. The scale at the bottom is in millimeters.*

Charles E. Lane

and subtropical seas that are the home of its owner. Occasionally, the shells are carried by the Gulf Stream to the beaches of Britain and other European countries, as far north as the Faroes. They are so numerous at times that they produce windrows of thousands of shells. Collectors know them as "little post horns" for their resemblance to the horns sounded on the old-fashioned horse coaches of Europe.

Spirula is usually about 3 inches long, half of this length being the tentacles. The body is cylindrical and whitish, with rust-red or brown markings. When frightened, it can withdraw its head and tentacles entirely into its mantle cavity, closing the end for protection. One of the few deep-sea animals that can withstand being brought to the surface, it has been kept alive for moderately long periods of time in shipboard aquaria, and its actions observed. The scientists on the *Dana,* who were among the first to capture *Spirula* in the early part of this century, watched it swim around with swift, jerky rushes, squirting water from its little siphon. It remained in a vertical position and fluttered two fins near the hind end of the body—the uppermost end—to help its swimming movements. Biologists of the *Galathea* Expedition caught twenty-six in a single haul off Natal in 1950 and kept one alive long enough to take motion pictures of it in a tank; then it succumbed to the heat of the photographic lamp.

The depths usually occupied by *Spirula* are about 300 to 1,600 feet, but it sometimes descends to 3,000 feet. It seems to make considerable vertical migrations, as much as half a mile, aided by the buoyancy of its gas-filled shell. How this shell manages to remain unbroken through the great changes in pressure at various depths is a mystery. Apparently the gas pressure can be altered only in the last shell chamber, since all the others are sealed. Anton Bruun, who was an expert on these little creatures, thought that air might be able to pass through the pores of the shell, but he was unsure of this.

Spirula has a small beadlike light organ at its upper (posterior) end that shines steadily with a pale, yellowish-green light. In this it is unlike the luminescent organs of many other deep-sea creatures, which flicker on and off.

The other line of evolution from *Piloceras* was divided again, leading to the ammonites and the nautiloids. The ammonites are animals with elegant coiled shells (their name comes from the Egyptian god Ammon, whose head was a ram's). The ammonites flourished in enormous variety in the late Mesozoic Era. Some had shells only a quarter of an inch across, but others carried great cartwheels as much as 15 feet in diameter. There are 6,000 species of fossil ammonites already known, and certainly there were many more in the late Mesozoic, when they were at their height. These shelled creatures were then among the grandest of living things, but—alas for pride!— they became extinct in the Cretaceous Period, some 100 million years ago. Their fossils resemble the petrified remains of coiled snakes, and they are sometimes called "serpent stones" by collectors.

The nautiloids also had coiled shells, chambered like those of the ammonites. Like the latter, they have nearly all disappeared, but one genus, *Nautilus,* persists as a living fossil. This can be regarded as a deep-sea animal, since it is caught as deep as 2,000 feet in the ocean. It is restricted to the South Pacific, where it scuttles along the bottom of the sea in search of shrimps and other dwellers of the sand and mud. Its handsome white and brown shell is a collector's item.

The best known of the six surviving species of the pearly nautilus (not to be confused with the little octopus *Argonauta,* the paper nautilus, whose female constructs a delicate shell-like egg case) is *Nautilus pompilius.* Its first shell is about the size of a pea, but as it grows, it successively builds a bigger and bigger chamber to its house, sealing off the last one with a pearly cross-wall but leaving a small opening between the two rooms. Eventually there may be as many as thirty-six of these chambers. The inner ones are filled with gas; this makes the shell semibuoyant, allowing the nautilus to adjust to various depths and to swim above the bottom in search of food. There are two series of arms, 42 in the outer group and a variable number in the inner; females have about 94 arms altogether, males about 60. The numerous arms make the animal look like "a shell with something like a cauliflower sticking out of it."

FIGURE 12 – 7. *Living fossil* Vampyroteuthis infernalis *is neither octopus nor squid. It has good eyesight, unusual for deep-sea animals that live in darkness. This specimen is an adult. When very young, it had a pair of paddle-shaped fins near the end of its body; later a second pair (seen in the picture) developed, and after a while the first fins disappeared.*

Valdivia Expedition

"THE LITTLE BLACK OCTOPUS"

The abyssal cephalopod, *Vampyroteuthis infernalis,* is an honest-to-goodness living fossil. It was first described in 1903 by Carl Chun, the famous German marine biologist, who called it an octopus. But then it was discovered to have an extra pair of arms, to make a total of ten, and this seemed to make it a squid. This did not appear to be right either, however, since it is the second pair of arms that is elongated and not the fourth pair, as is the case in the squids; further, the modified arms are not studded with suckers for catching prey, as in the squids, but are slender and mobile and fit into special pockets in the web that joins the arms.

This odd creature, neither octopus nor squid, turns out to be most closely related to an ancient group, the Mesoteuthoidea. All the representatives of this once-large tribe disappeared from the seas millions of years ago, during the Cretaceous Period. *Vampyroteuthis infernalis* is the sole species still living, and it occupies a whole order of the cephalopods by itself.

Perhaps *Vampyroteuthis* outlived its relatives by retreating to the deep sea, where it now lives exclusively. It seldom rises above a 3,000-foot depth and is commonest from there down to twice that depth. The *Galathea* found it in the Kermadec Deep at 9,850 feet.

At the depth where this animal is usually caught, no glimmerings of light remain, and even at its nearest approach to the surface only the faintest possible sparks of greenish light can penetrate. In view of this it may seem odd that *Vampyroteuthis* has efficient eyes of large size. They must either pick up light far dimmer than the human eye can, or they must be used to see the luminescent creatures of the deep sea. The octopus-squid is itself highly luminous, and good eyesight would therefore be an advantage in identifying friends and prospective mates.

The extra arms of *Vampyroteuthis* are unique among cephalopods. They are long and slender and tuck into special pockets in the body. Nearly the whole central axis of the arm consists of a large nerve, and these organs are undoubtedly sensory. Their exact function is a mystery, but they may simply be feelers.

Vampyroteuthis is velvety jet black with purplish tones. The males attain lengths of only about 5 inches; the biggest female known was an 8½-inch individual caught off Durban by the *Galathea*. The eight arms are equipped with a single row of suckers and are joined by broad webs. Unlike some other deep-sea cephalopods, this one has no ink sac and produces no luminous secretion.

Near the apex of the body is a pair of paddle-shaped fins. These have an interesting larval history. In the very young animal they are almost at the end of the body, closer to the apex than a pair of prominent light organs. Then, as growth takes place, a second pair of fins develops, this time on the other side of the light organs. For a period in its life *Vampyroteuthis* has two pairs of fins. Now the first pair begins to disappear, and eventually the adult has only one pair again. This curious waxing and waning of the fins, so that in some individuals there are two pairs and in others only one pair in either of the two different positions, confused biologists. One consequence was that no fewer than eleven species were named. Now, as a result of the work of Dr. Grace Pickford, an American biologist, it is believed that all these are various stages of a single species.

There are three kinds of light organs on *Vampyroteuthis infernalis*. The first of these are simple pinpoints scattered over most of the body with the exception of the inner surface of the webs connecting the arms, and they are especially numerous around the funnel. On the back of the neck are two clusters of photophores of somewhat greater complexity. Most elaborate of all are two sizable light organs, one behind the base of each of the paired fins, equipped with "eyelids" that can be closed at will to shut off the light.

In common with nearly all other abyssal cephalopods, *Vampyroteuthis* is weak-muscled. Gelatinous tissue and oil droplets have invaded its muscles, and it seems certain that it is largely a drifter rather than a vigorous swimmer. This likelihood is made stronger by the high development of the static organ, which controls balance, and by the great webs that join the arms and form a parachute around the mouth.

Vampyroteuthis was so unusual, even among the scores of curious animals brought to the surface by the research ship *Dana* as she dragged the deep oceans of the world, that all the scientists and even the sailors were familiar with "the little black octopus." William Beebe must have been struck by some of the same emotions that caused Carl Chun to give the animal its dramatic name, for in *Arcturus Adventure* Beebe says he captured "a very small but very terrible octopus, black as night, with ivory white jaws and blood red eyes. This came along, half swimming, half sidling, its eight cupped arms all joined together by an ebony web."

Grace Pickford considers the discovery of this archaic creature less spectacular than that of the coelacanth, but still one of the more exciting discoveries of the twentieth century.

What is the evidence that the deep sea is a general haven for animals that flourished in very ancient times but withdrew to the abyss because they could not compete successfully in the heavily populated areas near shore? Certainly it seems quite clear that the proportion of ancient animals to those of recent origin is much greater in the deep sea than in shallow areas. Soviet scientists, reporting in the latter part of 1960, estimated that approximately 16 per cent of the species of animals caught in depths over 13,000 feet were archaic, while only 0.00005 per cent of those from the shelf areas were ancient in form. Then there are the spectacular captures of *Latimeria, Neopilina, Spirula,* and *Vampyroteuthis.* Still, these are the exceptions rather than the rule. One cannot assume that the deep sea is so full of ancient animals that if we look long enough we will find representatives of all the old groups of animals known at present only from their fossil remains in rocks.

Nonetheless, there are genuine living fossils in the deep sea. In all probability, they result from the slower rate of evolution in the relatively unchanging conditions in the abyss, where alterations in environment take place exceedingly slowly. The much smaller number of species found in deep water than in shallow contributes to this slower rate of differentiation too: while changes are taking place in the structure of the animals that inhabit the shallows, the deep-sea creatures are faced not only with less opportunity but with less necessity for change.

13   The Mark of the Abyss

Our environment molds us: we are what we see, what we feel, what we hear, what we eat. The same is true, of course, of other animals in other environments. The fantastic shapes and colors and details of anatomy of the denizens of the deep sea are part of evolution's effort to get them in tune with their incredibly harsh environment. Let us review for a moment the conditions that abyssal animals face. First, their living space is cold. Temperatures fall quickly from the surface down to the thermocline (the water layer of sharp temperature and density change) and from there on steadily but more slowly; in the great depths the temperature is from 38° F. to below the freezing point of fresh water. The depths are dimly lit at best; deep-sea creatures are surrounded by perpetual, absolute darkness. Pressures are enormous, with the animals of the abyss being subjected to a crushing burden of 7 or 8 tons per square inch. Food, important enough to any creature even when it is plentiful, becomes a dominating fact of life when it is scarce—and it is painfully scarce in the deep sea. Of the four major elements of the environment, neither pressure nor cold has as much effect on the deep-sea creatures as might be expected; it is the scarcity of food and the lack of light that most influence the form and behavior of the inhabitants of the abyss.

PRESSURE AND COLD

How can it be that the frightful pressures of the abyss do not squeeze the life out of the animals there? For the same reason that land animals can live at the bottom of a sea of air under pressures of 14.7 pounds per square inch: the tissues are surrounded and pervaded with water, which makes up most of their substance. Water is almost incompressible, and its cushion protects the living material from harm. It is when the fish has an air-filled swim bladder that quick changes in the depth, and therefore in pressure, can cause trouble. Fish brought rapidly to the surface even from moderate depths may have their air bladders forced out of their mouths. The fish must maintain the air bladder at a certain volume in order to maintain buoyancy in the water; to counterbalance the increased pressure of the depths, air must be forced into the bladder by secretion from the blood. Under ordinary circumstances, when the fish swims upward at a normal rate, air can be gradually withdrawn again, so that the pressure is kept adjusted. But if a fish is snatched upward so rapidly that its body is unable to make this adjustment, the internal pressure of the air swells the bladder, which may be ejected as when a balloon, blown up while being held tightly inside cupped hands, forces its way through an opening between the fingers.

Considering the harshness of their surroundings, we might expect the fishes of the deep sea to be tough, stringy creatures, with muscles like tanned leather and with thick, strong bones. Quite the opposite is true. The flesh of many deep-sea fishes and squids resembles the substance of a jellyfish. One group of deep-sea fishes, the stomiatoids, are particularly noted for their gelatinous flesh, and one small member of the angler fishes, *Haplophryne mollis,* has an envelope of gelatin surrounding its body, covering even its fins and eyes.

FIGURE 13 – 1. *Deep-sea codlike fish* Mora mora, *after being dragged up from a depth of 3,000 feet. As a fish swims upward at a normal speed, the air is gradually withdrawn from its bladder. But when the fish is dragged up quickly, the bladder expands and is forced out of its mouth.*

J. Murray and J. Hjort, *The Depths of the Ocean*

It is living in the still waters of the ocean depths where there are no water movements except for feeble currents that makes it possible for fish to dispense with cumbersome skeletons. The skeletal structures of other deep-sea creatures are reduced too, and for the same reasons. Since deep-sea snails, mussels, and sponges do not need to withstand pounding by waves on the beach or tossing by currents in shallower waters, they save themselves the trouble of constructing heavy shells or spicules. This reduction in the bulk of skeletal material is the result of the scarcity of calcium, which is essential to the construction of bony and most shelly skeletal material and which exists only in small amounts near the bottom of the ocean. Vitamin D is another essential for bone-building, and the sunlight necessary for its manufacture is absent in the deep sea. The consequence is that many deep-sea fishes might be said to have rickets. But where they live, rickets is no handicap.

The increase in the amount of gelatinous material in the body of mid-water deep-sea animals and the reduction in the mass of their bones have the effect of making the creatures more buoyant. This is of great importance to their survival: it conserves energy that would otherwise be dissipated in attempts to keep from sinking into the inky depths.

But the weak bones and the gelatinous padding markedly reduce the agility of these fishes. Many of them—remember the deep-sea anglers and the gulpers—are relatively sluggish compared to the darting creatures of the sunlit upper waters.

THE LILLIPUTIANS AND THE GIANTS

When someone first sees photographs or drawings of deep-sea fishes, without any clues as to their actual size, he often assumes that they are huge creatures. Perhaps this is the inevitable reaction to such ugliness and strangeness. In most cases, as we have seen, these fishes are small—and it is just as well. Imagine how terrifying the viper fish would be if it were 20 feet long, or the damage that a deep-sea angler could do if it were as big as a wolf! Fortunately, the little viper is frightening only to shrimps, since it is some 4 inches long, and anglers often measure only 1 or 2 inches and frighten almost nothing.

The pioneer Norwegian oceanographer Johan Hjort called deep-sea fishes "Lilliputian." It is probably the scarcity of food that accounts for smallness of most fishes in the depths. In the face of this

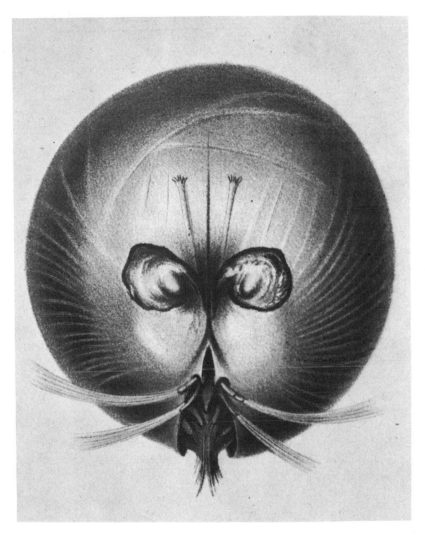

FIGURE 13–2. *Orange-red ostracod* Gigantocypris agassizii, *a relative of the shrimp. Only the size of a cherry, it is a giant compared to its shallow-water brothers.*

Valdivia Expedition

scarcity, there are a few deep-dwelling fishes that are successful enough foragers to support fairly large bodies. One of the gulpers reaches about 2 feet in length, another at least 6 feet; one of the angler fishes grows to about a yard. But of course there is a good chance that there are more big fish in the deep sea than we now realize—that we have simply never caught them because the nets used for very deep hauls are not designed to catch fast-moving fish.

By contrast with the deep-sea fishes, in several groups of invertebrates (animals without backbones) it is the largest individuals that inhabit the deep sea. For example, living in the depths there is an orange-red ostracod (a small bivalved relative of the shrimp) that grows as big as a cherry. Since most ostracods are tiny—even microscopic—this fellow is a veritable giant and bears the name *Gigantocypris agassizi*. The biggest sea urchins found anywhere are in the deep waters, their leathery bodies being a foot thick. A giant crab, orange in color and 6 feet across from the tip of one leg to another, inhabits the deep western Pacific. Another big crab is shown in Figure 6-7. Isopods on land include the pill bugs encountered in damp ground, and these are little fellows about a quarter of an inch long. The isopods in the ocean are usually in this range of size too, so the deep-sea species, *Bathynomus doederleini* and *Bathynomus giganteus*, look decidely outsized at 6 to 8 inches. The hydras are ordinarily insignificant-looking creatures, perhaps a few inches high, resembling flowers more than the animals that they really are. The grandly named "imperial wax flower hydra," of the deep sea off Japan, has a stature —some 8 feet—to go with its appellation. Some sea pens, colonial animals whose size is ordinarily a foot or so, grow to about 8 feet in the great depths. And we have already described the enormous squids that live in deep water.

THE EFFECT OF DARKNESS

The progressive loss of light from surface to bottom in the sea has a profound effect on marine animals, principally in their colors and in the development of their eyes. Most fishes living near the surface of the ocean are silvery or light-colored underneath, and darker —usually bluish or greenish—on top. In the well-lit upper regions of the ocean a predator underneath a fish has a harder time making out the form of its intended victim if the underside of the latter is light-colored, blending with the light diffusing from above. To a predator peering down, a dark back is less easily seen against the comparative gloom of the depths.

Like most adult fishes in shallow waters, many of the planktonic animals are light-colored, either transparent or faintly red or blue. The arrowworms, the copepods, and many of the other swarming crustaceans and molluscs thus escape the notice of their enemies either by allowing the light to pass right through them or by bouncing it back like a mirror. Dark-colored creatures would stand little chance of survival here; they would be picked off like the red-coated British soldiers among the green forests of the New World.

But the opposite is true as the light begins to fail. The commandos of World War II who attacked inside enemy territory in night raids were careful to don dark clothing and blacken their faces. As the biologist in his bathysphere goes down through the water, the colors of the animals he sees become darker.

Beginning about 500 feet down, the fishes are gray or silvery. From this depth to 1,600 feet or so, there is still enough light to be reflected off silvery flanks, and many hatchet fishes have this coloration. Pale brown or gray is the color of the lantern fish and some other groups. In the next region the fishes become dark brown or black, and the crustaceans are red, violet, and brown. Some of the fishes in this region are red too, and some of the black fishes have a violet- or copper-colored cast. Generally the vertebrates seem to have adopted sober browns and sables, while the shrimps and their allies are more gay.

If most fishes seek protection from their enemies by assuming dark garb, how can some fishes and many invertebrates get away with bright red coloration? The answer lies in the characteristics of light. The long waves of light at the red and orange end of the spectrum are absorbed first by water, so that none of those waves penetrate deeply into the sea. An object appears a certain color to our eyes because it is bouncing waves of that color back to us. Since there is no long-wave light in the depths of the ocean, a red animal cannot reflect light and appears black to potential enemies.

This phenomenon is vividly described by William Beebe. Gazing entranced out the window of his bathysphere, "I happened to glance at the large deep-sea shrimp I had brought along in its bottle for experiment. It was no longer scarlet, but a deep black, with an orange tone." He looked at the colored illustrations of scarlet shrimps in Murray and Hjort's book *The Depths of the Ocean*, and the pictures appeared not red but "as black as night." This was in less than 50 feet, in the clear waters off Bermuda, but the red rays had already been filtered out.

It is interesting to compare the adaptations of deep-sea animals with those of cave-dwelling fishes. In both cases light is absent, but the cave dwellers typically are bleached, while the marine creatures usually exhibit vivid—if dark—coloration. The reason for this difference is not clear, any more than the reason for the tendency of the fish to favor black or other somber pigments while the crustaceans and other invertebrates produce pigments on the red and orange side.

FISH EYES

The nature of light in the deep sea determines the character of the eyes of animals there. As the amount of light diminishes steadily with depth, the usefulness of eyes would also seem to decrease; we would not be surprised if these organs got weaker until, at the point where all light is absorbed, the animals become sightless. This is what happens in cave-dwelling fishes; but just as abyssal oceanic fishes do not become bleached like cave denizens, they do not ordinarily lose their eyesight either. Among the few blind fishes is the whalefish *Ditropichthys storeri*. A strange little deep-sea fish, *Bathymicrops regis*, has vestigial eyes covered with scales, while a relative, *Ipnops*, has no trace of eyes. The blind invertebrates are mostly creatures that spend their time in the mud of the ocean bottom, and this, more than the absence of light in the water, seems to account for their lack of sight. But the deep-sea swimming octopus, *Cirrothauma murrayi* (Figure 8-5), is blind.

Living alongside creatures that have lost the use of their eyes are many more with functional, and in some cases very well-developed, organs of vision. Among the deep-sea fishes with good eyes are the viper fish (the stomiatoids), hatchet fish, rat-tails, and many other groups. The big crystal-clear amphipod *Cystosoma neptuni* has enormous red eyes that almost cover an outsized head.

Nature is usually too thrifty to permit such complex structures as eyes to be developed if they are not going to be used, so we are forced to the conclusion that the deep-sea creatures that have eyes use them despite the inky blackness of their surroundings. As we have already seen, there are two reasons for the occurrence of eyes in deep-sea animals. First of all, some fish do not remain in the dark

FIGURE 13-3. *Amphi-pod* Cystosoma neptuni, *whose enormous red eyes are adapted to the exceedingly dim light of the deep sea.*

J. Murray and J. Hjort,
The Depths of the Ocean

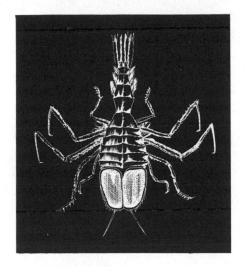

regions all the time but migrate up to feed or spawn. Second, there is another kind of light besides that produced by the sun: the light generated by the creatures themselves. Both of these factors will be discussed in detail.

The eyes of deep-sea fishes are so constructed that they do not see sharp images. Probably their range of vision is limited. On the other hand, from short distances many deep-sea fishes' eyes can detect objects in very dim light indeed—light that would be useless to our own.

Many deep-sea fishes increase the efficiency of their eyes by enlarging them. But the mere fact of enlargement would not be enough: if the diameter of an eye is doubled, its area is quadrupled, so that, even with a larger opening, the same amount of light strikes a given area of the sensitive screen called the retina—the "film," if the eye is regarded as a camera—and the fish is no better off than before. But if the eye is enlarged and the pupil enlarged even more, the amount of light striking an area of the retina is increased and dimmer images can be perceived. This is what happens in the eyes of many deep-sea fishes.

The shape of most vertebrate eyes is nearly spherical, and so are those of most deep-sea fishes. But in some the eyes have deviated and are tubular in shape. These are often called "telescopic" eyes, but the word applies only to the shape of the eyes. Far from being

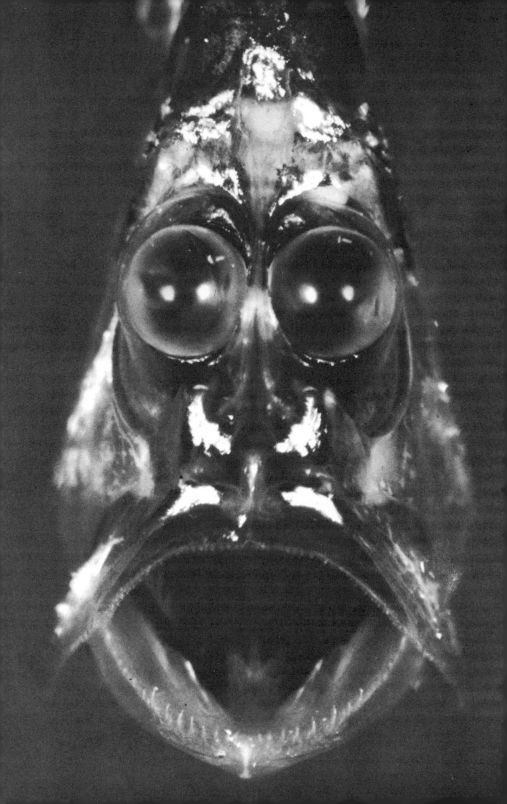

able to bring distant objects into better view, telescopic eyes are notably shortsighted. The hatchet fish seen in Figure 13-4 also possesses telescopic eyes which bulge very prominently upward.

The tubular eye has two retinas. The principal one is on the floor of the eye, opposite the lens. An accessory retina is located on the inside wall of the eye. The fish apparently makes use of this double retina to compensate for its virtual inability to focus on objects as they approach. Our eyes do this by changing the shape of the lens, but a fish's eye lacks the necessary muscles. The fishes with tubular eyes perceive distant objects with the accessory retina; as an object comes closer, its image is sensed instead by the main retina.

The principal advantage of tubular eyes, occurring in at least nine families of deep-sea fishes, is to provide binocular vision. Dr. G. L. Walls, who made exhaustive studies of the vertebrate eye, stated the problem of deep-sea fishes vividly when he wrote: "If the reader will imagine trying to estimate the distance of a faint dot of light in a dark room, with one eye closed, he will appreciate the value of having bearings on such a stimulus from two angles at once."

We tend to assume that all creatures have the same skills as ourselves, and we take for granted our own binocular vision, with its ability to give us a stereoscopic image. But consider the ordinary fish, such as a trout. Its eyes are on opposite sides of its head and it sees two completely separate images, both flat. Two eyes must be directed at the same scene, from very slightly different aspects, to bring depth into the picture and increase the usefulness and meaning of what is seen. Most of us have seen motion pictures taken with two slightly offset cameras, and realize how the three-dimensional picture differs from the more common one.

Dr. Johan Hjort is one of many students of deep-sea fishes who have sought to understand the relation between the depth at which fish dwell and the development of their eyes. "Nothing," he admitted, "has appeared more hopeless in biological oceanography than the attempt to explain the connection between the development of the eyes and the intensity of light at different depths in the ocean. In a

FIGURE 13-4. *Deep-sea hatchet fish* Argyropelecus, *which has enormous eyes with enlarged pupils designed for seeing dim objects in the zone where sunlight still penetrates feebly. Another source of deep-sea light is the phosphorescence of abyssal creatures.*

Photo by Paul A. Zahl, National Geographic Society

trawling from abyssal depths in the ocean we may find fishes with large eyes along with others with very small eyes or totally blind." Hjort did show, however, that there is some relationship between depth and eye development. He proposed a two-zone breakdown. In the "twilight" region from 500 to 1,600 feet, he pointed out, the eyes of the creatures are large, as though straining to make use of the feeble light that persists there. Below that depth the eyes get smaller and less versatile, tending to become mere light receptors. It is as though the eyes of the deepest fishes finally give up the struggle, decrease in size, and in some cases—although in surprisingly few—actually disappear.

SOUND AND HEARING

Inefficient as some of them may be, eyes are probably much more important to deep-sea animals in interpreting their environment than are their ears. This is in spite of the fact that the reputedly "silent world" of the sea is actually filled with the most varied and fascinating sounds. Here are some of the noises that investigators have reported: boops, honks, cackles, moans, shrieks, and groans; squawks, squeals, crackles, and whistles; ticking, mewing, clicking, grunting, hammering, and croaking; drumming, clucking, and sizzling; noises like a pneumatic drill, a deep-toned organ, jingling bells; sounds like an enormous harp, like coal rolling down a chute, like chains being dragged over the floor!

And all this is from a region inhabited almost entirely by creatures without vocal cords. But even without vocal cords many marine animals "talk." It is an amusing experience to listen to recordings, made under water, of a group of porpoises or whales carrying on a "conversation." In this case the sounds are mostly whistles and squeals, but the pitch and modulation vary in a most entertaining manner. The rest of the sound-makers produce their racket in two principal ways—by rubbing parts of the skeleton together or by vibrating the swim bladder. We know a good deal about the sound production of shallow-water fishes. Many of them, such as the file fish, make sounds by rubbing spines of their fins against other parts of their skeletons. Jacks, triggerfish, the ocean sunfish, and others produce sound that seems to come from vocal cords—but is actually made by rubbing together teeth located deep in their throats. Spiny lobsters can make a loud noise by rubbing their long antennae against their horny armor. Snapping shrimps perpetually "crack their knuckles" by clicking the two joints of the large claw that is their chief weapon. When a lot of snapping shrimps meet in convention the din is deafening—and it has been heard under water as far away as 2½ miles!

Of much wider occurrence are fishes that use the swim bladder to make sounds, either as a resonator to amplify sounds produced by the teeth in the throat or as the primary organ of sound. In the latter case slowly contracting muscles called the drumming muscles, attached to the backbone or skull, set the gas-filled bladder into resonant vibrations at about 24 per second.

The drum fish are the best known of the sound producers. They include the common weakfish or sea trouts, the red and black drums, the croakers and others, and their ability to produce sound is truly remarkable. Other noisy fishes include the gurnards and sea robins, the sea perches, the toadfish, and the singing midshipman *Porichthys notatus*.

FIGURE 1 3 – 5 . *Snapping shrimp* Alpheus formosus. *These little animals are responsible for a great deal of crackling and sizzling noise in the shallow sea near shore. They crack their knuckles by clicking the two joints of the large claw. A group of snapping shrimps sounds like an immense underwater frying pan.*
Bulletin of the Vanderbilt Marine Museum

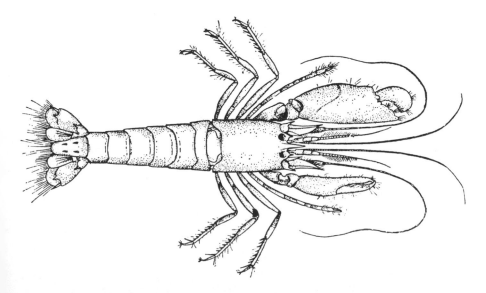

Some deep-sea animals apparently produce sounds, since hydrophones dangled in deep water off Bermuda picked up ". . . strange mewing sounds, shrieks and ghostly moans." There was no way of telling what creatures produced this racket, and because of the great difficulty in making observations, almost none of the still rare sounds recorded from the deep sea have even partially been linked to their source. But Dr. N. B. Marshall of the British Museum recounts an interesting experience by American scientists which suggests strongly that one kind of loud noise from the abyss was made by a fish. The research ship *Atlantis* was working in very deep water (about 16,800 feet) 170 miles north of Puerto Rico, on March 7, 1949. Over the side of the ship a hydrophone was recording underwater sounds. Among these were loud calls, followed at an interval of about 1½ seconds by a fainter call, which was judged to be an echo from the deep-sea floor. Calculating from the speed of sound in water, it was determined how much farther to the ship the echo had to travel than did the sound directly from the source, and from this it was concluded that the source was 11,500 feet deep. Whales do not dive as deep as this, and crustaceans do not emit loud calls, so the noisy creature of the abyss could, apparently, only have been a fish.

We know even less about the purpose of the sounds. By analogy with shallow-water animals, it would seem likely that sounds some-

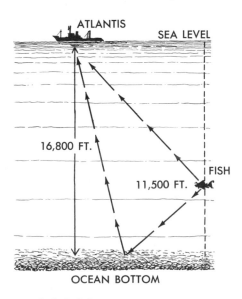

FIGURE 13–6. *Sound from what was thought to be a fish swimming at enormous depth was recorded by the research ship* Atlantis. *A loud call was followed by a fainter echo about 1½ seconds later. The fish, with no landmarks to assist it in fixing its position, may have been echo-sounding to determine its distance from the bottom.*

times provide a signal for rallying the clan, especially at spawning time. The fact, for example, that in the red drum only the males are able to emit sound suggests that there is some sexual significance to this skill. It is also possible that some fishes, such as the one heard off Puerto Rico, may actually "echo-sound," that is, send impulses to the bottom and time their echo so that the fishes can orient themselves in surroundings entirely devoid of physical points of reference.

SMELL AND TOUCH

In humans the sense of smell is very subordinate to sight and other senses, but we know that in dogs and many other animals this sense is an important link with the surrounding world. Might deep-sea animals share this characteristic, since their reliance on sight is so greatly reduced by the dimness of their environment? The fact is that smell does not serve a major purpose for any but a few deep-sea animals, any more than it does for shallow-water creatures. Some fishes do appear to locate their prey by its odor. The eel is famous for the acuity of its olfactory system, and the salmon has been shown to find its home stream by smell, but these may be somewhat exceptional. The comparative subordination of smell is indicated by the small size of the nose in most fishes. It is usually a double pit on the snout: water enters one opening and flows out the other, and during its passage chemicals in it stimulate sensory folds containing nerve endings.

Touch is another minor sense in fish, although it may be more important in some deep-sea animals than is generally true in their shallow-dwelling brothers. A few crustaceans and fishes have notable sets of feelers. Witness the greatly extended antennae and legs of some prawns and crabs, many with large numbers of sensory hairs. We have seen that some fishes, including the stomiatoid *Grammatostomias,* have ridiculously elongated barbels, many times the length of the body. Such creatures as these seem to use their organs of touch to compensate for poor eyesight, much as a blind man employs a cane to help him get around.

THE LATERAL-LINE SYSTEM

Up to now we have been on familiar ground, making direct comparisons between our own senses and the sensory organs of the deep-sea animals. But we launch into strange areas when we consider the lateral-line system, something entirely lacking in our own bodies and those of terrestrial animals in general. It gives the fish a "sixth sense" in the truest meaning of the phrase, something like touch, a little like hearing, a little like radio receiving.

The lateral line along the flank is a plainly visible feature of many fishes, but the branches of the system on the head, which are of the greatest importance in some fishes, are more apt to be missed in a casual examination. These lines represent rows of the individual receptor organs that make up the sensory system. They consist, typically, of delicate cushions of cells, each with a hairlike element protruding above the level of the skin surface. They are sensitive to low-frequency vibrations—acoustic waves, or "silent sound." Water currents, sometimes of extremely small strength, move the hairs and stimulate the nerves, signaling the fish that something is moving nearby.

Impulses sufficient to stimulate the lateral-line system are set up when an animal in the vicinity moves its tail or fins or changes speed. These acoustic waves travel at great speed, nearly 5,000 feet a second. This stimulus sets the lateral-line alarm system jangling, and the fish is alerted to the presence of its swimming neighbor. This may be a potential meal or an enemy; in either case its presence may literally be a matter of survival to the fish lying in wait.

The organ serves for some fishes as an "echo-ranging" apparatus. Energy waves are set up by the movement of the fish itself through the water, and move outward in all directions. When they meet an object in the water they are reflected. The time required for the wave to go out and come back is a measure of the distance to the reflecting object. The lateral-line system can record the energy produced by the fish's own bow wave, reflected after striking an obstruction ahead, and this may warn the fish that it is in danger of a collision.

The lateral-line organ can be astonishingly sensitive. Experiments with blinded minnows show that fishes conditioned to associate small disturbances in the water with the presence of food can locate a hairlike glass rod being moved very slightly in the water at some distance, especially if the movement is near its head or tail. It has been calculated that some of the bigger fishes with especially sensitive lateral-line systems may be able to detect fair-sized prey at distances as great as 60 feet or more.

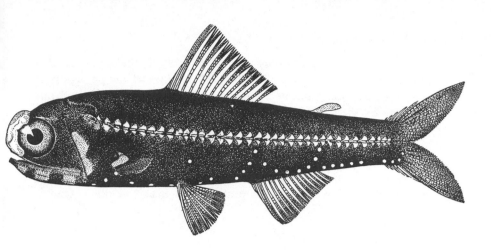

FIGURE 13–7. *Lantern fish* Aethoprora effulgens. *Its conspicuous lateral line consists of delicate organs that respond to sound waves in the water. The movement of a nearby animal sets the lateral-line alarm system jangling, and the lantern fish is alerted to the presence of a potential meal or enemy.*

G. B. Goode and T. H. Bean, *Oceanic Ichthyology*

Many of the deep-sea fishes have this faculty of "distant touch," and those with the poorest sight seem to have been compensated with the most sensitive lateral-line organs. While most fishes from mid-depths, in the twilight zone of the ocean, are equipped with relatively poor lateral-line systems, the lantern fish have extremely well-developed organs, especially on their heads. As we go deeper into the abyss, more groups of fishes seem to depend on the lateral line for contact with their environment. The whalefish and some of the swallowers have highly sensitive organs. It may be significant that some of the blind fishes, such as many of the brotulids, have the most sensitive lateral-line systems of all.

The rat-tails have the most sensitive and elaborate lateral-line system of all the fishes living on the deep-sea floor. In particular the big head of the fish contains wide canals and very large sensory cell cushions, with hairlike and club-shaped papillae, or protuberances, as the sensitive receptors. As the fish swims smoothly and silently through the inky blackness, these acutely responsive organs pick up the faint stirrings in the water set up by the tiny breathing or feeding currents of creatures in the mud. Sometimes a sponge harboring little

[269]

shrimps or other creatures may give them away by the water currents it passes through its pores. The glasslike spicules of the sponge may protect it from harm, but its associated animals will be picked off one at a time by the fish.

THE SWIM BLADDER

Other structures besides the sense organs have developed in the body of fish to keep them in harmony with their surroundings. One of

FIGURE 13 – 8. *Air bladders of three rat-tailed fishes, and, below, an exterior view of a rat-tail. The air bladder (or swim bladder) is a hollow organ near the top of the body, between the alimentary canal and backbone. "Air," actually a mixture of gases, buoys up the fish. The dark winglike structures are drumming muscles that make the bladder produce sounds.*

N. B. Marshall, *Aspects of Deep Sea Biology*

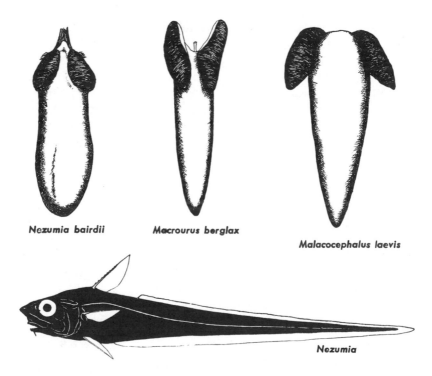

Nezumia bairdii   *Macrourus berglax*

Malacocephalus laevis

Nezumia

the most important is the swim bladder or air bladder, to which we have already alluded as an accessory sound-producing organ. Its main function, however, is to keep the fish approximately "weightless" and so help it float easily in the water.

The air bladder is a hollow organ lying near the top of the body cavity, between the alimentary canal, or gut, and the backbone. The air bladder is derived from the gut, and in some fishes a connection persists between the two organs through which "air"—actually a mixture of various gases—can pass. In other fishes this connecting tube is lost, and air is added to the bladder from the blood as needed; gases in solution there are delivered to the bladder by a rich supply of blood vessels. The gases are originally drawn into the blood from the water.

The swim bladder works by adjusting the specific gravity of the fish—the relation between its weight and volume. When Archimedes, sitting in his bath, cried "Eureka!" it was because he had realized in a sudden flash of genius that water buoyed up objects in proportion to their volume, with the buoying effect being equal to the weight of the water displaced. A boat floats because it weighs less than the water it displaces. The body tissues of a fish are slightly denser—heavier for a given volume—than water, and without some help the animal would sink. This help is provided by a hollow area in the body, which decreases the fish's average density so that the water volume it occupies has a weight equal to that of the fish. This hollow area is the swim bladder. It occupies about 5 per cent of the volume of the fish in sea-water species. For fresh-water species the buoyant effect of the water is less, the fish needs more help, and the volume of the bladder is 7 to 10 per cent. Because the density of water varies slightly with depth, and, more important, because there is a marked variation in pressure with depth, the volume of the swim bladder must be adjustable if the delicate equilibrium between fish and water is to be maintained. If a fish moves, as some of them can, from about 165 feet to 1,600 feet in only 2 hours or so, the pressure increases from 88 pounds to 750 pounds per square inch. This would squeeze the air—and thus the bladder—to about one tenth its original size, and the fish must work very hard to prevent this by secreting gases into the bladder. The journey upward from deep into shallow water presents the opposite problem, and the fish corrects for this by resorbing gases into the blood in a special region of the bladder called the "oval."

Sydney Hickson, writing in 1893, thought that deep-sea fishes might "tumble upward," forced to the surface by a blown-up swim bladder, if they were foolish enough to rise too quickly in the excitement of chasing a meal, for example. Actually, however, the fish would almost certainly be able to detect trouble in time to check its headlong rush to a lower-pressure region, and the oval would absorb gas quickly enough to allow rapid but controlled ascents. The occasional deep-water fish found on the surface with its swim bladder everted (like *Mora* in Figure 13-1) is probably either sick or has been dragged there by a fisherman or some other predator.

Almost all fresh-water fishes have swim bladders, and in most of them it is connected to the outside through the gut. With this mechanism fishes that live near the surface can make quick adjustments in the pressure of the air bladder by actually gulping air or by releasing gases into the gut. Some oceanic fishes that swim in the top layers of the water, including the salmons and the many members of the herring family, are able to do this. Some fishes, like the tarpon, go further, using this device to supply themselves with oxygen; the air bladder functions as a lung. The deeper oceanic fishes have no use for this ability to suck in atmospheric gases. It would take them too long to swim to the surface; moreover, they could not draw in enough air to be useful after it is reduced in volume by the great pressures they encounter when they return to the depths. So these fishes depend on their ability to secrete gases into the bladder from the blood. How they do this, expanding the bladder against the enormous pressures in the depths, is one of the many unexplained mysteries of the physiology of deep-sea fishes.

Many fishes living to depths of about 1,600 feet have exceedingly well-developed swim bladders, and this is reflected in their ability to swim easily and gracefully through the water, taking full advantage of their weightlessness. The hatchet fish and the lantern fish are good examples of this group. Below this depth we find that many fishes have lost the swim bladder, which may disappear entirely or may be transformed into an organ for the storage of fat. This group of fishes includes the gulpers and some of the angler fish. The swim bladder is a handicap to fishes in the great depths because of the energy expended in changing the volume of the organ in opposition to great pressure. The loss of the bladder was probably one adaptation that made possible the successful occupation of the abyss by the fish.

THE BIG EATERS

Deep-sea animals get their food in two principal ways. One depends largely on vertical migrations, since food is carried from the lighted upper areas by animals that temporarily invade this region and then descend to supply food for populations at the next level beneath. The chain through which food passes from the surface areas of the sea to animals in the depths starts when the swarming grazers in the sunlit regions feed on the diatoms and flagellates. Then these creatures—the copepods, euphausids, the many kinds of larvae, the salps and numerous other animals—serve as tasty morsels for the jellyfish, the squids, the prawns, the arrowworms, and the little fish. Some of these small predators live all their lives in the upper reaches of the sea, but others, especially the squids, shrimps, and fish, have many species that spend the day in deep waters and come to the surface at sunset to feed. When these mid-water creatures retreat to the dim depths again they carry with them bellies loaded with the substance so liberally created in the upper regions of the sea, and if they become the victims of the animals occupying the middle layers, the energy stored by the diatoms and flagellates is passed along to the angler fish, the squids, and the other inhabitants of the dark areas where no primary food production can take place. Gradually, in steps, food is provided even for fishes in the abyss.

The second method by which deep-sea animals get food is more chancy than the one just described. It involves the drift of dead and dying animals and plants down through the water column. Individuals of both plants and animals in the plankton have developed various devices to prevent themselves from sinking, but sometimes these do not function perfectly and there is a slow sinking of individuals. In any case death comes at last to all of the animals, so that part of the mass of organic material that the plankton represents finally filters down in a slow and trancelike snowfall to the hungry maws of the deep-sea creatures. As they sink through thousands of feet of water the food scraps are snapped up by ravenous passers-by, becoming fewer and fewer with every fathom of depth. Finally, in the abyss, there are exceedingly few bits of food left for the animals at the end of the line.

We are not surprised when a wolf eats a deer, but we would be astounded if the deer were swallowed whole. Many of the fishes gulp down animals more than twice as big as themselves, the reason for this greediness being the compelling necessity of making use of every-thing that comes along in the way of food—almost regardless of size. Meals are few in the deep sea and must be accepted when they appear; large digestive capacity is a price of survival. The fish do not become lethargic after their outsized meals the way a boa constrictor does; their digestion is rapid and they are soon searching for another meal.

The most notable of the "big eaters" are the gulpers, the deep-sea anglers, the giant-tails, and the deep-sea perches. One of the deep-sea perches, *Chiasmodus niger,* cannibalizes individuals of its own species and swallows fish as much as twice its own size. A tale of gluttony in the angler fish, *Linophryne lucifer,* was told by George Brown Goode and Tarlton H. Bean, who wrote of a 2-inch speci-men caught by a Captain P. Andresen in 1877, northwest of Ma-deira: "He was capturing turtle in his boat; there was a heavy swell but the water was smooth. After a time he caught sight of this little fish, which lay on the surface quite alive but almost motionless, which was not surprising when it was discovered that it had just swallowed a fish longer than itself. It did not lie on its side but was unable to swim away. By getting the bailer under it he was able to get it out with ease, and in order to keep it fresh he gave up his search for turtle and rowed to the ship, where he placed it in spirits for preservation."

Even the greediest wolf could never alter its jaws enough to

F I G U R E 1 3 – 9 . *Deep-sea perch* Chiasmodus niger, *its belly enormously distended with a fish bigger than itself. The upper jaw is hinged to the front of the skull, and it can swing wide to swallow large prey.*

G. B. Goode and T. H. Bean, *Oceanic Ichthyology*

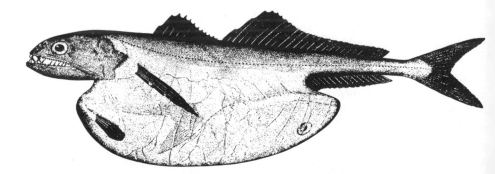

accommodate a deer, because, as in ourselves and all other higher vertebrates, its upper jaw is fastened tightly to its skull, and its lower jaw is hinged firmly to the back part of the upper jaw. In fish, however, the upper jaw is hinged to the skull only at the front; the hind part swings from a special device called the "suspensorium." The deep-sea fishes—and strangely, only the deep-sea fishes—have seized upon this loophole in their anatomy to develop their marvelously distensible jaws.

The mouth of the gulper is enormous, but its ability to engulf huge prey depends upon the elongation of the suspensorium and the jaws. The latter are several times as long as the skull, so that as the mouth opens it gapes far past the backbone, as though the fish were opening up its chest as well as its mouth. The bones that suspend the jaws act as a series of levers, hinged to unfold on one another as the fish prepares "to welcome little fishes in." As the great mouth opens it forces the backbone to bend backwards on itself in a sharp fold. When this awesome maw drops open to receive its prey, needle-sharp teeth grip the food and help stuff it down.

Some stomiatoids add a lower jaw like a steel trap to the mechanism owned by the gulper, and also increase the size of the mouth in effect by throwing the head back into the neck. Their backbone, where it attaches to the skull, is cartilaginous and sometimes also bent into an S-shape, both adaptations helping to absorb the shock of the shifting head and the struggling prey. With the upper jaw thus prepared, the lower jaw is dropped far down to produce a widely gaping mouth. Now, as the prey comes into range, the lower jaw snaps shut with lightning speed and the food is pulled into the mouth on huge canine teeth and, despite a desperate struggle, is swallowed. Some stomiatoids of the genus *Malacosteus* have gone one step further in efficiency, having discarded the walls and floors of their mouths. This makes possible even more rapid snapping forward of the lower jaw, since there is no water resistance against the inside of the mouth to slow the action. Small prey can, to be sure, escape through the openings in the skeletal mouth, but bigger animals have little chance.

THE REPRODUCTIVE PROCESS

The perpetuation of the species poses special problems to deep-sea creatures. In most shallow-water fishes, reproduction is a matter of mass spawning of great schools, and the individuals in these schools

FIGURE 13-10. *Deep-sea stomiatoid* Chauliodus *swallowing a lantern fish. As* Chauliodus *throws back its head, its jaws gape astonishingly to receive the prey, which is gripped by the long, curving teeth.*

N. B. Marshall, *Aspects of Deep Sea Biology*

are able to see each other in the lighted waters. By contrast, the solitary habits of many deep-sea fishes and the darkness of their environment make a meeting of the sexes in the deep sea a sometime thing. It is not hard to imagine isolated fish roaming for long periods without encountering members of their own kind. And in the gloom prospective mates might pass near each other without realizing it, like lovers in a suspense movie.

To solve these problems deep-sea fishes have adopted several structural and behavioral devices. They carry lights whose patterns identify kind and sex. Some males undergo early sexual development so that spawning can take place before the numbers in a new generation have been reduced. And there is the ultimate adaptation: the permanent joining of the two sexes once a male and female angler fish have found each other.

The distribution, color, and flashing sequence of the deep-sea fishes' lights are quite characteristic of a species; a fish is surely even better able to use them to identify its own kind than the ichthyologist separating species in his laboratory. In addition, the photophores on some kinds of fishes are different in the male from the female, and this difference may very well be a means of sex recognition.

PARASITIC MALES

In 1922, when small fish were first noticed attached to the bellies of female angler fish, they were thought to be the young of the bigger fish. In 1925 Tate Regan, Director of the British Museum, found a male attached to a large angler caught off Iceland. He suddenly realized the truth—that the males of angler fish are parasites on their mates—and searching back through the specimens caught by the *Dana* he found other fish with attached miniature males. Now it is clear that many species of the deep-sea anglers have this amazing adaptation.

In the first days of their life the male anglers resemble the females closely, and it is difficult to tell the sexes apart at this point— except that on the forehead of the female there is the unmistakable beginning of the fishing line. The eggs that produce the little anglers are probably spawned in the deep sea and float upward near the surface. In the upper levels, where light and food are abundant, angler larvae hatch, and feed on copepods and other small plankton organisms. The tiny anglers are peculiar in having a transparent envelope of gelatin under colorless skin. This probably functions as a flotation device, helping to prevent the fish from sinking too soon into the deep sea. It also may cut down predation by increasing the size of the larvae and thus reducing the numbers of small planktonic predators that can swallow them.

Larval life for the anglers is probably over in less than two months. Gradually the two sexes take on the differences that will be so striking in adulthood. The female assumes a more globular shape, while the male develops into a slender creature without the outsized head. The skin of both sexes darkens. In the female the illicium develops, while in the male curious denticles like "buck teeth," which will later be used to grip the female in a permanent marriage grasp, begin to make their appearance. The envelope of gelatin becomes much reduced and there follows the rapid descent into the great depths which will be their permanent home from now on.

At a young age—much younger than is the case with the female —the males are ready for reproduction. One important consequence of this earlier maturity is that the number of males seeking a mate will greatly outnumber the mature females, since merciless attrition will have decimated the ranks of the latter.

The eyes of the male continue to grow and assume a slightly tubular shape, pointing upward toward the fading light. These eyes, and all the other resources at the command of the diminutive male, including high ability as swimmers, are now used with as little delay as possible in seeking a female. The males range rapidly through the ocean, alert for the faint flashes of the light organs of prospective mates. It seems probable that male angler fish also seek mates by sniffing them down. In relation to their size, the male anglers have the largest olfactory organs of any vertebrate in the animal kingdom. For the male to put this to use in tracking a mate, the female must have some characteristic odor; as she progresses slowly through the dark waters 6,000 feet below the surface, she presumably leaves behind a faint but fascinating spoor.

The unseemly haste on the part of the male is prompted by the angler's difficulty in finding a mate in the vast lightless tracts of the deep sea, where chance encounters are so rare. If a female is located, permanent troth is plighted, with the male attaching itself to her body for the rest of their lives.

Having found the female, the male takes no chances of having to go through the search again—for, if he fails to find a mate within a few months, he dies. With specialized little teeth he grips the surface of the female's body in a viselike hold. His position on the female is apparently of no consequence; males have been found attached to the belly, the head, and the gill covers of the females.

Now remarkable changes take place in the body of the male. Since he will henceforth obtain his nourishment from the female, his mouth and alimentary tract degenerate, the lips and mouth tissues of the male fusing with the body skin of the female; small openings are retained at each side of the mouth to admit water for the male's respiratory purposes. The blood streams of the two individuals are connected. The large, bowl-shaped eyes that served the male during its brief period of bachelorhood now degenerate, and the smooth skin becomes spiny. The testes develop and the male takes on its extraordinary existence as little more than a reproductive organ of the female.

The male angler fish continues to grow after attachment, but he never rivals the female in size; the very largest male of *Ceratias holboelli* found was only a third as long as its mate. Usually they are much smaller than this, some only one twenty-fifth as large. One female caught near Iceland was 40 inches long, while its attached male measured only 4 inches; another female was 2½ inches long— but its hitchhiking consort was only two fifths of an inch. The differences in weight are much greater still, with attached male individuals of the genus *Cryptopsaras* weighing only one twenty-thousandth as much as the female. Comparing the largest known female of *Ceratias holboelli* with the smallest known male, the former is half a million times as heavy as the latter! The result is that, while the male is a true parasite, gaining all its sustenance through the placenta-like connection with the blood stream of the female, its requirements are

F I G U R E 1 3 – 1 1 . *Female angler fish* Linophryne argyresca *and a parasitic male. Huge by comparison with her tiny spouse, the female angler is only 3¼ inches long. The male, attached permanently to the belly of the female, gets his sustenance from her.*

Dana Expeditions

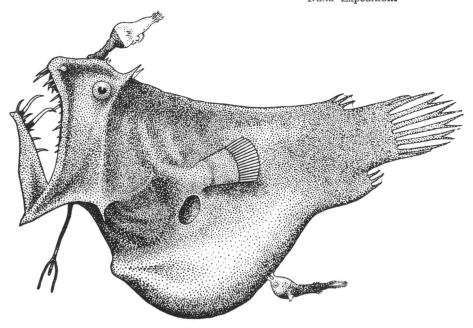

modest and it does not put undue strain on its mate to keep it alive.

Idiacanthus is another fish with curious adaptations for reproduction in the deep sea. Again the male is a runt compared to his spouse. While females reach about a foot in length, the males do not exceed half an inch. The male seems to retain some larval characteristics all his life; in some respects he never grows up. He lacks the formidable teeth of the female and has no chin barbel, and instead of being black in color he is pale. Both sexes have rows of light organs along their sides, and the male has a large red or green luminous organ on his cheek. The male sex organs begin to swell even before the little fish has finished the postlarval period. Bent from an early age on reproduction, he swims forth in search of a mate. Perhaps he uses the light from the big organ on his cheek to signal the females of his kind and to cajole them into joining him. His digestive system is very underdeveloped except for the liver, on which the little male seems to subsist during adulthood. Reproduction is his whole life; he spawns only once and then he dies.

14 *Living Light*

Cold and darkness are associated with death, warmth and light with life. Deep-sea animals accept cold as the condition of life and largely adapt to the absence of the sun's illumination. But, as if reluctant to settle for inky blackness, they have made a brave attempt to light their forbidding environment. They do it as miners do who work buried in the earth: they carry their own lights on their heads—and on every other conceivable part of their bodies, including their eyes and tails and the insides of their mouths! The light they shed is living light.

Such "biological luminescence" is seen in some kinds of plants and in most of the great subdivisions of the animals. Many bacteria (which are plants) can emit light; they often team up with animals, which collect them in certain areas of their bodies and make them shine on command. Some fungi produce light too, including toadstools and simpler kinds that emit a ghostly glow on rotting logs. Among the animals that exhibit luminescence are one-celled radiolarians and flagellates, sponges, corals (especially the soft corals like sea fans), hydroids and jellyfish, the comb jellies, many crustaceans, worms, some clams and snails, many squids and fishes. On land, centipedes, millipedes, and many insects can luminesce. For some reason no species emitting light is found in any of the higher vertebrate classes. The occurrence of luminescence is sporadic, often with one species producing light and a closely related form lacking this skill. Strangely, the only major habitat that is without self-lighting organisms (or almost so) is fresh water, where only one snail and a glowworm are luminescent. There is no good explanation of this anomaly, especially since luminescent species in the sea have closely related nonluminous relatives in fresh water.

The most familiar and best studied of luminescent creatures are the fireflies, which are not flies at all but a group of wondrously endowed beetles. They have charmed people with their nocturnal pyrotechnics since time began, and fascinated scientists have studied for many years the mystery of their light-production mechanism. The firefly's larva, the glowworm, shares the adult's ability to emit light. Dr. William D. McElroy and Dr. Howard H. Seliger of Johns Hopkins University recounted in *Scientific American* how glow-worms light up caves: "It is a spectacular sight to see the glowworms that live in caves in New Zealand, the most famous being at Waitomo, about 200 miles north of Wellington. The ceilings of these caves are covered with thousands of glowing larvae, and from each is suspended a long luminescent thread that apparently serves to catch food particles or small insects. If one talks loudly, or if the wall of the cave is tapped sharply, the larvae turn off their lights virtually as one. After a brief period, the lights come on again, tentatively at first and then more boldly, until the whole ceiling is once again ablaze."

Another beetle with spectacular lights lives in Central and South America: *Phrixothrix,* whose larva is called the "railroad worm." It has eleven pairs of green lights along the sides of its body that flash when it crawls, and two bright red spots on the head that look like headlights. The whole effect is that of a tiny railroad train.

"PHOSPHORESCENCE" AT SEA

It is in the sea that luminescence blossoms in its full glory. Few people who live near the ocean have failed to be charmed by what ordinarily goes by the name "phosphorescence." Sometimes this merely takes the form of a glowing bow wave in front of a skiff rowed across a bay at night, while the oar strokes make fiery swirls in the water. At other times more spectacular displays present themselves, the very surface of the sea appearing to be alight. In his book *Aspects of Deep Sea Biology,* Dr. N. B. Marshall says that in tropical seas in summer the glow from luminescence "may be so intense as to outshine the light from the stars." He quotes a merchant skipper's description of light seen on one such occasion: "It was first sighted as a line of phosphorescent water stretching across the horizon ahead from east to west. As the ship approached the area, it presented a curious scintillating effect. On passing through, it was found to be a belt about a half a mile in width . . . the effect at close quarters was as though thousands of powerful beams of light directed upwards from under water, each illuminating a patch of some twenty to thirty square yards of sea surface, were being switched on and off alternately, independently of each other. If any one of the patches were watched, the intervals of light and darkness were found to be of surprising regularity, about one to one and a half seconds."

Charles Darwin describes a brilliant display of luminescence in South American waters during the voyage of the *Beagle:* "While sailing a little south of the Plata on one very dark night, the sea presented a wonderful and most beautiful spectacle. There was a fresh breeze, and every part of the surface, which during the day is seen as foam, now glowed with a pale light. The vessel drove before her bows two billows of liquid phosphorus, and in her wake she was followed by a milky train. As far as the eye reached, the crest of every wave was bright."

The fiery displays occur in the warm months. In the other months of the year the little creatures that produce the luminescence often disappear from the open sea. In certain sheltered bays, however, there may be permanent colonies. One such "fire lake" near Nassau in the Bahamas used to be a showplace for tourists. Two other famed fire lakes occur in the Caribbean Sea, one being Oyster Bay near Falmouth, Jamaica, and the other a small bay on the south coast of Puerto Rico. In these bays every fish that makes the slightest movement is outlined with white fire, and every wave looks as if it were aflame. All these areas have similar physical characteristics: they are salt-water bodies of small size surrounded by mangroves. The conditions that make for continued luminescence appear to be delicately adjusted: when the opening from the Bahamian bay to the sea was widened, its brilliant display was ended.

The "burning of the sea" has always intrigued observers, and a great many quaint and ingenious theories as to its cause were put forth before the truth—which is really much stranger than the fiction —was eventually discovered. In *The Open Sea: The World of Plankton,* Sir Alister Hardy reviews some of these theories: "Tachard, an ecclesiastic, expressed the opinion in 1686 that the ocean absorbed the light of the sun by day and emitted it again at night, and our great chemist, Robert Boyle, in the same period believed that the light of the sea was caused by friction, either between the waves and the atmosphere or by the waves striking an object like a ship. Benjamin Franklin at first thought that it must be an electrical phenomenon, taking place between the particles of water and those of salt; later, however, he gave up this view as a result of an experiment he made in 1750 in which he found that sea water ceased to sparkle after being kept for a time. In the same year two Venetian naturalists, Professor Vianelli and Dr. Grixellini, discovered that phosphorescence in the Adriatic was due to the organism *Noctiluca;* and a little later Spallanzani observed the luminous jellyfish *Pelagia* in the Mediterranean, although in fact he was only confirming what was already known to Pliny."

THE CHEMISTRY OF LIGHT

Thus it was discovered that what is commonly called phosphorescence has nothing to do with phosphorus but is due to the amazing ability of living creatures to produce light. Careful studies on fish and other animals, but especially on fireflies, have revealed that light production by living creatures is a complex chemical phenomenon; bioluminescence is "chemiluminescence." Inside the cells of the light-producing organ, or in some cases in the areas between the cells, a chemical called "luciferin" combines with oxygen, yielding oxyluciferin and water. In the process energy is released—in the form of light. The reaction cannot take place unless a third chemical, luciferase, is present to stimulate the change. This catalytic, or "helping-hand," chemical is called an enzyme, and it does its job without itself being altered. The conversion of luciferin to oxyluciferin is not a simple oxidation reaction. Instead, a whole series of rapid steps take place, and a sequence of chemicals is formed and quickly altered to the next phase. The aspect of the chemical reaction of greatest significance to us here is the fact that light energy is released. The reaction is a reversible one; the oxyluciferin can combine with water and thereby be "reduced" to the original form, luciferin, ready to produce light the next time it is triggered.

It is only very recently that the chemical nature of luciferin and luciferase has been fully understood, although as early as 1887 a French scientist, Raphael Dubois, expressed the opinion that luciferase was an enzyme. He worked with the boring clam *Pholas dactylus.* Dubois found that extracts of the clam would fluoresce only if they were not heated, and that it required two substances to create

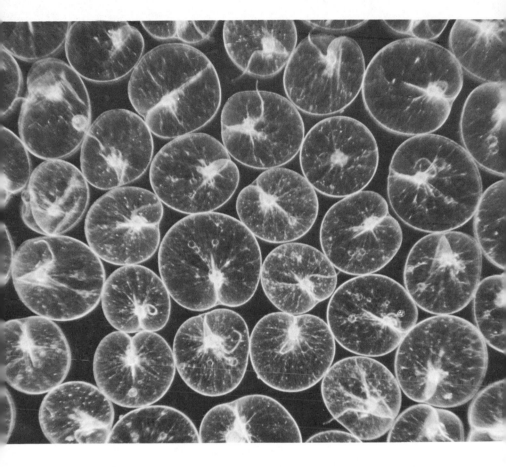

FIGURE 14 − 1 . *The night light* Noctiluca scintillans, *photographed alive. These one-celled dinoflagellates are one of the commonest sources of luminescence in the sea. They are excited into flashing by the movement of a fish or a boat.*

Douglas P. Wilson

the light. It was he who named them "luciferin," from Lucifer the "light bearer," and "luciferase," the "-ase" ending being chemist's shorthand denoting an enzyme. It is characteristic of this class of chemicals that they are destroyed by heat.

In the present century, Dr. E. Newton Harvey of Princeton University spent most of his active life on the study of luminescence. He showed that there were several kinds of luciferin-luciferase reactions in different animals, and that they were clearly enzyme reactions. In the late 1950's McElroy and Seliger isolated firefly luciferin in the laboratory, determined its complicated chemical structure, and created it synthetically. They have done the same for firefly luciferase and have found that its structure is exceedingly complex. It is a protein whose molecule contains about 1,000 subunits, making it larger than any other protein whose structure is known. Japanese scientists have determined the structure of the luciferin of *Cypridina,* the little crustacean famous for light production, and found it similar to that of the firefly luciferin but not identical.

Dr. Harvey had come to the conclusion many years ago that the light-producing chemicals of different animals were not exactly the same. He could produce light by mixing luciferin and luciferase if both were extracted from the same animal, but not if they came from different species unless they were very closely related. It may be that every luminescing creature has light-producing chemicals with slight differences in structure.

One of the most interesting and significant facts about living light is its efficiency. Manmade light is accompanied by a good deal of heat, as anyone knows who has tried to unscrew a light bulb that has been burning for any length of time or has sat under floodlights while his picture was being taken. Since the object is to produce illumination, the heat is a waste product, dissipating some of the electrical energy and reducing the efficiency of the light source. In most kinds of manmade light more than half the energy is lost in the form of heat, and sometimes the loss is considerably over this fraction. It had been suspected that in some forms of living light, such as that of fireflies and *Cypridina,* only as little as 1 per cent of the energy was wasted as heat. Judging by recent research, even this estimate may be too conservative. The Johns Hopkins scientists have found in their experiments that each molecule of pure luciferin oxidized results in the release of exactly one quantum or unit of light.

FIGURE 14 – 2 . *Living ctenophores* (Pleurobrachia pileus), *called comb jellies because of the "combs" that run from top to bottom. Their brightly lit tentacles can be retracted (specimen on the upper right).*

Douglas P. Wilson

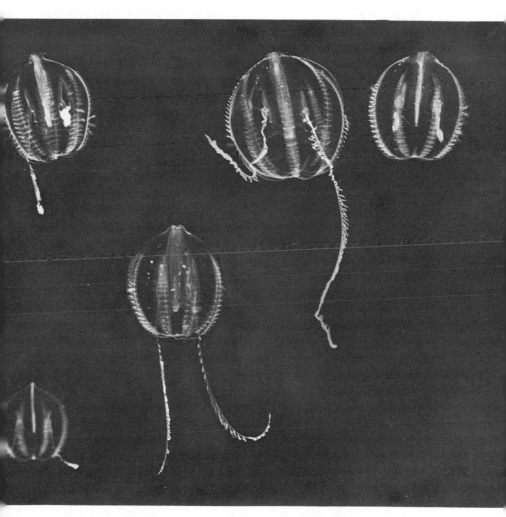

This suggests that the light-production mechanism of the firefly is 100 per cent efficient: there is no waste heat, and the phrase "cold light," which had long been applied to bioluminescence, is strictly accurate. The study of the cold light produced by living things is of practical value as well as of great theoretical interest, for, if we can unlock the secret of the high efficiency of these tiny living machines, we might learn to make much better use of some of our energy sources.

The luminous displays on the surface of the ocean are most often due to the presence of myriads of one-celled planktonic creatures, members of the phylum Protozoa. Perhaps the most famous of these is the one mentioned above—*Noctiluca,* a relatively large species about the size of a pinhead. *Noctiluca* and other luminous protozoans produce a steady glow and are likely to be responsible for the illumination that outlines fish and waves and oars in fire. The more intermittent scintillating displays like the one described by the ship captain quoted earlier are produced by ctenophores (small globular "jellyfishes") and tunicates. Occasionally luminous fishes, crustaceans, squids, and other animals show their lights at the surface, but more often the larger creatures flash only in the dark depths of the ocean.

There are two distinct sources of luminescence in the fish and other large animals. The light may be emitted by self-luminous organs of the creature itself or by colonies of luminous bacteria.

LUMINOUS BACTERIA

Some marine bacteria are brilliantly luminescent, and occasionally they may independently be the cause of displays of light in the sea. Bacteria are responsible for the glow often seen in the dark on unrefrigerated meat and fish. The bacteria that cover the ghostly meat or fish are harmless but vigorous; cultured on a nutrient jelly, they grow into splotchy colonies of light. The chemist Robert Boyle was able to show that such bacteria must have air to produce their glow. Raphael Dubois created a spectacular display at the turn of the century with cultures of luminous bacteria. His account is quoted by McElroy and Seliger: "In 1900, at the Palace of Optics, at the International Exposition in Paris, I have been able to illuminate, as from the clearest light of the moon, a vast chamber, using large glass flasks of 25-liter capacity—containing very brilliant photo-bacteria. . . . In the evening as soon as one entered the chamber one could read and see all the people in the room."

The most interesting occurrence of luminous bacteria is in symbiosis, a mutually agreeable "living together" on the bodies of other animals. Symbiotic bacteria derive nourishment from their hosts and in return act as the light-producing elements of many creatures, especially prawns, squids, and fish. The bacteria are organized into compact colonies. Since these patches of bacteria always occur in the same place on the host animal—as we shall see, the pattern of the light organs of any kind is constant within a species—the bacteria must be passed along as part of the legacy of heredity from one generation to the next. We still do not know how this is done. Inoculation may take place by the bacteria being carried on the egg, or the young animal may be "infected" by contact with others of its kind that already have colonies of the symbiotic bacteria.

THE PHOTOPHORES

Whether bacteria are involved in the production of light or the animal itself has light-emitting tissues, the light is produced in special organs called photophores. These light organs vary greatly in size, shape, and color. Most often cup-shaped, they may also take the form of a flask, a paper bag, or a shell. In structure they bear a distinct resemblance to eyes, and in fact they were so regarded by early naturalists. There is really nothing very mysterious about this resemblance, of course, since it reflects the fact that organs receiving or transmitting light must have certain components. Depending on the stage of their complexity, photophores may have focusing lenses, reflectors formed of masses of interwoven glossy fibers, linings of black or red opaque pigment to isolate the organ from the other parts of the animal, layers of cells to produce the light, a diaphragm to control the amount of light emitted, muscles to work this diaphragm, and color cells to give a characteristic hue to the organ. Light organs are always well supplied with blood vessels but usually not especially well endowed with nerves; in this latter respect they differ widely from eyes.

In bacteria and fungi, and in the simpler animals, the color of the light is usually blue or yellow. The most common colors in fish, crustaceans, and other higher groups are bluish-green, reddish, and white, but many other shades have been observed. An angler fish that Beebe saw through the porthole of the bathysphere had a lemon-colored light, while that of another angler was purplish. The ventral light organs of the hatchet fish throw beams of greenish-yellow. Other colors mentioned include violet, ruby red, green, purple-orange, brilliant white, and many more.

The position of photophores on the body is equally varied, and they can sometimes be found in the most incredible places. In general they are located on the flanks and underside of the animal rather than on the upper areas and are not scattered randomly but usually arranged in rows and patterns on the skin. Quite commonly there is a row of prominent photophores along the side of the fish like the portholes of a ship. This is the case in the singing midshipman *Porichthys,* although the light organs must have reminded whoever named it of the brass buttons on a junior naval officer's uniform. *Porichthys notatus,* from California waters, has 840 variously sized photophores in complex patterns all over the body, with numerous minute organs scattered among the more prominent ones. In other species the number may be in the thousands.

Photophores may occur, as we have seen, on the ends of the illicia, or fishing rods, of the deep-sea anglers and on the ends of the chin barbels of the stomiatoids. They may also occur on the ends of long tentacles, as in many squids. Sometimes they are located deep in the body, buried near the digestive tract. In the prawn *Sergestes* there are luminous liver tubules, and on the roof of the gill chamber there is an elongated photophore with a horny lens. A common location, amazingly, is on the eye: *Systellaspis affinis,* a deep-sea prawn, has about 125 light organs on the eyestalk, and many squids have photophores in this position.

CTENOPHORES

Many groups of animals possess luminescent representatives, and in some the proportion of light bearers is high. But it is only in the phylum Ctenophora (Figure 14-2) that all members of the group are luminescent. The ctenophores are responsible for many of the most spectacular displays of light in the sea, often shooting their beams like flares as they put on scintillating pyrotechnics. As in most organisms except the bacteria, the production of light in the ctenophores is not continuous; it comes in short flashes, either according to a fixed time pattern or only when the animal is disturbed.

Recently I encountered a brilliant display put on by the ctenophore *Mnemiopsis gardeni*. In the course of research on the biology of shrimps, four of us were fishing for them with a small-meshed net stretched across a canal leading from a brackish lake to the sea. The tide was running in slowly from Florida Bay, at the foot of the Everglades near Flamingo, and we were not getting many shrimps. The net was filling rapidly with ctenophores, however, and when they struck the wall of netting, they glowed and winked in an enchanting manner. Sometimes, just before we raised the net, we saw many hundreds flashing and gleaming in the dark water. Every half-hour, when we emptied the net to collect the shrimps, the ctenophores formed into a nearly spherical mass as they were lifted into the boat, flaring out in a brilliant viridian green. With each bump of the net they glowed anew, and as they were spilled into a tub in the boat a great wash of color flowed over the captured animals. Even after they had been dumped on the ground at the canal's edge so that we could search through the slippery mass for the few shrimps, the ctenophores gleamed one last time.

THE COLONIALS

Many colonial animals such as the sea pens, sea whips, and "soft corals" emit brilliant and beautiful light when they are stimulated. Sea pens and gorgonians (horny corals) are described as giving out "a lambent white light so brilliant that the hands of a watch showed quite distinctly." Wyville Thomson described the contents of a deep-sea net in the following terms: "The trawl seemed to have gone over a regular field of a delicate, simple Gorgonid . . . ; the stems, which were from eighteen inches to two feet in length, were coiled in great hanks around the trawl beam and engaged in masses in the net; and as they showed a most vivid phosphorescence of a pale lilac color, their immense number suggested a wonderful state of things beneath—animated cornfields waving gently in a slow tidal current and glowing with a soft diffused light, scintillating and sparkling at the slightest touch, and now and again breaking into long avenues of vivid light indicating the path of fishes or other wandering denizens of their enchanted region."

But for all the beauty of the ctenophores, and of the gorgonians that excited Thomson to such poetic enthusiasm, it is the fish, the squids, and the crustaceans that have the most interesting, varied, and lovely patterns of light.

LUMINOUS FISHES

Much of our knowledge of luminescence in marine animals comes, as might be expected, from research on shallow-water forms, which are more accessible to study than those from the deep sea. Harvey worked with two such fishes from Indonesian waters, *Anomalops* and *Photoblepharon*. Each has a conspicuous white or cream-colored organ under the eye, oval in shape and about as long as the diameter of the eye. In the eighteenth century these organs were thought to serve as shields to protect the eye of the fish from coral or from the powerful rays of the tropical sun. Actually, they are photophores, and Harvey discovered that their light comes from a colony of bacteria living in the organ. A layer of black tissue behind each organ protects the fish from the effects of the luminescence.

Since the light is bacterial in origin it is emitted continuously, and the fishes have developed mechanical ways of turning the light on and off. Curiously, the method is quite different in the two genera, despite the fact that the fishes are closely related. *Anomalops* rotates the organ downward until the light is obscured by a black envelope of tissue; *Photoblepharon* draws a fold of black tissue on the lower edge of the organ socket up over the light like an inverted eyelid. While *Anomalops* constantly flicks its light on for 10 seconds and off for 5, *Photoblepharon* keeps its light shining most of the time. The natives of the Banda Islands, where these two fishes live, say that the illumination is used to attract prey and confuse enemies.

These natives have found an interesting use for the light organs: They cut them out of the fish and use them for bait in nighttime fishing. Because the light comes from bacteria and does not depend on the vital processes of the fish to keep it glowing, it serves for a long time as an effective and unique lure. Luminous baits are also prepared by fishermen of Cezimbra in Portugal. They rub pieces of dogfish flesh in a thick, yellow liquid exuded from the luminous glands of one of the rat-tails, *Malacocephalus laevis*. This bait then luminesces "with a light like that of the blue sky" that lasts several hours.

During World War II, Japanese soldiers found an excellent

source of light in the remains of the ostracod *Cypridina*. Ostracods are tiny crustaceans related to the shrimps, and many of the marine forms are luminous, emitting tiny puffs of bluish light as they swim through the water. Since they produce a copious luminescent secretion and are easy to obtain in considerable numbers, they have been the object of much of the biochemical study of luminescence. They can be preserved dry for long periods—at least up to 30 years—without losing their ability to luminesce. Japanese army officers, when too close to the enemy to risk a bright light, would place small amounts of dried ostracods in the palms of their hands, moisten the material and read their dispatches by the ghostly light of the long-dead *Cypridina*.

Even some sharks display luminescence, although not so commonly as the bony fishes. The earliest record we have of luminescence in sharks was set down in 1840 by a biologist named Bennet, who watched a live shark in an aquarium. "When the larger specimen, taken at night, was removed to a dark apartment, it afforded a very extraordinary spectacle. The entire inferior surface of the body and head emitted a vivid and greenish phosphorescent gleam, imparting to the creature, by its own light, a truly ghastly and terrific appearance." Another species of shark is described as having a white luminescence, and a third as emitting a greenish-blue light.

FIGURE 14-3. *Stomiatoid fish* Idiacanthus panamensis, *about 6 inches long. A luminous organ glows on the end of the chin barbel and many photophores, alternating with bluish-white patches, stud its body. Swimming perhaps 5,000 feet deep in the daytime, this fish comes almost to the surface at night.*

Dana Expeditions

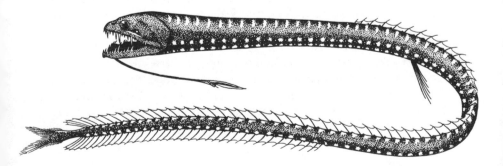

Among fishes the most spectacular displays are produced by some of the deep-sea species. Consider, for example, the little lantern fish *Myctophum coccoi*, which Beebe described. The body was covered with small, round luminous organs. There were thirty-two lights on each side of the lower body, extending from the tip of the lower jaw to the base of the tail. About twelve lateral lights were strung along the head and body, and a series of three to six median lights near the base of the tail. The colors were mostly pale green, blue, yellow, and reddish. The luminescent decorations of another fish, *Idiacanthus fasciola*, are blue-violet, those of the ventral surface being set in gilt frames.

By no means all species of the deep-sea fishes carry torches. At some depths as few as 10 per cent are luminescent, and the maximum seems to be about two thirds. But in terms of the numbers of individuals, a large proportion possess light organs. The *Valdivia* expedition found that 78 to 88 per cent of the individuals caught below 1,300 feet were luminescent. Beebe fished a small area south of Bermuda far more intensively than any deep area has ever been worked, even to this day. In his years of sampling he captured 115,747 individual fish, mostly from depths between 1,000 and 1,300 feet, and 96.5 per cent of them had light organs.

It is curious that the frequency of occurrence of light organs does not seem to increase with depth, since the fish in the gloomiest surroundings would seem to need illumination most. The highest incidence is in the mid-depths, between about 1,000 and 8,000 feet. Below 8,000 feet, fish, squids, and shrimps alike seem to abandon themselves to perpetual and utter gloom. We have as yet no explanation of why they seem to have given up.

THE SQUIDS

It would appear that the prize for luminescent glory must go to the squids. Jean Baptiste Verany, a French naturalist of the early part of the nineteenth century, was captivated by the beautiful light display of a Mediterranean squid, *Histioteuthis bonelliana;* his reaction was the stronger because apparently no one before him had recorded the colored lights of these animals. Securing a specimen from fishermen in Nice he put it in a tank of water. "At that moment I enjoyed the astonishing spectacle of the brilliant spots—sometimes it was a ray of sapphire blue which blinded me; sometimes of opalescent topaz yellow—at other times these two rich colors mingled their magnificent rays. During the night these opalescent spots emitted a phosphorescent brilliance which rendered this mollusc one of the most splendid of Nature's products."

As more marine explorers dive deeper into the ocean in the next few years, they may see the kind of spectacle described by the German marine biologist Carl Chun, who described a live deep-sea squid, which he later named *Lycoteuthis diadema,* that came to the deck of his vessel: "Among all the marvels of coloration which the animals of the sea exhibited to us, nothing can be even distantly compared with the hues of these light organs. One would think the body was adorned with a diadem of brilliant gems. The middle organs of the eye shone with ultramarine light, the lateral ones with a pearly sheen. Those toward the front of the lower surface gave out a ruby-red light, while those behind were snow-white or pearly, except the median one, which was sky blue. It was indeed a glorious spectacle."

The position of many of the light organs on the eyeball is one of the most amazing things about the squids. Perhaps they serve as searchlights, illuminating the area ahead of the animal. Sexual display and signals to their kind are probable additional functions of the photophores, which frequently appear as white buttons on the arms, head, and mantle, as well as on the eyes. There are usually more light organs on the underside of the squid, and they are apt to be crowded under the eyes. In some species there are even photophores within the mantle cavity, and these shine through when the mantle is transparent. Strong lights that flash on and off are sometimes placed on the ends of the arms—perhaps signaling to friend or foe or perhaps illuminating the dark waters.

The Japanese firefly squid, *Watasenia scintillans,* has lights on the ends of some of its tentacles. These are brighter than the photophores on other parts of the body, and they flash periodically like those of the firefly itself. The Japanese biologist Shozaburo Watase (for whom the squid was named) says that its photophores "shine with a brilliant light like that of the stars in heaven." These light organs can be covered by pigment cells (chromatophores) in the skin of the squid and the light thus extinguished. Then (to quote Watase again), "When the animal is about to produce the light the membranes [pigment cells] covering the spots will concentrate and remove themselves, thus opening a way for the light. The light is so brilliant that it seems like a sunbeam shot through a tiny hole in the window curtain."

It would appear that one of the best tricks of the squids, that of producing a smoke screen or a substitute shape in the water with dark ink, would be lost in the deep sea. But deep-water species have adapted to the darkness by producing luminous "ink" rather than the somber sepia of the animals in the lighted shallows. When they are disturbed they pour from their funnels a cloud of light. This confuses the predator, which lunges at the bright shape while the squid shoots off with powerful squirts of its jet system. The luminous discharge is commonly blue-green or bright blue in color.

THE USES OF LIGHT

A very large number and variety of creatures can produce light, and most of them have evolved complex light organs. Moreover, this ability and the special organs that go with it have evolved independently in different orders of animals and by many parallel lines of development. All this suggests that bioluminescence must have some useful purpose—some survival value. While this may seem reasonable, and some safe guesses can be made as to the nature of these uses, they are actually far from obvious, and the proof of their validity is often hard to come by.

LIGHT AS A TRAP

It seems almost certain that many deep-sea creatures use light as a lure, although there has not yet been an actual observation of such behavior. We are all familiar with the attracting quality of light: the way a lamp fascinates a moth is proverbial, and many insect traps are based on this reaction. Marine animals seem to be even more strongly attracted to light than the moths and other insects. A light on a boat at night may attract flying fish and squids so strongly that they fling themselves from the water and land on deck. In tropical waters, a light reflected from a white sail is a good trap for flying fish —and a sailor had better not get his head in the way or he might get a fish full in the eye. "Night-lighting" is a standard method scientists use to collect marine animals of many kinds.

FIGURE 14 – 4. *Fabulous black angler fish* Galatheathauma *axeli, which has its lighted bait placed conveniently in its mouth. This fish was dredged from 11,778 feet off tropical West Africa; unlike most deep-sea anglers, it lives on the bottom. It is about a foot and a half long—big for anglers in the abyss. Only one specimen has ever been captured.*

Poul H. Winther, *Galathea* Expedition

It is difficult to believe that the luminescent organs of some fishes of the abyss and perhaps of some crustaceans and squids are not used for this same purpose. A shallow-water angler, the goosefish *Lophius piscatorius,* has often been seen to use its illicium (the light at the end of the long first fin ray) as a lure. Although no deep-water angler has been seen fishing, all but two of the more than eighty species have lighted illicia so placed and so fashioned that any other function but as bait is unthinkable.

The viper fish, *Chauliodus,* probably uses the elongated second dorsal fin ray in the same manner. Here the "fishing rod" is long enough to dangle in front of the mouth, making it all the easier for the predator to move in on a victim beguiled by the light (Figure 13-10). In the stomiatoids another structure, the barbel, is lighted, and the German biologist August Brauer was the first of many to suggest that these, too, are lures (Figures 10-6 and 10-8). Since the barbel is attached to the chin, it brings the prey into position for a quick snap of the fish's mouth.

Photophores may even occur inside the mouth. What guile and efficiency this represents. The victim is lured by the fascination of a flashing light right into the maw of the predator. *Chauliodus* has about 350 photophores scattered over the roof of its mouth, which is thereby brilliantly illuminated. Little shrimps and fish, the prey of the viper fish, may be attracted when the lighted mouth is opened so the fish can draw water over its gills, and *"Chauliodus* feeds as it breathes." The lantern fish *Neoscopelus* has a considerable number of light organs on its tongue, and Brauer believed these too are to attract food animals into the mouth. The remarkable angler fish *Galatheathauma* has its lighted bait attached inside its mouth instead of dangling from its head!

LIGHT TO SEE BY

The second likely use of living light in the deep sea is simply to illuminate the surroundings. Some deep-sea animals appear to use the more powerful of their photophores as searchlights. An account of how one stomiatoid used its light for illumination is given by biologist E. R. Gunther in the log of the research ship *William Scoresby.* The ship was working near the South Sandwich Islands, at the edge of the antarctic pack ice, when an eellike stomiatoid with silvery scales was seen feeding on krill, which support so much of the life of far southern seas. Says Gunther: "From a pair of luminous organs in the orbital [eye] region, the fish (which was 9 to 12 inches in length) emitted a beam, of varying intensity, of strong blue light which shone directly forwards for a distance of about two feet. The fish had the habit of lurking at a depth of 2–6 feet below the surface, poised at an angle of about 35–40° from the horizontal—this gave the beam an upward tilt: occasionally the fish swam round and with a quick action snapped at the cloud of krill above it. In its manner of lurking and of snapping prey it resembled the freshwater pike." (From *Discovery Report*, 1950.)

In most deep-sea fishes the photophores are crowded on the flanks and undersides. This concentration casts a glow downward, and presumably illuminates the area below the fish. The stomiatoids, for example, have two rows of their most powerful lights along each side of the belly, as well as two rows along the underpart of the head (Figures 10-6 and 10-8). Lantern fishes also cast a sheet of light downward. Watching the lantern fish *Gonichthys coccoi* in an aquarium, William Beebe "saw good-sized copepods and other organisms come close, within range of the ventral light, then turn and swim still closer to the fish, whereupon the fish twisted around and seized several of the small beings."

The euphausids (of which the krill of the Antarctic is a famous example) all have photophores, sometimes in rich profusion over the body. Some biologists believe that euphausids use the more powerful of these lights to illuminate their prey. When a euphausid flexes its abdomen, all the lights converge on an area in front of it. Bright lights are also located in the eyes of some species, and Dr. N. B. Marshall suggests that they might be turned on at an appropriate time when prey is nearby: "When a copepod has been located by the antennae and the euphausid draws nearer, are the eye lights turned on and the eyes rotated so that the beams from the two lights shine forward? When the prey is eventually illuminated do the eyes and the beams follow it until it is seized?" These are still open questions, but they may soon be answered, since these animals are not difficult to keep alive in captivity.

SIGNAL LIGHTS

The lights of euphausids may also serve as flashing signals between the individuals of a swarm, and this represents a third probable function of photophores. We know that many animals use light signals as part of their reproductive behavior, to attract mates and stimulate spawning. Certainly the flashing lights of fireflies, which have been easier to observe than the fish, have a sexual function. The pinpricks of firefly flashes over a meadow may seem to be haphazard, but in fact there is a well-defined pattern to the flashes emitted by an individual insect. Professor J. B. Buck, who studied fireflies intensively, said he could distinguish among a dozen species flying together in a darkened field by timing the flash pattern.

At dusk the two sexes emerge from the grass. The male flies through the air a short distance from the ground, winking his light; the female (which may be wingless) sits on the ground and responds in turn. The better to see and be seen, she may climb to the top of a blade of grass. The light flashes of the male are always of the same length and always the same time apart, and the female always waits exactly the same length of time before answering with her own flash; in *Photinus* the delay is about 2 seconds. She never flashes except in response to his signal. The male may circle and the codes be exchanged again—as many as five or ten times. At length the two meet, and mating takes place. In certain species of fireflies in Burma and Thailand, all the males apparently sit on one tree and all the females on another; the males flash simultaneously, lighting their tree, while at a distance all the females glow lovingly after the proper maidenly interval, illuminating another tree for an instant.

Equally unmistakable uses of light in mating occur in some marine animals. A spectacular example is the curious mating ceremony of the fireworm *Odontosyllis enopla,* which is common off Bermuda and elsewhere in the Atlantic. Columbus, when approaching the Bahamas on his first voyage in 1492, reported seeing lights in the water that looked like moving candles, and these may have been fireworms. Two or three days after the full moon the females rise from their burrows in the bottom of the sea to swim in small circles at the surface, glowing brightly green. Males in the vicinity that see the circles of light swim toward them, emitting flashes of light themselves as they go. If no male appears, the female turns off her lights and stops swimming in a circle for a while, then in a few moments lights up again and resumes her dance. If the female darkens while the male is approaching, he will stop swimming toward her, wander aimlessly, and only advance again when she luminesces once more. Several males may be attracted to a single female, and the whole group then swims in a fiery circle. Eggs and sperm are emitted simultaneously, and the fertilized eggs float off to create the next generation of worms.

In many creatures the light patterns are as complicated as the brilliant signs in Times Square, with some groups of lights flashing independently of other groups, some flashing rhythmically, with several colors sparkling a message to the aquatic audience. The deep-sea prawn *Sergestes prehensilis* (Figure 7-6) has about 150 greenish-yellow photophores that flash on and off very rapidly. According to one Japanese observer, "Each single photophore lighted up for nearly a moment only, and as soon as a light disappeared, another appeared a short distance behind in rapid succession, so that there were scarcely ever observed more than one light alive at a time. It took one to two seconds from start to finish of a single series of illuminations of the above sort. At other times only a limited number of photophores in a certain body region were observed to light up simultaneously, this time the lights remained steadily alive for several seconds." (Quoted in Harvey's *Bioluminescence.*)

A common characteristic of many animals is the difference in color between the two sexes. This is especially prevalent among birds and insects. In some shallow-water fishes, too, the males are different from the females in color pattern; the same distinction is achieved in some deep-sea species by differences in lighting. For example, in the lantern fish the number, size, and position of the photophores on the tail are different in the two sexes. The male has the larger and more powerful photophores, and they are placed on the top of the tail; the smaller, dimmer light organs of the female are placed on the underside. Dr. Harvey has suggested that even when the spatial arrangement of photophores is the same in the two sexes, the pattern in which they are flashed may be different. Beebe, watching *Gonichthys coccoi* in the laboratory, found that it was possible "to distinguish . . . the sexes by the upward or downward direction of the tail lights. I have never seen the latter illuminations given out by fish swimming alone in the aquarium . . . it is very evident that the caudal flashes have some sexual significance."

In *Idiacanthus,* another deep-sea fish (Figure 14-3), the male has a relatively enormous light just behind the eye, while the female has a smaller one. The female is about 10 inches long, enormously larger than the male, which averages about an inch. It may be that the bright light on the cheek of the male guides the more powerfully swimming female to her mate and so increases the chance of successful spawning.

Colored-light signaling may even provide a deep-sea parallel to the practice, common in many primitive human communities, of advertising the specific sexual status of an individual. The girls of a tribe may arrange their hair, put marks on their skin, or wear certain colors of flowers to announce whether they are marriageable, married, or widowed. The deep-sea squids may do something of the same kind. Their complex patterns of many-colored lights make possible code signals in virtually endless variety; even the color of an individual photophore may be subject to change by drawing over the organ a colored filter consisting of a fold of skin. With this virtuoso ability to flash signals the squids may announce to interested observers their sexual abilities and inclinations: whether they are mature, whether they are ready to mate or have recently mated.

Just as colors serve as recognition signals for schools of surface fishes, the photophores probably assume this function for deep-sea fishes. As we saw in the case of the lantern fish, light patterns often differ in some respects as between the sexes of a species, but in other respects a pattern is often "species-specific"—that is, common to all its members and only to them. Beebe said he "could tell at a

FIGURE 14 – 5 . *Patterns of light organs on five species of lantern fishes; the main lights are along their lower flanks. An expert can tell from the arrangement of lights alone which of some 170 species he is looking at.*

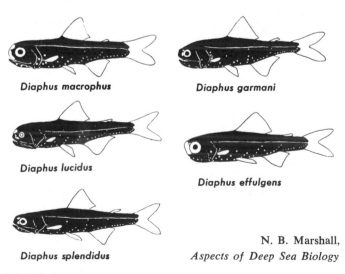

Diaphus macrophus

Diaphus garmani

Diaphus lucidus

Diaphus effulgens

Diaphus splendidus

N. B. Marshall,
Aspects of Deep Sea Biology

glance what and how many of each species were represented in a new catch, solely from their luminous hieroglyphics. When several fish were swimming about, their side port-holes were almost always alight, and it seems reasonable that they may serve as recognition signs, enabling members of a school to keep together, and to show stray individuals the way to safety." Deep-sea prawns and euphausids are usually brilliantly luminescent, and they occur in enormous swarms; here, too, one function of the lights may be to maintain the schools by showing straying members where the crowd is. Dr. R. Dennell, who has made recent studies of the biology of crustaceans, estimates that the green light of prawn photophores may be visible to other prawns at distances as great as 330 feet.

DEFENSE SYSTEM

A final possible function of luminescence is defense: rather than seeking a meal, to avoid becoming one. There is considerable evidence that deep-sea creatures use lights to distract or blind an attacker. In his book *Half Mile Down,* writing again about *Gonichthys coccoi,* Beebe said, "Although it is very evident that the caudal flashes have some sexual significance, yet another important function seems to be that of obliteration. . . . When the ventral lights die out they do so gradually, so that the eye holds the image of the fish for a time after their disappearance, but the eye is so blinded by the sudden flare of the tail lights that when they are as instantly quenched there follow several seconds when our retina can make no use of the faint diffusion of remaining light but becomes quite blinded. A better method of defense and escape would be difficult to imagine."

It would appear that the extraordinary ability of many deep-sea prawns to "explode" with light serves a similar purpose. This trick confused Beebe during his early dives off Bermuda. In his first dive in 1932, he looked out the quartz window of the bathysphere and "watched one gorgeous light as big as a ten cent piece coming steadily toward me, until, without the slightest warning, it seemed to explode, so that I jerked my head backward from the window. What had happened was that the organism had struck against the outer surface of the glass and was stimulated to a hundred brilliant points instead of one. Instead of all these vanishing as does correspondingly excited phosphorescence at the surface, every light persisted strongly, as the creature writhed and twisted to the left, still glowing, and vanished without my being able to tell even its phylum." (From *Half Mile Down*.) This puzzling kind of experience was repeated several times. On another dive, "at 1,200 feet an explosion occurred, not at the window but a few feet away, so baffling that I decided to watch intently for repetitions. . . . Suddenly in the distance a strong glow shot forth, covering a space of perhaps eight inches. Not even the wildest guess would help with such an occurrence." And again, "Three times, at different levels, creatures had struck against the glass, and, utterly meaningless as it sounds, exploded there, so abruptly that we instinctively jerked back our heads." Then, finally the explanation for these queer exploding lights was obtained on the last deep dive, at 1,680 feet, when "I saw some creature, several inches long, dart toward the window, turn sideways—and explode. This time my eyes were focussed and my mind ready, and at the flash, which was so strong that it illuminated my face and the inner sill of the window, I saw the great red shrimp and the outpouring fluid of flame . . . ; all the previous 'explosions' against the glass became intelligible."

It would appear that some fishes react to luminous light from another creature by a defensive flash of their own, probably followed by a quick retreat. Beebe watched the lantern fish *Myctophum affine* in a darkened aquarium. At intervals the entire fish was lighted; then it would become completely dark and light up again at about 15-second intervals. "I happened to lift my wrist-watch with its dully luminous dial close to the fish and it reacted at once, giving out two strong discharges of the caudal glands. Concealing the watch and later displaying the light of its face resulted in an instant reaction. This happened eight times. I then flashed my much stronger flashlight with no result. For five minutes I alternated the two artificial sources of illumination with identical results, the fish reacting vigorously to the watch dial, but paying no attention to the electric torch." (From *Half Mile Down*.) In the sea it is possible that the fish would have dived for safety after flashing its strong light in the eyes of the luminescent intruder.

Certain deep-sea prawns, including *Hoplophorus spinicauda* and *Systellaspis debilis,* can do the same, emitting the fiery cloud from glands near their mouths. Beebe observed from the bathysphere that two distinct kinds of illuminated clouds are produced by the deep-sea shrimps. One of these diffuses into a small cloud of uniformly glowing mist, while the other is dimmer but is interspersed with brilliant

stars and pinheads, "for all the world like a diminutive roman candle.
. . . For an instant the shrimp would be outlined in its own light—
vivid scarlet body, black eyes, long rostrum—and then would vanish,
leaving behind it the confused glow of fluid." The light died out
slowly, leaving a grayish mass of fluid that puzzled Beebe until he
finally discovered that it was the burned-out luminous cloud produced by the shrimps.

A group of deep-sea herringlike fishes, the Searsidae, may also
be capable of emitting a luminous cloud, again perhaps as a method
of confusing an enemy. In the skin of the shoulder they have a gland
filled with soft tissues and lined with black-pigmented cells. It opens
through a tube pointing backward and upward, and is thought to
secrete luminous mucus. Dr. Bertelsen, the authority on deep-sea
anglers, describes odd structures on the females, in front of the dorsal
fin. They are lined with pigment and open to the outside. When
pressed they emitted "a whitish-yellow secretion which poured into
the water like a thread and quickly broke and spread into numerous
points of light, which could be observed scattered over the basin for
about a minute."

One of the marine worms, *Acholoe,* goes further: it sacrifices
part of itself in order that another part—the one containing the reproductive organs that ensure the perpetuation of the race—can
survive. If the worm is cut in two by the bite of a predator, the front
part, containing the main organs, remains dark and swims away, later
to regenerate another rear part; the original hind end meanwhile
gleams brightly to hold the attention of the attacker.

THE EVOLUTION OF LUMINESCENCE

Why, one might ask, does illumination have survival value? If
the animals quietly went about their business in complete darkness,
would they not have a better chance of survival? This would certainly be true for any individual animal, but the life of the individual
is not what counts in the evolutionary history of a species; survival is
assured if enough individuals escape to reproduce their kind. The
race therefore trades the dangers of light for its advantages, and
counts on the mass fecundity of all its members to keep its population in existence.

This statement, like so many regarding the changes that accom-
pany evolution, may sound as though animals have consciously

weighed the pros and cons of carrying light organs, and have decided whether or not to have them. Of course nothing of the sort is the case. An animal either possesses photophores or not, depending on its evolutionary history. If a survival value attaches to their possession—if, in other words, they turn out to be useful—then the light organs, which to be sure developed fortuitously by mutation, persist in the species because the animals that have them multiply.

The Johns Hopkins investigators McElroy and Seliger believe that the production of light in living creatures probably arose as a chemical accident in the course of evolution. They point out that the earliest forms of life are thought to have developed in the absence of oxygen, which was missing from the primordial seas. Later, when the sun began to decompose water vapor to produce oxygen, and the earliest plants produced it by photosynthesis, this gas would have been toxic to the cells of the primitive creatures. The easiest way to get rid of unwanted oxygen would be to reduce it to form water, using the organic compounds of the cell. This would result in a release of energy—if in the form of light, the organism would be potentially luminescent. Later developments of body machinery in the more highly organized creatures would make it possible for them to use the oxygen, and the need to free the cells of this gas would disappear. But the mechanism, including the by-product light, would persist. While some light-producing organisms would never put the light to work (it seems unlikely that the luminescence of bacteria or fungi has any function), in other creatures light would prove useful and would result in higher survival of those possessing it. Special organs would be developed, including eyelike organs to produce and concentrate the light. This hypothesis could explain why luminescence has developed in a parallel fashion in many different kinds of creatures instead of from a single, light-bearing ancestor.

15 *To the Deep Sea and Back*

Animals move in mysterious ways, and men have always been fascinated by their migrations. How do ants march miles through jungles in orderly array? How do bees guide one another to nectar? How do birds travel 12,000 miles from winter homes to summer breeding grounds, moving on precise schedules and with unerring accuracy? How, above all, can one explain the movements of fish through the unvarying sea, without even the sun and the stars to guide them? Consider the journey of the salmon to its ancestral spawning site in a small creek 1,000 miles from the ocean, or the incredible migration of baby eels from deep in the Sargasso Sea to a pond in Denmark. Consider, too, the vertical migrations, the diurnal rise and fall of millions of planktonic creatures. In recent years much has been learned about the migrations of deep-sea animals, but many questions are still unanswered.

Fish are in nearly constant motion, often over considerable distances. Not all these movements are migrations. The term should be reserved for regular journeys in specific directions, ordinarily motivated by either feeding or reproduction needs. Nor are true migrations necessarily over great distances. The herring shoals, those "armies in the sea," change their encampments with the seasons, making a few notably long journeys (for example, between Norway and Iceland) but for the most part staying near home. The important little flatfish of Europe, the plaice, finds the North Sea and nearby waters congenial, and seldom wanders far afield. And tagging experiments in Florida waters by scientists from the Institute of Marine Science reveal that many local species travel only short distances. The black mullet, *Mugil cephalus,* and the spotted weakfish, *Cynoscion nebulosus,* seldom travel more than 30 miles from home base, even after many months.

THE HOMING SALMON

But if the common migratory behavior of fish is of this character, there are some remarkable exceptions, and one of the most extraordinary is the behavior of those great wanderers, the Pacific salmon. The salmon's life starts in one of the cold rivers of the Pacific Northwest. Born from an egg buried deep in the gravel bed of a swift, clear stream, the little salmon emerges to spend the first few months of its life as a fresh-water creature. Then the little fish, an inch or so in length, are carried in great hordes downstream to the Pacific Ocean. Little is seen of them for the next several years, while they range over the continental shelf and beyond, eating little shrimps, arrowworms, and small fish. Salmon from Oregon, Washington, and southern British Columbia range up to Alaska, and out to sea. Salmon from Alaskan waters may swim far out toward the Asian coast. In mid-Pacific they meet and mingle with their kind from Asian rivers. (This international comity is being copied by the citizens of Japan, Canada, and the United States, whose scientists have joined in the International North Pacific Fisheries Commission. Its principal job is to protect the common interests of the men who fish for salmon. All the wisdom and patience the Commission members can summon will be required in years to come, because eager Japanese fishermen are now pursuing the salmon to the mid-Pacific and are catching fish spawned in American streams, to the alarm of fishery scientists on this continent.)

After from two to five or six years in the sea (depending on the species), the salmon prepare to return to fresh water. They collect into shining, silvery schools and head for their native streams. With very few errors (perhaps one in 10,000, or 100,000) they swim up the very stream they descended years before. The "home-stream theory" is no longer a theory but an established fact. Furthermore, it is certain that the fish make their way back not only to the identical river but even to the small tributary stream in which they were spawned.

How they get back over hundreds of miles of apparently trackless ocean to the entrance of their native river is shrouded in mystery. There are many records of tagged salmon whose journeys were 1,000 miles or longer. One king salmon, *Oncorhynchus tshawytsha,* was tagged off the west coast of the Queen Charlotte Islands in British Columbia and recovered 870 miles away, near Marshfield, Oregon, 60 days later.

Having won its way to the entrance of the river, the salmon must still fight upstream past the teeth of seals, the nets of fishermen, and the spears of Indians, past falls and rapids, to the spawning ground. This river journey is often as much as 1,000 miles long; in the Yukon River it can be 2,000. As a result of the work of Arthur Hasler and Warren Wisby, we know more about the mechanism of this migration than we do of the ocean trip. Apparently it is by means of an astonishingly acute sense of smell that the salmon picks its own particular tributary from among the many streams of the river system. At every branch of the river where the fish is confronted with a choice as to which way to turn, it is guided by the characteristic odor from the gravel and humus of its birthplace. What an amazing apparatus the salmon nose must be to pick out this faint and subtle exhalation from the rich smells of hundreds upon hundreds of other small tributaries. But this is what it does, unerringly. Eventually it reaches the spawning ground, digs its nest, and deposits its eggs. Having finally paid its debt to its kind by initiating the next generation, it dies. No Pacific salmon ever survives to see its offspring or to make another journey into the dark green waters of the Pacific.

In this failure to survive spawning, the Pacific salmon differs from its Atlantic counterpart, *Salmo salar,* which may spawn two or even three times. The Atlantic salmon can make long migrations too. One individual tagged at Loch na Croic in Scotland was recaptured in Eqaluq Fjord, Greenland. This migration of 1,730 miles across the North Atlantic was accomplished in a little under 11 months.

MYSTERY OF THE EELS

At one stage in its life the eel (*Anguilla*) is a deep-sea fish. This statement may be hard to accept by anyone who has seen eels in tiny brooks far from the sea, in lakes, stagnant ponds, or even irrigation ditches. There is some evidence that some eels can slither through the grass from one ditch to another, where the only water is the dampness of the dew. This remarkable fish can actually use atmospheric air for its oxygen supply in brief periods of emergency, absorbing the oxygen through damp gills. Yet, despite its unquestionable link with the land and with fresh waters, the eel is an authentic creature of the ocean depths. Its dual character has been established only in recent decades, and the facts of its discovery are almost as wonderful as the eel's life cycle itself.

The rivers and other fresh waters of Europe swarm with eels. Few species of fishes are more abundant—or more distinctive; "eellike" evokes an immediate image for almost anyone. Eels have long snakelike bodies, almost round in cross-section. This shape serves the fish's habit of living in the soft bottom of a river, wriggling in and out of the mud or insinuating itself into holes and crevices. Its head is long and bullet-shaped, with big eyes and a long cleft of mouth armed with pointed teeth. The dorsal and anal fins combine with the tail fin to form a continuous fringe around the hind part of the body. The body is slimy and apparently scaleless; in truth, there are tiny oblong scales embedded in the skin. Big females (which grow a foot or so larger than the biggest males) attain a length of 5 feet and a weight of 20 pounds.

Yet, despite its familiarity, the eel has always seemed strange, mysterious, and somewhat malevolent. Something approaching witchcraft was often attributed to this snakelike, naked, slippery fish. It loved the dark, burying itself in the mud during the day. Its power to adapt to sweet water or foul, to running water or still, to fresh water or salt seemed preternatural; its ability to invade the land seemed most unfishlike. Its blood and slime were said to cause serious and even fatal infections when rubbed into cuts on human hands. Even its omnivorous appetite was looked on with suspicion and fear —especially since it was said that the eel had a special liking for dead human flesh.

But all of these mysterious attributes of the eel faded into the background compared with the puzzle of its reproduction. Unlike proper animals, it seemed to have no young. No eggs or larval eels were ever found in the fresh waters, and the adults did not seem to develop ripe reproductive organs. Some fantastic explanations were proposed to account for this. Aristotle suggested that eels were generated out of the mud in which they were so commonly found. The "May dew" was said by medieval men to be transformed into eels. A very common superstition—and one that some people still retain—

was that when horsehairs fell into a rain barrel they first produced a horsehair worm (*Gordius*) and then an eel.

Careful observers finally noticed that adult eels migrated out of the rivers into the ocean when they reached a large size and that tiny, transparent "glass" eels moved into fresh water in prodigious numbers at certain other times of the year. It was assumed—incorrectly—that the little glass eels were the offspring of adults that had gone to sea the season before and had spawned in the ocean near shore.

THE LOST LARVA

Meanwhile, naturalists were busy collecting marine animals and describing and naming them. One of the most interesting was a transparent, leaflike creature about 2 inches long. Called *Leptocephalus* ("puny-headed"), from the small size of its head, it was regarded as the adult stage of some strange sea animal. Then two Italian biologists, Giovanni Battista Grassi and Salvatore Calandruccio, made an exciting discovery. After catching a little leaf-shaped *Leptocephalus* in their plankton net they watched it in an aquarium until, to their amazement, it was transmuted into a baby glass eel. Science was on the track of a remarkable story.

In 1905 the International Council for the Exploration of the Sea assigned to Denmark's Carlsbad Laboratory the problem of solving the life history of the eel. Dr. Johannes Schmidt was given this difficult task, and the brilliant way in which he solved it has established his permanent fame in the annals of science.

When adult eels leave the rivers they disappear, never to be seen again. The possibility of discovering their migration route and tracing them to their spawning ground seemed hopeless in the vast reaches of the ocean, and Schmidt wisely did not try to do this. Instead, he decided to attack the problem from the opposite direction: to collect the larvae. Following the lead of Grassi and Calandruccio, the Dane began his observations in the Mediterranean, hoping to find eels spawning there. Two fruitless years were spent in this effort. Next, Schmidt began the prodigious job of straining the waters of the Atlantic Ocean for the young stages of the eel. This monumental task was made possible by the cooperation of sea captains who towed plankton nets during their regular transatlantic passages. By this means, and with research vessels to fill in gaps, the Atlantic was crisscrossed and thousands of young eels were strained from the water. Some of these larvae were big—about to metamorphose, or transform into glass eels—or were actually in the process of metamorphosis; others were smaller; still others were so small that larvae of their size had never before been encountered by scientists.

And there was a remarkable pattern of size distribution. The largest and presumably oldest larvae were caught near the continent of Europe, while larvae collected farther to the west were progressively smaller. The smallest leptocephali were found, astonishingly, thousands of miles from Europe, near the coast of North America. Gradually closing in on the spawning eel, the scientists tracked it finally to an area near the northern edge of the Sargasso Sea, that giant, storied eddy of the western Atlantic Ocean. Here, eggs and

FIGURE 15 – 1 . *Distribution of the European eel. The tiny transparent eels that swarm into the rivers of Europe were spawned in the Sargasso Sea. As adults they will make the long voyage back—2,000 to 3,000 miles—to the Sargasso to spawn more eels. The heavy marking along the European coast indicates the countries where adult eels occur in fresh water; contour lines show areas where larvae of various sizes (10, 15, 25, and 45 millimeters) are found. Letters "ul" show the zone where the largest larvae occur.*

Johannes Schmidt

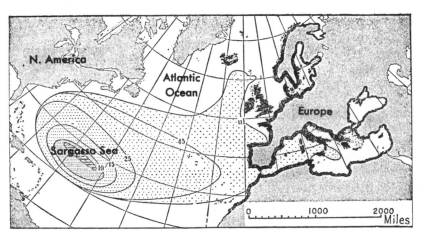

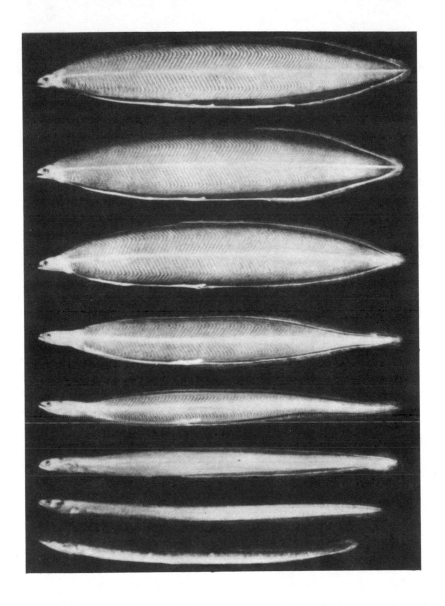

FIGURE 15‑2. *Development stages of the European eel* Anguilla vulgaris. *At the top is the young larva with its transparent leaflike body and the tiny head for which it is called leptocephalus. Each succeeding stage shows an older but smaller larva, until it becomes a young eel, or elver.*

Johannes Schmidt

newly hatched larvae were caught in plankton nets at depths of about 350 to 550 feet, sometimes considerably deeper. Here, then, was the spawning ground of the eels of Europe, in a place the wildest guess could never hit upon. Deep in the ocean southwest of Bermuda the adult eels end their migration, spawn, and die.

But how do these tiny, transparent, completely helpless baby eels find their way back to the rivers of Europe? It is not easy, as it turns out; nor is it accomplished quickly. Three years of extremely hazardous travel are involved, the motive power being supplied not by the small fish but by that great wheel of current, the Gulf Stream.

And so it transpires that the swarming glass eels that crowd into the rivers of Dr. Schmidt's Denmark and the other countries of Europe are three years old, and that their parents had left their freshwater home long ago to travel 2,000 to 3,500 miles to fulfill their destiny. The vast numbers of eels—probably running into the thousands of millions—are a tribute to the fecundity of the adults, since undoubtedly only a tiny fraction of those spawned escape the jaws of hungry predators during the long sea voyage, and are carried at last by a favorable current to the mouth of a friendly river.

But the remarkable story is not yet complete. North America has its eels, too, and they, too, are born in the deep sea. Their spawning ground is right next to that of the European eel; in fact, the two grounds overlap. But North America is considerably closer to the Sargasso Sea than Europe is; surely the American eels must spend less time en route to their rivers than the three years required by the Old World eel larvae. Would not the American larvae be too small and immature to be ready for transformation into glass eels and for the subsequent ascent of the rivers? This problem has been neatly solved: the American eel's larval life is telescoped just enough so that the small eel is ready to metamorphose precisely at the time it arrives off the American Atlantic coast. But then how do the larvae sort themselves out, one group taking the European route and the other drifting properly to the shores of America? This is an unsolved mystery. It is possible that no such sorting takes place, and if baby eels get into the wrong set of currents they perish. Only the strength of their enormous numbers keeps the population going.

The final question is how the adult eel finds its way from the estuary of some river in Europe, across endless miles of dark and trackless ocean, to the depths of the Sargasso Sea where it spawns.

The eel's method of navigation is inexplicable in the present state of our knowledge. Can we obtain any clues from the migrations of other famous wanderers, the birds, which may travel as much as 12,000 miles from their summer to winter homes?

HOW BIRDS DO IT

Many possible navigation systems have been put forward to explain the remarkable bird migrations, but until recent years none of the theories has stood the test of critical scrutiny. Even now there are wide gaps in our knowledge, but these promise to be filled, perhaps at a rapid rate, since there is a fever of activity in research on bird navigation.

Migrations of some species of birds appear to depend on their having been over the route before, presumably led the first time by older birds. They have extremely acute vision and excellent visual memory, and some of them need no further clues to negotiate a route a second time. As an illustration of the visual memory of some birds, B. F. Skinner trained pigeons to peck at certain places on an aerial photograph. Four years later the birds pecked at the same positions. Racing and homing pigeons are trained to the countryside before they can return to their own loft. They lose their way easily if the ground is covered with snow or if the visibility is bad.

But there are numerous instances in which birds make their first migrations without the benefit of experienced companions. Young petrels (*Oceanites*) are deserted by their parents a week or so before the fledglings leave the land. The young birds then proceed to travel solitarily out to sea, and have often been observed making their way over the ocean far from land; they can receive no guidance from the adults, which by this time are much farther at sea. Young gannets (*Morus*) are abandoned in the same way by their parents, and make their first flight alone. They overtake the adult migrating bands, or at least pass through their wintering range, and proceed to a juvenile nursery or wintering ground that they have never seen before, sometimes hundreds of miles beyond that of the adults.

Some ingenious theories depend on the birds' possession of special sense organs. For example, one European ornithologist has suggested that birds can detect infrared rays and are guided by them. There seems to be no evidence to support this. Other workers have supposed that the lines of magnetic force around the earth supply directional clues and that birds have receptor organs sensitive to them.

In 1946 G. Isling suggested that birds could perceive mechanical forces resulting from the rotation of the earth—meaning the Coriolis force, described in the chapter on currents, to the right in the Northern Hemisphere and to the left in the Southern Hemisphere. The force is zero at the equator and at a maximum at the poles. The lines of equal Coriolis force are thus parallel to the lines of latitude, and the bird is supposed to determine its latitude by measuring the intensity of the Coriolis force. But the magnitude of this force is so small that most scientists doubt that the birds have sensory receptors so delicately attuned that they can measure it. The semicircular canals of the ear have been suggested as the sense organs involved, but birds' ears are less highly developed than those of mammals, and nothing in their structure suggests the level of sensitivity to mechanical forces implied in this theory.

SUN FOLLOWERS

As recently as 1949 the German investigator Gustav Kramer noticed that starlings kept in a cage at migration time stayed mostly in the part of the cage pointing toward their normal migration route. If the birds could see the sun they maintained their proper bearings. If the apparent direction of the sun was altered with mirrors, the birds were fooled into changing their orientation. They were unable to get their bearings when the sun was hidden by clouds. These experiments and others that followed made it clear that some migrants use the position of the sun to find their location, making continual corrections for the sun's movement through the sky. In order to do this they must have a remarkably precise internal clock.

Evidence of this internal time sense is abundant. Starlings were trained in natural daylight to orient themselves in a fixed direction. Then they were confined in a room with artificial light, where "day" began and ended 6 hours later than the natural one. Returning to natural daylight, they oriented themselves in a direction 90° to the right of the previous one. In one of these experiments a starling that could see only patches of sky above the horizon but could not see the sun, continued to orient itself in the proper direction for migration. Observations in the field with Manx shearwaters confirm the necessity of a clear sky for the efficient use of this skill.

THE EELS AGAIN

If the migration of birds seems difficult to understand, the journey of the eels seems impossible. They are certainly not following a route they have learned. It could not be the reverse of their larval journey: that would be the long way around, against the powerful current of the Gulf Stream. More likely they take a shorter route across the sea or use the southern arc of the great Atlantic current and approach the spawning ground from the south and east. As Sir Alister Hardy points out, if they go straight across to the Sargasso Sea from the mouth of whatever river they leave as maturing adults, eels from Scotland would have to go some 2,000 miles due southwest, while those from Spain would go south-by-west. Furthermore, eels from the North Sea and others from the Mediterranean must first find their way out of their landlocked seas before they can strike out on the long ocean trip that other eels begin immediately after leaving their rivers.

Whether the adult eels swim to the spawning grounds straight across the ocean or go south and then west with the current system, they have never gone that way before; even if they had, there are no "landmarks" to recognize. Changes in the salinity of the water have been suggested as the guide, but there is no steady change that would make this explanation acceptable. Sun or night-sky navigation like that used by the birds seems impossible. Currents could not serve as signposts, since there is no steady current pattern—and in any case how would the fish detect that it was in a particular current when there is no reference point like the bank or bottom of a stream by which to judge its relative movement? The temperature requirements of the spawning eels may seem to offer the most likely explanation, since the eels appear to spawn in a narrow range of temperature found only in the Sargasso Sea between 55° and 68° West latitude. But this explanation is incomplete, too, since it is impossible to imagine how the eel could detect the slight changes in temperature that mark its proper course across the ocean.

An English biologist, Dr. D. W. Tucker, suggested in 1959 that perhaps the mystery could be solved by supposing that the European eels never do get to the Sargasso Sea; that, in fact, they never spawn! In putting forth this theory he stated that the condition of European eels as they leave their rivers is much poorer than that of American eels, and that the Europeans have much farther to travel to the

[317]

Sargasso Sea. The European eels had progressed so far in maturity, with accompanying loss of vitality, that Tucker thought it doubtful they could survive the long journey across the ocean. He suggested that all the eel larvae in the Sargasso Sea are produced by American adults, and that the young eels that eventually swarm up the rivers of Europe are the survivors of the young produced by American eels and accidentally swept eastward by the Gulf Stream. The larvae that had reached the American coast from the same spawning would eventually produce all the spawners for Europe and America in some future generation. Most biologists acquainted with the problem believe that Tucker's arguments are less convincing than the older ones of Schmidt.

UP AND DOWN MIGRATION

The eel is a bridge between animals that take part in horizontal migrations in the sea and those that move vertically, for the eel does both. Vertical migration is remarkable in its universality; practically all deep-sea and mid-depth marine creatures exhibit these movements: single-celled protozoans, jellyfish, arrowworms, true worms, copepods, euphausids, prawns, shrimps, squids, and many pelagic fishes. There must be an advantage to the animal in this periodic shift in habitat, and a profound biological significance is certainly attached to such widespread behavior.

The definition of plankton as "drifters" is not strictly accurate, for many animals of the plankton can propel themselves—especially up and down. Their vertical migrations are diurnal—closely tied to the alternation of day and night. One planktonic creature that carries out daily vertical migrations is the copepod *Calanus finmarchicus*. During the day *Calanus* dwells some 350 feet deep in the ocean; at night it is found at the surface. They do not rise in a compact body; a few pioneers precede the main body and some stragglers are left behind. As the light fades progressively, more and more of the population move toward the surface. They sink again as the sun rises the next day, and most of the group is deep in the ocean when the sun once more reaches the zenith. Light seems to be the principal factor governing the copepods' vertical movements, but it is clearly not the only one. There is a good deal of variation in behavior, related to the individual's stage of maturity and the temperature or pressure

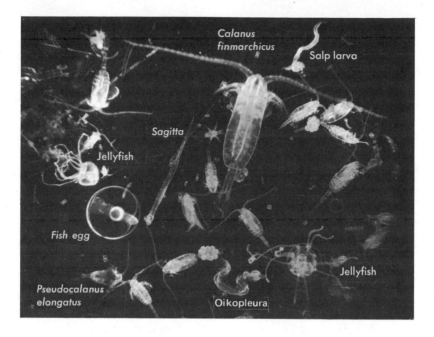

FIGURE 15–3. *The planktonic copepod* Calanus fin-marchicus, *although the largest animal in the picture, is only ⅛ of an inch long. It spends the day 350 feet deep in the ocean. As daylight fades, it swims to the surface, remaining there until dawn approaches. The smallest change in the amount of light makes* Calanus *change its position. Other animals in the photograph include smaller copepods* Pseudo-calanus elongatus, *a round gadoid egg, arrowworm* Sagitta, *a tunicate larva, and two small jellyfish with long tentacles.*

Douglas P. Wilson

of the water. Furthermore, it is not clear whether the animals react to the absolute amount of light or to the rate of change of illumination.

Many species of euphausids and prawns, crustacean cousins of *Calanus,* exhibit stronger vertical migrations than their smaller relatives. Some of them move upward 1,200 feet or more and then swim back into the depths again. For example, several species of euphausids appear in enormous swarms at the surface of antarctic seas at night.

Perhaps it is in pursuit of euphausids or other delicacies that many kinds of fishes also exhibit vertical migrations, or perhaps they, too, are caught up in the same great universal rhythm as the invertebrates. In any case many deep-sea fishes also migrate vertically. The lantern fish and stomiatoids, for example, can be regarded as fishes of the surface as well as of the depths, since they appear to spend part of each 24-hour period in the upper layers of the sea. Some fishes are thought to be able to make amazing daily vertical journeys of as much as 1,000 feet—perhaps more. One of the principal problems faced by these animals in their migrations must be the change in pressure, which could be as much as 40 to 50 atmospheres in the course of a dive or an ascent. In addition, a fish whose daylight hours might be spent in cold water of about 50° F. may swim into tepid tropical surface waters in the 70's or 80's—the equivalent of a surface journey from the coast of Iceland to the equator.

The instrument that has provided most of our information on vertical migrations is the echo sounder, a device developed before World War II as an aid to navigation. It operates by sending high-frequency sound waves down into the water and recording them when they bounce back from the bottom of the sea. On some of these machines the contour of the bottom is traced on a moving piece of paper. The task of the old-time sailor who cast a heavy lead line at the bow of a ship groping cautiously over a shoaling shore line is now done by an electronic brain that takes thousands of soundings while a man could take one, and all this while the ship is proceeding at full speed.

Unexpected complications often arise with the use of new tools, and, during World War II, echo sounders often showed a "bottom" where it was known that none existed. It developed that the dark patches that looked like a second bottom were really schools of fish in mid-water. A valuable new tool was handed to the commercial fisherman, who could now locate fish that gave no hint of their presence at the surface. Herring and sardine fishermen, among others, now employ echo sounders as a matter of course.

Biologists testing these machines on research boats and physicists working with the United States Navy to develop antisubmarine defenses sometimes saw on their echo charts marks that were neither fish nor submarines. Instead of the sharp, clear echo characteristic of that from a submarine, or the somewhat less precise but still definite mark made by a school of fish, the new echoes were diffuse and soft.

THE DSL

The "phantom bottoms" were double or triple on many occasions and occurred in moderately deep to very deep water. They were named "deep scattering layers," or DSLs, since they scattered the sound waves sent down by the echo sounder.

Presently, another remarkable characteristic of the DSLs was noticed: they moved up and down in the water. There was a distinct pattern to these movements, the layers being closer to the surface at night and at their deepest during daylight. In fact, they behaved like the pelagic animals of the sea, with typical diurnal movements toward the surface and back to the depths. It is logical to assume that the deep scattering layers are produced by vertically migrating creatures of the sea.

At first it was suggested that squids or concentrations of big fish were responsible for the DSLs, and fishermen became excited at the prospect of rich hauls with nets placed at the proper depths. But fish of even moderate size make a darker, firmer mark on the echogram than those of the DSLs. Not infrequently, black marks do appear in the paler wash of the DSL echoes, moving up in the water column with them. These are apparently fish in pursuit of whatever is in the deep scattering layer. Curiously, while some of these marks can be traced upward in company with the DSLs, they seem never to go down again. Instead, they seem to appear suddenly in deep water, later to repeat the upward climb.

Then it was suggested, on the basis of occasional catches of euphausids at depths where a deep scattering layer was detected, that these little crustaceans, an inch or so long (Figure 7-5), were the cause of the phenomenon. Lantern fish (Figure 10-11) and other small deep-swimming species were also suggested as candidates because they, too, were sometimes captured under suitable conditions. It now appears that these two groups, along with another small kind of crustacean, the sergestid shrimps (Figure 7-6), are the main causes of the DSLs. Each of these three kinds of creatures seems to account for one of the three layers that typically appear. Echo traces made in the daytime show the shallowest layer at about 700 to 950

[321]

feet, the depth varying with the location in the ocean. The second layer is at about 1,400 feet and the deepest at 1,700 to 1,900—sometimes as much as 2,400. As night approaches, the three layers all move toward the surface of the sea, merging into a single band as much as 500 feet thick. With the return of daylight, the three bands sort themselves out again. The first to leave is apparently that representing the lantern fish. At the appropriate interval the euphausids peel off and begin their dive. Finally the sergestids move down, and the typical pattern of the daylight hours is established. The three layers of the DSLs never cross; they know their position in life. Recent work also points to siphonophores as one of the causes of sound scattering. Scientists working with the bathyscaph *Trieste* have caught these little colonial "jellyfishes" under circumstances which associate them with the deep scattering layer. Probably the presence of an air bubble in the top of the colony is responsible for most of the reflection of the sound.

The little creatures of the deep scattering layers are remarkably sensitive to light. Long before dawn, with the first faint glimmerings of light, they begin their downward migration. They move rapidly, as fast as 25 feet per minute, and, by the time the sun rises, they are more than halfway to their daylight position. In another hour they have arrived. Their sensitivity to illumination is revealed when they move downward during moonlight nights, and when they rise slightly as a cloud passes over the sun during the day. Dr. Robert S. Dietz of the United States Navy Electronics Laboratory in San Diego recounts how he followed DSLs all the way from California to the Antarctic. The echoes stayed between 900 and 2,100 feet during the day. At night the creatures came to the surface in the characteristic manner, but the time of their upward migration changed gradually with the increasingly earlier sunrise and later sunset as the southern continent was approached. At last, as the length of the night was reduced to a mere 4 hours near the Antarctic Continent, the little creatures apparently gave up the struggle and the migrations broke up in confusion.

The illusion of swarming life suggested by the breadth and darkness of some echograms implies that whatever seems to be cluttering up the water so heavily should be caught in profusion. But plankton nets and mid-water trawls catch only an occasional creature, and attempts to photograph the DSL organisms or pick them up on underwater television have met with little success. Observations by divers in bathyscaphs now make it clear that the little shrimps and fish are considerable distances apart. Eric G. Barnham of the Navy Electronics Laboratory has, however, seen moderate to heavy concentrations in a few of his dives. Once he entered a zone of sergestid prawns between 1,200 and 1,500 feet—so many that he could not count them. In the 200 feet below this he saw numerous lantern fish, and after a gap he saw heavy concentrations from 2,150 to 2,300 feet of what appeared also to be lantern fish. He sometimes counted eight within view at a time, and finally saw so many that he lost his tally. This was an unusual case, however; in other dives he saw far fewer organisms. The DSL creatures are not likely to support any new commercial fishery.

Not all vertically migrating animals come so near the surface. One group observed by French oceanographers in the Antarctic varied their position from a minimum depth of about 900 feet to a maximum of about 1,700; a deeper group shifted from about 1,400 to 3,000 feet. Although neither came anywhere near the surface, both groups moved in a rhythm related to the time of day and therefore probably to the amount of light. The surprising fact is that nearly all animals in the plankton and the fishes associated with them exhibit the same kind of patterned rhythm.

Why do they do it? The most obvious possibility is that the food supply is better in the surface layers. So far so good, but why do the animals not stay in the richer upper layers? Since they surface only at night it might be assumed that the creatures retreat to the dim depths at daylight for safety. If that is the case, however, why do so many of the creatures involved in these migrations blaze with bioluminescent light as though straining to draw attention to themselves?

Sir Alister Hardy believes that the important advantage of vertical migration is actually horizontal movement. By moving vertically in the water through a few hundred feet, the planktonic animals descend to currents that move at different rates of speed from the surface layers and often in different directions. Rising again to feed, they may find themselves transported a mile or more from where they were a day before. In this way a weakly swimming organism acquires extensive powers of movement by hitching rides on various currents.

THE GIANT DIVER

From the small creatures of the plankton, almost entirely at the mercy of the currents, it is an enormous jump to the mighty whales, proud masters of their own destiny with power to roam the oceans of the world from pole to pole and from top to bottom. But the whales, in common with the little animals of the deep scattering layers, make frequent excursions from the surface to the deep sea and back again. The vertical movements of the whales are not controlled by the waxing and waning of light, as are those of the crustaceans and fish of the DSLs, but they are motivated by the same need: the pursuit of food.

The baleen, or whalebone whales—those that feed on the plankton—commonly make dives as deep as 300 feet. They need not go any deeper since the krill, their favorite diet, are concentrated in the top 30 feet of the sea and thin out rapidly as the water deepens. The toothed whales, on the other hand, feed on squids and fish and pursue them to much greater depths, sometimes all the way into the deep sea. The hunger drive makes the sperm whale and some of its relatives the champions of all the diving animals of the ocean: they can plunge more than half a mile deep.

If they must go so far and work so hard for their food, why are the whales so immense? Would it not be easier for them to accomplish their extraordinary feats of vertical movement if they were smaller? The fact is that the strength to swim rapidly and dive deeply can best be provided by great muscle mass in relation to surface area. Hence, the larger the animal the faster it can swim and the deeper it can dive. A blue whale can cruise at 14 or 15 knots and go as fast as 20 knots in bursts, developing about 500 horsepower for the higher speed.

The whales are the biggest animals of the earth—now or in the past. They are far larger than the biggest modern land animals, the elephants, and larger even than the dinosaurs of ancient times, whose size is often used as the standard for vastness. The biggest of all is the blue whale. The British *Discovery* Expedition in the Antarctic measured a blue whale 98½ feet long, while the Norwegians are said to have measured one over 100 feet. The biggest blue whale ever weighed, an 89-footer, grossed over 120 tons—the weight of forty Indian elephants. *Brontosaurus,* the great dinosaur of the Mesozoic Era, reached only 80 feet and must have weighed much less than the whale, since much of its length was taken up by a slim neck and a still slimmer tail.

The reason whales have to keep diving and returning to the surface, of course, is that they must break water at frequent intervals to breathe air into their lungs. Whales are not fish but mammals that long ago returned to the sea, whence their ancient ancestors had come millennia before. The whales have resumed the streamlined fishlike shape of those ancestors; they have regained fins (paddlelike flippers, which do not propel the whale but act as steering planes), lost their hind legs (which are represented only by vestiges deep inside the body), and gained a tail (which is now horizontal, instead of vertical as in the fish). They have cut themselves off completely from the land, feeding, reproducing, living, and dying wholly in the sea. In this divorcement from terrestrial life they have proceeded farther than the seals, which spend enough time on land to mate and produce their young.

The whales breathe by breaking the surface and taking air into their lungs through nostrils on the top of their heads. After a dive the whale rises to the surface, "blows," inhales, sinks vertically, and rises again. At last, after repeated breaths, the big animal rolls forward so that its dorsal fin appears, followed by the great tail flukes as the whale dives down once more.

The blow of a whale was once thought to be water, but this is clearly not so. Later it was said to be condensed vapor, like a man's breath on a cold day. But the fact that whales could be seen blowing in tropical oceans where the air is hot seemed to throw an impossible difficulty on the condensed-vapor theory. Two British biologists, Dr. F. C. Fraser and Mr. P. E. Purves of the Natural History Museum in London, suggested that this difficulty could be avoided by supposing that the visible part of the whale's blow is a foamy mucus that mixes with water to produce an oily emulsion. However, the Dutch whale authority, Dr. E. J. Slijper, believes that the blow of the whale *is* composed of water vapor, and offers an explanation for its visibility in hot climates. He says that the air expelled by the whale is under pressure, that by the way the whale whistles when it blows "we could tell that the air was being forced through a narrow opening under great pressure, subsequently to undergo great expansion outside." When air expands suddenly it is cooled; the cooling condenses water vapor as visible droplets, or "steam."

The behavior of the whale in diving to great depths poses a

number of questions concerning the physiological mechanisms by which this is accomplished. The first of these is how the whale withstands the enormous pressures involved. With every increase in depth of a little over 30 feet, the pressure increases one atmosphere, or about 15 pounds per square inch. Thus at 300 feet, the depth to which the baleen whale dives, the pressure is 150 pounds per square inch. It is crushingly greater in the regions reached by some of the other whales. The most accurate evidence of the depth of dives comes from observations of Dr. P. F. Scholander, a Norwegian physiologist now working in the United States. He attached manometers, or pressure gauges, to the harpoons used to shoot fin whales (*Balaenoptera physalus*) and recorded maximum pressures that indicated depths of 276, 342, 444, 756, and 1,164 feet. There is less direct evidence on the depths attained by sperm whales (*Physeter catodon*), yet it is virtually certain that among diving animals they are unexcelled. The sperm whale feeds to a considerable extent on various species of squids, whose normal depth range is known. Dr. Robert Clarke, a British whale biologist, has recovered bottom-living creatures from the stomachs of sperm whales—such prey as spider crabs, octopuses, and skates. This bottom-feeding habit is confirmed by records of sperm whales that became entangled in submarine telephone cables; apparently they had fallen foul of the cables as they scoured the bottom. Dr. Bruce Heezen of the Lamont Geological Observatory of Columbia University has listed fourteen cases of sperm whales caught in cables. Six were from around 3,000 feet, and one was from 3,720 feet—seven tenths of a mile. One of the best-known cases was a whale brought up from 3,000 feet by the cable ship *All America* off the coast of Colombia in April, 1932. The victim was a 45-foot male sperm whale, and the cable had twisted around its lower jaw, one flipper, and the tail. Other such cases have been reported off the Pacific coast of the United States, off Brazil and other parts of South America, off Nova Scotia, and in the Persian Gulf. Heezen suggests that the whales become entangled as they plow the bottom with their lower jaw in search of food; perhaps they mistake the cable for the arm of an octopus or squid.

At 3,000 feet, where the whale recovered by the *All America* was trapped, the animal had been subjected to a pressure of 1,500 pounds per square inch; at the record 3,720 feet the pressure would be a crushing 1,860 pounds per square inch. Why is the whale not squeezed to death under these conditions?

The tissues of the whale are mostly water, and the semifluid substance of the cells, like all liquids, is nearly incompressible. Covered with an elastic membrane, the individual cell responds only slightly to the pressure of the ocean water. But any air-filled spaces inside the whale's body would be affected by water pressure. Some fish, it will be recalled, have air bladders; these would be squeezed to a fraction of their original volume as the fish descended into the depths were it not that the fish secretes gases to increase the pressure inside the air bladder as the external pressure mounts.

The whale, of course, has lungs—and no machinery for adding gases there during a dive. It must depend on other devices. Its lung capacity is small compared with that of terrestrial mammals, being only about half the relative size. In addition, the position of the lungs and the diaphragm, and the more flexible structure of the rib cage as compared with those of other mammals, all make it possible for the thoracic cavity to be reduced in size as the lungs are compressed without damage to the whale.

HOLD YOUR BREATH

The second physiological question is how the whale secures enough oxygen. Most human beings run out of oxygen and must pop to the surface to breathe in much less than a minute; highly trained human divers can last 2½ minutes. A muskrat can stay submerged for a maximum of 12 minutes, and so can a porpoise. A duck has been timed at 17 minutes, a gray seal and a ringed seal at 20 minutes.

Whales stay submerged for much longer intervals. The fin whale can stay under for 30 minutes, the blue whale for 50 minutes, and the Greenland whale for 60 minutes. The titleholders are the sperm whale, which can stay under for 90 minutes, and the bottle-nosed whale, which has been timed at 2 hours.

A series of interacting physiological adaptations make long submergence possible. Although a whale's rate of oxygen consumption at the surface is about the same as that of land mammals, its oxygen needs are greatly reduced when it submerges. This is associated with a marked slowing of the heartbeat. In general, the smaller the animal, the faster is the heartbeat; in cats, it is 150 per minute, in hedgehogs 300, in mice 650. Big animals have much slower heartbeats, with horses averaging 40 per minute and elephants 30; the average for man is about 70. Given this relationship, the heartbeat of a whale should be very slow even at the surface. The big whales have not had their pulses taken satisfactorily as yet because of obvious practical difficulties. A fin whale stranded at Cape Cod in 1949 stayed alive for 24 hours, during which time J. Kanwisher of the Woods Hole Marine Biological Laboratory measured its pulse at 25 per minute. Stranded animals have a heartbeat at least three times as fast as the normal rate of those swimming at the surface, so presumably the fin whale and similar large species have rates of 5 to 8 in water. Dr. Paul Dudley White, the famous Boston heart specialist who attended President Eisenhower at the time of his celebrated heart attack in 1955, has measured the heartbeat of one of the smallest whales, the beluga (*Delphinapterus leucas*). This species reaches a length of about 18 feet, and the individual studied by Dr. White weighed about 2,260 pounds. On the basis of this size, the beluga would be expected to have a heartbeat of about 35 per minute while the animal is breathing air. Dr. White measured the beat while the whale was submerged, and it was a steady 16 to 17. This slower rate under water is typical of diving animals. For example, the beat of a beaver falls from a regular 140 at the surface to 10 when submerged, a penguin's from 240 to 20, a seal's from 120 to 10. Dolphins have been measured at 110 beats per minute out of water and 50 submerged. The slowing of the heartbeat under water is thought to be possible because the flow of blood to a considerable part of the body is cut off, the muscles and some other parts of the body being forced to do without blood so that the oxygen supply can go to the brain and the heart, which cannot be deprived of oxygen except for very brief periods. The brain suffers serious damage when the blood supply is cut off for 3 to 5 minutes; the heart can last without damage somewhat longer. By contrast the muscles of the arms and legs can survive some hours of deprivation. In diving whales a ring muscle around a great artery just in front of the diaphragm cuts off the blood supply except to the vital organs, which must have oxygen.

The muscles therefore accumulate an "oxygen debt." Lacking oxygen, the muscle cells store waste materials, chiefly lactic acid, that are disposed of by oxidation. When it finally surfaces, the whale pays off its oxygen debt as it blows and breathes repeatedly, coming up for air as often as six times a minute. After shallow dives the breathing rate may average only once every two to three minutes.

The whale's relatively small lung capacity would seem to aggravate the animal's problem of securing enough oxygen for a long dive. But the whale makes the most of its equipment. Our own lungs are usually only about half filled, and when we breathe we exchange only a fraction of the gases in our lungs, about 10 to 20 per cent. But the whales fill their lungs to capacity before a dive and exchange 80 to 90 per cent of the gases contained in them.

The next factor that helps the whale survive under water is its manner of storing oxygen. In human beings about 34 per cent of the oxygen is stored in the lungs, 41 per cent in the blood, 13 per cent in the muscles, and the remaining 12 per cent in other tissues. The whale's lungs are less important in oxygen storage, and the muscles assume a much more important role. Only 9 per cent of the oxygen is stored in the lungs of the fin whale, 41 per cent in the blood, another 41 per cent in the muscles, and 9 per cent in other tissues. Some whales store a full 50 per cent of the oxygen in their muscles. In muscle the oxygen is stored in "muscle hemoglobin," or myoglobin. The amount of myoglobin in the muscles of diving animals is very high. In seals there are 7,715 milligrams per 100 grams of muscle, compared to 1,084 in 100 grams of beef muscle. The richness of myoglobin in whale muscle accounts for the very dark, almost black, color of the flesh.

The blood corpuscles of diving animals have, moreover, an extraordinary size and capacity to carry oxygen. The hemoglobin of seals has an oxygen capacity of 1.78 milliliters per gram, compared to 1.23 milliliters per gram for man, and whale hemoglobin is similar to that of the seal.

Air-breathing mammals are impelled to breathe, not when oxygen stores run low, but when the concentration of carbon dioxide builds up. The feeling of suffocation is the result of carbon dioxide acting on the breathing center of the brain, and the reaction is to gasp air. In whales the effect of carbon dioxide on the breathing center is greatly reduced, so that they are not forced to breathe until the concentration of this gas is far beyond what would produce suffocation in a land animal.

The final question is how the whale escapes "the bends," the dreaded caisson disease that afflicts a man who ascends from deep water with incautious speed. The cause is blockage of the blood vessels by bubbles of gas—mostly nitrogen—coming out of solution in the victim's blood and other body fluids.

Look closely at the next bottle of soda water you open. There are no bubbles until the cap pops off. The soda water was bottled under pressure for the specific purpose of forcing carbon dioxide gas into solution; when you open the bottle, reducing the pressure, the gas bubbles out of solution. Similarly, the blood of a diver is under pressure and becomes "supersaturated" with gas. Unless he comes to the surface again very slowly, and thus reduces the pressure gradually so that the lungs can reabsorb the excess nitrogen and void it in breathing, he is in danger of it bubbling out of his blood and causing the bends.

A whale does not bother to inch its way to the surface after one of its spectacular dives, but zooms upward at will. How does it escape the bends?

Fraser and Purves of the London Natural History Museum believe, as mentioned earlier, that the spume shot out by a blowing whale consists of finely divided oil droplets, produced from a fatty foam from the lungs and trachea of the whale. Since fat has five or six times as much affinity for nitrogen as water, they postulate that the oil in the lungs takes up the dangerous nitrogen and thus keeps it from entering into the blood stream. A great deal of oil would have to be available; the English scientists suggest that it comes from the spermaceti reservoir. This is the store of liquid oil that was largely responsible for the beginning of the whaling industry, since it was the main product sought by the old-time whalers out of New England. Its function in the physiology of the whale has always been a mystery.

There are difficulties in the way of this fascinating theory. It appears that fatty foam in the small passages of the lung would interfere with oxygen transfer. Unless a great deal of the foam were ejected at each blow, the fat would become saturated and lose its affinity for nitrogen. Further, there does not seem to be any duct leading from the spermaceti organ to the lung cavity that would allow the oil to be supplied to the lungs and trachea as suggested. Finally, the spermaceti material is primarily a wax rather than an oil.

The more traditional and perhaps more satisfactory explanation of how a whale avoids the bends is that it simply does not receive so much nitrogen that the concentration can become dangerously high. It will be recalled that the lung capacity of the whale is small compared to that of terrestrial animals, so that the whale carries comparatively little air with it when it dives. Then, unlike the human diver, who has compressed air pumped to him continuously, the whale receives no more. It is the great volume of nitrogen forced into the blood as long as the human diver is submerged that later bursts out of solution as bubbles. Even the smaller amount of nitrogen in the lungs of a whale may enter the blood to a lesser extent than in humans, since the great pressure of the water may partially collapse the lungs, squeezing the spongy compartments where the gas exchange takes place and forcing the air into the more rigid windpipe and nasal passages, where much less absorption of nitrogen can take place.

But it is clear that because of the great difficulties involved in gathering information about the whale, whose size makes it an exceedingly inconvenient experimental animal, we are still groping for trustworthy data on the adaptations that enable it to be a true deep-sea animal while remaining an air-breathing mammal. It is equally clear that we are far from understanding fully the motivations and the mechanisms that control a great many other creatures' movements to the deep sea and back.

16 Uses of the Sea

"If there be one thing of which it seems safe to predict that our knowledge has almost reached its limit, it is that of life at great sea-depths."

This injudicious statement was made in 1904 by Frank Thomas Bullen, an English naturalist. He could hardly have made it at a worse time. Twentieth-century technology was gathering its forces to push back the limit of knowledge of life in the ocean far beyond the point where Bullen found it. He based his opinion not only on the great difficulties attending the study of the abyss but on the "exceedingly important" factor that "such investigation can promise no great commercial or even scientific gain. Its pursuit can at best be only rewarded by the acquisition of much curious, out-of-the-way knowledge of a side of life." In the very year that Bullen's book was published, the marine nations of Europe had recognized the commercial and scientific importance of knowledge of the sea acutely enough to sink national differences and form the International Council for the Exploration of the Sea, beginning fruitful studies that continue to this day, proving beyond doubt that research on life in the sea—deep and shallow—has vast utility and scientific interest. Dr. Bullen's misjudgment emphasizes once more the danger of underestimating the capacity of man to push forward the bounds of knowledge. The only safe ground for prophets seems to be to say flatly that science has no foreseeable limits, that no secrets are sure to withstand the persistent, probing mind of man.

And oceanography has one of the most exciting futures of any branch of science. If "past is prologue" in the sea, then the astonishing glimpses that science has snatched in recent years promise revelations of vast and cosmic proportions. We are offered nothing less than the keys to the enigmas of the origin of the earth, the solar system, the very universe itself. The sea is the cradle of life; where else are we as likely to solve the puzzle of the origin of life? The prospects for pure scientific investigation of the oceans are limitless, as will be seen in the next chapter. The questions are many, and the answers will come slowly. In the meantime science and technology will turn to the sea in an effort to solve some of mankind's urgent practical problems. One such problem is to satisfy the hunger that grips a large proportion of the people of the earth.

HUNGRY WORLD

The population of the earth is increasing at a frightening rate, and, instead of catching up with hunger through improved agriculture and technology, we are falling farther behind. Two thirds of the world's people are sometimes hungry, and many over half suffer from malnutrition. The shocking truth is that the specter of hunger —which has pursued man since, stone club in hand, he grimly stalked a three-toed horse in prehistoric times—is closer on our heels today than ever before; more people are starving, and the hungry of the world are hungrier today than ever before.

The population of the United States is expected to increase from 213 million to 263 million by the end of the century. If each of these future citizens eats as much fish as today's Americans, the 10 billion pounds of seafood used here annually will have to be increased to about 12¼ billion pounds. And if other kinds of protein foods are in short supply and we turn to the sea to take up the slack, yet another billion pounds of seafood may have to be provided. And this is in America, mind you, the best-fed country on earth, where malnutrition is exceptional and starvation a rarity. Think of the vast new demands on the sea from other parts of the world, where starvation is often the daily companion of more than half the people.

Can the sea provide more food? It should. Scientists have discovered in recent years that the ocean is about as fertile, acre for acre, as the land. Someday it may become a major supplier of man's food, instead of yielding a mere 3 per cent.

FOOD FROM THE SEA

At present the upper layers of the sea hold greater promise of adding to the world's food supply than the deep sea, but the latter has rich potential. There are, at considerable depths, some populations of edible animals which are large enough to produce sizable catches; they await only fishing methods cheap enough to show the fisherman a profit. Many of the deep-sea prawns, for example, are as large as or larger than their shallow-water cousins, and quite as tasty. Most of them are at present too difficult and too expensive to catch commercially. One deep-sea variety, however, comes into waters almost shallow enough for commercial fishing even today. This is the large red shrimp *Hymenopenaeus,* which was fished as deep as 1,650 feet by a few boats off Florida and the Mississippi delta area beginning in 1956, and sold under the appealing name "royal red shrimp." The fishing has been sporadic, and few boats have engaged in it, but royal red shrimp may some day be among a number of deep-sea creatures pursued on a commercial basis.

The large populations of squid-eating whales (and the even larger numbers that existed in the past before overfishing depleted them) attest to the swarming numbers of squids that must live in the deep sea. We are not making use of this resource to an extent approaching its potential. Only the Japanese catch any considerable quantity of squids—about 650,000 tons a year—but even they catch only the shallow-water forms. It may be that, if proper gear can be devised to outwit the squids, the world can look forward to a greatly increased source of food from these molluscs, especially those of the deep sea. Of course, this may not sound particularly attractive to some people—especially to those of northern European origin, who usually do not share the whale's enthusiasm for squids. Their prejudice is largely unjustified, and is based on the appearance of the animal; most people who "do not like squid" have never tasted it. It is not easy to realize that squids and octopuses are related to oysters and clams and not essentially different from these highly regarded seafoods in flavor or texture. The breaking down of dietary prejudices may be more difficult than the development of gear to exploit deep-swimming populations of squids.

The fish, however, are the principal source of food from the sea at present, and if we ever probe into the deep sea to increase our food supply it will be fish that yield the biggest catches. Fishermen

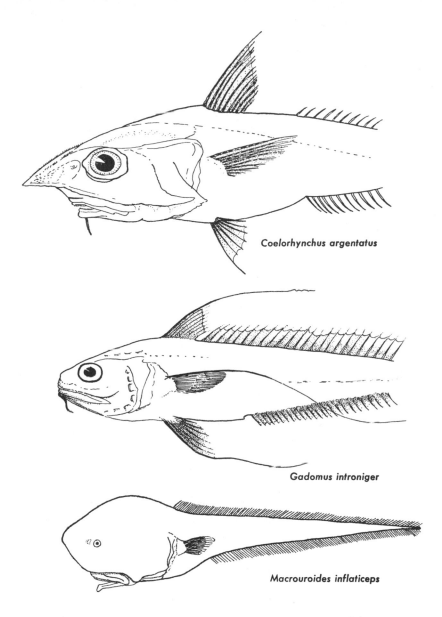

Coelorhynchus argentatus

Gadomus introniger

Macrouroides inflaticeps

FIGURE 16-1. *Rat-tails are the most numerous fishes on the edge of the deep sea, and they are good to eat. While new fishing gear may make rat-tails available for general consumption, their bizarre appearance may deter people from eating them.*

N. B. Marshall, *Aspects of Deep Sea Biology*

[335]

are already catching some fish in waters we have defined as being "deep-sea"—below about 650 feet. Some boats are capable of dragging trawls—large conical nets that operate on the sea bottom —at depths greater than this. The Soviets are said to have fished at least to 2,100 feet. Fishermen out of Eureka, California, routinely fish at 3,600-3,900 feet, and have made tests to 4,800 feet. They catch mostly Dover sole (*Microstomus pacificus*), black cod (*Anoplopoma fimbria*), and the thornyhead (*Sebastolobus alascanus*), but they are also landing some rattails, *Hemimacrus acrolopis* and *Shalinura pectoralis*. Better gear and more powerful boats will permit more deep-sea fishing, provided that the costs do not exceed the profits realized from such operations.

The fishes showing the greatest promise of commercially important catches from the deep-sea floor are the rat-tails. They are found in oceans over the whole world. One species, *Malacocephalus laevis,* was caught by the scientists of the *Galathea* Expedition in such far-flung places as the Skagerrak between Denmark and Norway, off East Africa, and off South Australia. Fishermen trawling deep for hake on the edge of the continental slope southwest of Ireland catch considerable quantities of two abundant rat-tails, *Malacocephalus laevis* and another species, *Coelorhynchus coelorhynchus,* and lesser quantities of a third rat-tail, *Bathygadus.* These occur together all the way from Ireland to Morocco, in depths of 650 feet to 2,600 feet, so they are sometimes within reach of deep-fishing trawls. Off the northeast coast of the United States another species, *Nezumia bairdii,* is very abundant on the continental slope deeper than 650 feet. The rat-tail is the most numerous of all the fishes on the edges of the deep sea, and its flesh is good.

Two things militate against extensive commercial exploitation of the rat-tails, in addition to the expense of fishing so deep: their name and—more important—their appearance. Of all the curious deep-sea creatures caught by the *Galathea,* one of the rat-tails attracted the most attention from visitors to the ship at various ports around the world. Newspapermen were more anxious to photograph this curious fish, a long-nosed *Trachyrhinchus,* than anything else— and it was very like some common rat-tails that would be prime candidates for commercial exploitation. But this kind of reaction can be dealt with. For many years the rosefish of the northeastern United States was thrown away as trash by trawlers out of Boston and Gloucester; the public was repelled by the name and the sharp-

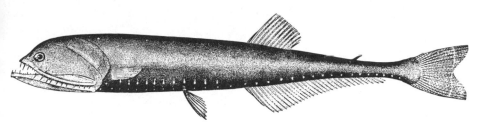

FIGURE 16-2. *Bristlemouth* Cyclothone elongata, *possibly the commonest fish in the ocean. Although it is only about 3 inches long, probably too small to be sold whole, it could make "fish flour."*

G. B. Goode and T. H. Bean, *Oceanic Ichthyology*

spined, odd appearance of these fishes. Then the name "ocean perch" was applied to them, and filleting techniques were developed so that the consumer only saw an attractive white piece of flesh. A new fishery began to flourish, and soon skyrocketed from nowhere to rank among the first ten in the United States. Now, far from throwing them away, fishermen avidly seek out ocean perch. An unfortunate common name can be changed, then, and an ugly exterior can be made graceful in an attractive package. The rat-tails are sometimes called grenadiers and this name might very well meet with approval from the public. If so, deep-fishing boats may find rich harvests in the nearly lightless waters of the continental slopes.

Some fishes swimming in mid-water, such as the tunas and the billfish (marlins and swordfish), are being caught commercially. The Japanese, the most skillful and persistent of the world's fishermen, lead in this kind of fishing. They float "long lines" in the deep ocean, with baited hooks dangling from supporting horizontal lines, sometimes miles in length. From the Japanese experience with such gear and from rare but exciting experience with sport and commercial gear in other parts of the world, we know that there are schools of very large marlins and tunas and other fishes at considerable depths. Perhaps methods yet to be developed will allow us to make better use of these in the future.

In most years, the herrings, sardines, menhaden, and other members of the clupeoid family provide by far the greatest commercial catch of fish in the world. Landings of these species reach the huge total of 30 to 47 billion pounds per year—and of course huge numbers escape the fishermen's nets. But it is quite possible that in total quantity a deep-sea fish surpasses the clupeoids; Dr. N. B. Marshall says that one of them, *Cyclothone* (bristlemouth), is the commonest fish in the ocean. If they prove to be too small to be handled on the market as food, they can provide the raw material for the manufacture of fish-protein concentrate, or "fish flour."

The lantern fish are often caught in scientists' nets, and in view of the relative inefficiency of such gears, these fishes, too, must be very numerous. They offer still better possibilities for commercial use than the bristlemouths, since they average 3 or 4 inches and reach a foot in maximum length, compared with a maximum of 3½ inches for the bristlemouths. If consumers prefer still bigger fish, then the lantern fish have great potential as sources of edible fish meal or fish flour. In addition, the lantern fishes are a source of oil. Besides the fat in their muscles and other parts of the body, their "air bladder," instead of being gas-filled, as in many other kinds of fishes, is full of oil.

The deep sea does not, of course, offer an immediate supply of additional food. The fact is that fish and shrimps and squids and the other animals of the deep sea may never be of much use to us if the cost of harvesting them is too great. Even shallow-water creatures of immense food value, like the swarming anchovies off California, are by no means being exploited to their full potential. Before they can be used, and before those from the deep sea can follow them to the tables of the world, several improvements in technology, marketing and distribution systems, and human taste will have to take place.

NEW TECHNOLOGY

At the present time the cost of fishing gear, boats, and labor is rising rapidly, while the returns from the sale of the fish are remaining nearly level. Many United States fisheries and those of some other countries as well are in economic trouble, and the downward trend of some of our fisheries will probably be reversed only by some really new methods of fishing. Advances may come in part from new and better ways of locating fish and of telling the skipper of the boat what kinds and what sizes are within reach of his gear. Sound will be one of the chief tools in solving this problem—both the sound made by the fish themselves and the sound bounced off their sides by echo-sounding fish-finders. Analysis of the bewildering variety of noises made by animals in the sea will allow us to locate useful species and track them down. Sonar will be improved so that sound waves can tell us where schools of fish are lurking, and pinpoint the species and size of the individuals in the schools. New fishing gear will include electrified nets that force fish of marketable size to swim within range of pumps that lift them aboard ship. Fish too small to be useful will be allowed to swim away, to grow and perhaps to be caught another day. Submarine trawlers and self-propelled trawl nets immune to heavy seas will replace some of the present boats that cannot function in stormy weather.

The science of hydrography, involving the description and the prediction of changes in the "climate and weather" of ocean waters, is certain to increase production of food from the sea. Changes in chemical and physical conditions in the sea must first be related accurately to fish behavior and abundance. Then, by learning to anticipate the changes, we will know what to expect in the way of abundance of fish. Ingenious new devices will be anchored permanently in the ocean and will continuously radio data to shore stations. Measurements of temperature, salinity, and other significant characteristics of the sea water, which up to now have been made with difficulty in a few areas at widely separate times, will be enormously increased and routinely available. The data will become so overwhelmingly numerous that present methods of recording and digesting them will be swamped. Then the marvelous new data-processing machines will be brought into play, and predictions of the abundance and location of valuable concentrations of fish will be supplied quickly to the fisherman of the future. In this manner the fisheries will at last approach the efficiency of many other industries.

RESTRICTIVE RULES

Commercial fishing in many countries, and particularly in the United States, is fettered with so many ill-conceived and unnecessary restrictions that the cost of producing fish is far higher than it needs to be. We seem to go out of our way to repress commercial fishing, forcing the use of inefficient gear and making it impossible to take full advantage of the bounty of the sea. This is a result of well-meaning but nonetheless mistaken concepts of conservation. Humans are characteristically nervous about the ability of living resources to maintain themselves in the face of human exploitation. There is some good reason for this attitude; such unhappy examples as the passenger pigeon and the bison are constant reminders of man's ability to squander his patrimony. But we have allowed the pendulum to swing too far. We have grimly outlawed those kinds of fishing gear capable of catching large quantities of fish quickly and cheaply.

The mere sight of large catches of fish in one place alarms some people and sends them hurrying to their lawmakers to prohibit the gear involved. They lose sight of the fact that the species may not require protection and that the failure to capture all the fish the population can yield in a particular year means that much food and wealth is lost forever. Or if the species is actually being overfished— a point that has not been established with certainty for many species —they fail to realize that a great number of small catches by many pieces of inefficient gear operated by many men scratching a meager living may deplete the stock as much as the few big catches by efficient gear operated at a profit. Political conflicts and competition between different kinds of gear and between commercial and sport fishermen complicate the problem.

The net result of these unfortunate policies is to make it as hard as possible to exploit the sea efficiently. No wonder the ocean has fallen behind in supplying any substantial share of the world's food. We cannot hope to increase greatly the amount of food from the sea with our present level of efficiency in methods or our present philosophy of exploitation of fish stocks. The fish are there, in enormous numbers. It only remains for mankind to be intelligent enough and daring enough to devise ways of making use of them.

ENERGY SOURCE

The sea has become an important source of energy. The oil and gas deposits under the sea may exceed those of the land. The U.S. Geological Survey estimates that off the United States coast, there are 660 to 780 billion barrels of oil and 1,640 to 2,220 trillion cubic feet of gas under the sea bottom from the three mile limit to a depth of 200 meters, with similar amounts between 200 and 2,500 meters. In 1974 oil from the ocean floor supplied 19 percent of the world's needs, and by 1980 this may rise to 30-40 percent. In the U.S. these resources are so valuable that they have been the subject of strenuous legal battles between the Federal government and some coastal states.

Fossil fuels, whether from the land or from beneath the sea, are disappearing rapidly. In a few decades perhaps, but certainly in a few centuries, supplies of oil and coal will have been used up and mankind will have turned to other ways of running its machines and warming its homes. Atomic reactors will be increasingly important. If thermonuclear power is ever practical—if the hydrogen fusion reaction of the H-bomb can be controlled—the sea is a source of vast quantities of the heavy hydrogen required to fuel it.

The immensely powerful pulsing of the tides has some possibilities as an energy source: a tide high enough can create a head of water to turn generator turbines. In one place in the world a plant has been built using the energy of the tides as the power source. This pioneer scheme is that of the Rance River, 2 miles upstream from Saint-Malo on the Brittany coast. In the Rance Estuary the maximum tidal range is 37 feet, and the water pours through a narrow opening into a 7¾-square-mile basin. A dam 2,329 feet long and 157 feet wide has been built, and the first power was produced in 1960. The project was scheduled for completion in 1963, at a cost of $100,000,000. The tidal power plant should save France a half-million tons of coal a year.

If the Rance scheme succeeds, the French will think seriously about an even bigger tidal-power proposal, which envisions enclosing the whole bay of Mont-Saint-Michel near the site of the Rance project. Here, 41-foot tides race into a basin of about 300 square miles. A 20-mile dam could produce about half of France's present electricity consumption.

If tides all over the world were harnessed, they could produce about 2 billion horsepower annually—about half the world's needs. Unfortunately, the areas of the world where conditions are favorable for obtaining power from the tides can be counted on two hands—with fingers to spare. The necessary conditions are a considerable tidal range and a large basin behind a narrow opening to the sea. Tides must rise not less than 10 feet to be useful as a power source, and 20 feet is a much more realistic minimum.

The few regions of the world seriously considered as potential tidal-power localities include Passamaquoddy Bay in the Bay of Fundy on the Maine-New Brunswick border. There is no area in the world where conditions are more favorable, and for nearly half a century serious efforts have been made to secure approval of plans to harness Passamaquoddy's tides. Early efforts failed when concern was expressed about the possible effects of a power project on the rich fisheries of the area. President Franklin D. Roosevelt, who spent his summers on Campobello Island in New Brunswick in the area of the great tides, tried to revive the scheme, but without success. In the late 1950's an International Joint Commission of Canada and the United States reviewed the idea once more. Advice to the Commission by fishery biologists was to the effect that the herring and most other fisheries of the area would not be adversely affected. Engineers reported the scheme feasible, and attached a price tag of from 566 million to 652 million dollars. Whether the "Quoddy Project" will ever come to fruition is still in doubt, but it holds more promise of economic feasibility than most other tidal-power schemes. Other possible areas include the Brittany coast area mentioned earlier, the Severn Estuary on the east coast of England, Cook Inlet in Alaska, and the San José and Deseada rivers in Argentina. There are at least two possible localities in the Soviet Union. One of these is Penzlinskaya Bay in Siberia, and the other near Kislogubsk on the shores of the White Sea. Work was reported to be actually under way at the latter site in 1963. Like the Rance River scheme, the Russian station at Kislogubsk is an experimental undertaking to investigate the feasibility of producing energy from the tides. Even if the tides in all the areas listed above were to be harnessed, they would produce, instead of half the world's power needs, perhaps 0.2 per cent. Some of these locations, moreover, are now so far from power-using areas as to be useless even if developed.

The tides are therefore not likely to be of widespread significance as a source of power. Of possible wider application would be a machine to take advantage of the temperature differences between the surface and the depths of the ocean. These differences can be as much as 40° to 50° F. in tropical oceans, and this is sufficiently great to drive turbines. An ordinary steam turbine produces power by burning fuel (oil, gas, coal) to produce steam at high pressure; this spins the turbines and generates electricity. The steam is passed to a condenser where lower temperatures change it back to water. A sea-water turbine could be set up in the tropics where the surface waters are about 80° F. This is too low a temperature to boil the water at atmospheric pressure, but not too low to boil it if the pressure is reduced in a partial vacuum. The 80° water would be run into a low-pressure boiler, forming steam whose pent-up energy would turn the turbines. Pressure reduction and condensation would be accomplished in condensers (also under partial vacuum) cooled by

[342]

40° water pumped from the deep sea. Someday our need for energy and the availability of more efficient machines may combine to make this source of energy economically feasible.

MINERALS FROM THE SEA

We are depleting other minerals besides oil, and we are already turning to the ocean to fill the gap. Magnesium and bromine are being extracted from sea water, where they are present in the form of salts. Sulfur and diamonds are already being mined from the sea bed. Recently researchers have seen indications of very large deposits of several kinds of useful metals on the sea bottom, some of which seem capable of supporting commercial mining operations. One sea-bottom mineral almost certain to have economic value is phosphorite, which has a ready market as an ingredient of fertilizer. Ores with a phosphorus content comparable to phosphorite mined in central Florida and Idaho are to be found on the ocean floor off Australia, Japan, South Africa, Spain, and both coasts of the United States. It could be used almost as it comes from the sea, needing only to be crushed.

In one area off southern California it is estimated that a billion tons of phosphorite exists over a 6,000 square mile area. Phosphorite occurs in the sea in the form of hard, glazed, light-brown nodules that look as though they were varnished. Though not a true resource of the deep sea since they are found only on offshore banks rising from the continental shelves, they may still lie in deep enough water that mining them will be expensive. This is the principal reason why large-scale economic operations have been delayed, despite excited predictions a few years ago that extensive operations would soon be under way. One attempt to establish phosphorite mining on Forty Mile Bank off San Diego ran into still another block—the existence of unexploded naval shells among the nodules, the area having been used as a firing range.

Some curious potato-sized lumps strewn over the bottom may be of even greater eventual importance. These are manganese nodules, and they occur in very large numbers on the floors of most of the oceans of the world, especially the Pacific. They are composed of as much as 36 percent manganese by weight, with important quantities of other valuable minerals including copper, nickel, and cobalt.

FIGURE 16–3. *Manganese lumps on the ocean floor.*
Official photograph, U.S. Navy

Major "ore provinces" in the North Atlantic occur on the Kelvin Seamount east of New York, the Blake Plateau off Florida and Georgia, and the Mid-Atlantic Ridge; in the South Atlantic on the Rio Grande Rise off the southern coast of Brazil and on the Angulhas Plateau south of the continent of Africa; in the Indian Ocean in the Madagascar Basin southeast of Madagascar and the Crozet Basin north of the Kerguelin Islands; and in the South Pacific on the Manihiki Plateau east of Samoa. None of these, however, is presently of economic interest because of low mineral content of the nodules. The only area of the ocean where the copper and nickel contents are high enough is a 500-mile band north of the equator, stretching from Central America to the Marshall Islands ($5° - 15°$ N and $120° - 160°$ W). Here the nodules lie half buried in siliceous (diatomaceous) ooze or in red clay. The nodules range from about one inch to nine inches in diameter; in a given area they are nearly uniform in size. They occur in patches, from several hundred yards to several miles in extent. The depths of the deposits vary from 12,000 to 18,000 feet, posing formidable engineering problems to the would-be miner.

But the returns may be worth the investment of a great deal of time and money to develop recovery and processing systems.

FIGURE 16–4. *Three kinds of machines that could be used to mine manganese nodules. At the left is a simple carpet-sweeper dredge that can be dragged over the ocean floor in shallow water. The middle dredge works like a vacuum cleaner, with motor and pump above the suction heads. Two floats, equipped with propellers, move the machine along the bottom. The right-hand dredge is a sleeker affair, with motor and pump inside a single float. Its suction heads sweep in circles.*
Scientific American

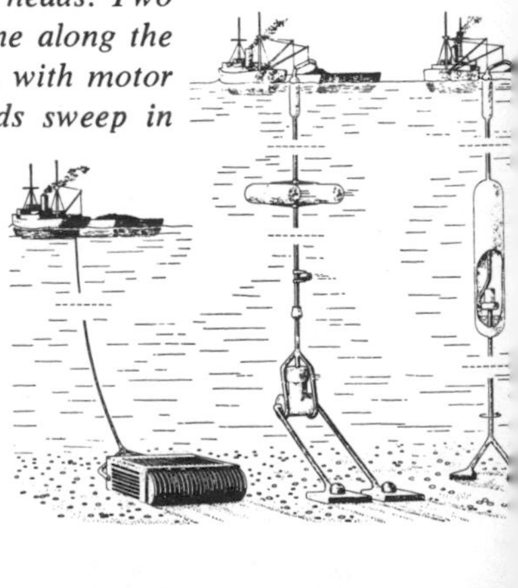

In the North Pacific ore province the mineral content of the nodules is about 1.2% nickel, 1.0% copper, 0.25% cobalt and 30% manganese, with traces of other minerals including molybdenum and zinc. One measure of the richness of these deposits is that some copper ores on land with less than 0.7% metal are being mined at a profit. The highest mineral content is exhibited in nodules occurring in the siliceous ooze—twice as high as those in red clay, and significantly higher than in nodules on the floor of other parts of the ocean. This difference may be due to the faster rate of sedimentation in the Atlantic Ocean as compared to that in the Pacific. The more sediment that is mixed into a nodule, the less metal. This relation stems from their method of formation. When cut in cross-section the nodules are seen to be composed of a series of onionlike layers. They are apparently formed when electrically charged particles of manganese dioxide and other minerals are accreted to some nucleus such as a piece of clay or pumice, or one of the shark's teeth or whale's earbones that occur so commonly on the sea floor as a result of their virtual insolubility. The growth of the layered nodule is exceedingly slow, varying from one millimeter (about one twenty-fifth of an inch) per thousand years to one millimeter per million years. Growth is faster if the surface of the nodule is free of sediment, since its electrical charge, and consequently its attracting property, is then greater. Since nodules are purer where there is less sediment mixed into them, the slower rate of sedimentation in the Pacific favors both the size and the quality of nodules.

The principal targets of mining will be copper and nickel, but the recovery of cobalt and manganese will add substantially to the value of the operations. Not only do some nodules have high metal contents, but they occur in stupendous quantities. Concentrations of 100,000 tons per square mile are common, and some deposits contain a minimum of ten billion tons of ore. The deposits of a single seamount could supply the world's manganese requirements for a hundred years.

Several countries have serious interest in mining deep-sea manganese nodules, with the United States at the forefront. The U.S. imports 10% of its copper, 88% of its nickel, and 100% of its manganese and cobalt—at an annual cost approaching a billion dollars. Four American companies have gone far enough in the development of mining and processing technology that they have announced readiness to begin commercial mining some time between

1977 and 1980, if national and international laws permit them to do so. These companies have mapped over 100 potential nodule mining sites in the Pacific, each with a life span of 20 years or more. These areas have at least two pounds per square foot (10 kg. per square meter) of nodules; mining operations are being designed to produce 5,000 metric tons of nodules per day.

Several types of deep-sea mining devices have been proposed (Figure 16-4), and over thirty U.S. patents had been granted by 1972 on equipment for mining or handling manganese nodules. Two general types of recovery machines have been successfully tested. These are mechanical dredges that dig the nodules off the sea bottom and lift them to the surface, and hydraulic devices which skim off the surface layer and lift it as a slurry.

One mechanical system is a continuous bucket dredge consisting of a long loop of rope hung over a platform at the sea surface. Attached to the rope are drag buckets which excavate nodules and mud from the sea floor; these are carried to the surface when the rope loop is rotated. In the hydraulic system slurry consisting of nodules, mud and water is forced to the surface either by the action of pumps in the pipeline, or by forcing air or water into the pipe under pressure. The nodule-gathering devices can either be towed or self-propelled by a power source on the sea floor.

Two nontechnical obstacles confront would-be deep-sea mining companies: the necessity to comply with environmental protection regulations, and the uncertainty about national and international legal constraints. The mining operations will disturb or kill plants and animals that may be associated with the manganese nodules, and will produce a plume of sediment that may have adverse effects on both bottom-living and midwater organisms some distance away. Since the number of living creatures is relatively small at the depths where mining will take place, and the amount of area affected will also be small, it is possible that little impact is to be expected. Nonetheless, this must be evaluated, and controls designed. United States Government agencies are conducting the necessary studies for this purpose.

It is unclear who owns the minerals of the sea floor beyond a depth of 200 meters (650 feet)—the limit of jurisdiction set forth in the 1958 U.N. Continental Shelf Treaty. At the 1974–1975 Law of the Sea Conference there developed general agreement that such resources are the property of all mankind. From that point, however, there is much dispute and uncertainty. Many countries insist that deep-sea mining must be controlled by an international agency, which will set performance and environmental regulations, and levy royalty or other charges. According to some proposals the income from these charges would be shared among the countries of the world, including those without coastlines. Meanwhile, United States companies are eager to begin commercial operations, and while U.S. Administration policy is to wait for decision from the Law of the Sea Conference, there are many, including influential members of Congress, who are impatient with the slow progress of international negotiations, and who want operations to begin. Such mining would be regulated by the U.S. Government, with suitable modification of the rules if the Law of the Sea Conference approves a treaty. It may well be that deep sea mining will be under way before this decade ends.

In 1974 and 1975 most of the nations of the world took part in the Third Law of the Sea Conference in Caracas and Geneva. It discussed the proposal that resources beyond the limits of national jurisdiction (which may turn out to be 200 miles, since that was one of the few things on which there was wide agreement) should be the common property of mankind. Nations or companies wanting to exploit these resources might have to obtain a permit from some global agency, and pay royalties to the community of nations for their harvest. Despite months of meetings, no treaty was agreed upon, and the Conference reconvened in the spring of 1976. Meanwhile, several countries are nearly ready to mine the deep ocean, including four companies in the United States, who expect to be capable of commercial operations by 1977–1980.

Diamonds hardly seem a marine resource, but they are now being "mined" from the ocean floor. Prospecting off the west coast of Africa began in October 1961, and results were so good that a sizable operation was launched. Production began in August 1962, and in the first three weeks 2,400 karats of diamonds worth about $100,800 were taken. The rich hauls were made in a stretch of coast 5 miles long by ½ mile wide, 60 miles north of the mouth of the Orange River in Southwest Africa. The Marine Diamond Corporation, the South African company engaged in this venture, has surveyed 720 miles of African coastline, and has located diamonds from the Olifants River in South Africa to Lüderitz in Southwest Africa. Soviet interests are also reported to be harvesting the gems off Africa. Lack of international rules governing exploitation of sea-bed resources beyond the limits of territorial waters may cause some severe legal problems before agreements can be worked out.

FIRST LINE OF DEFENSE

Yet another very practical reason for learning more about the sea is its importance to national defense. Security depends to an enormous degree on our ability to control the oceans, and this control will be impossible without intimate knowledge of the sea—knowledge that we are far from possessing now. The "capital ship" of the navies is no longer the battleship, as it was early in World War II, or the aircraft carrier, as it was by the end of that conflict, but the submarine. The development of atomic power has helped free the submarine of its dependence on the surface. The arrival of the era of the nuclear submarine was symbolized on January 21, 1954, when the United States launched the nuclear-powered atttack submarine *Nautilus*.

This dramatic event, the immense significance of which escaped the attention of much of the American public, went further than putting the submarine in the place of the aircraft carrier as the main unit of the navy. It pushed the navy itself into the dominant position among the three military services. No one would seriously suggest scrapping the air force or land-based ballistic missiles, but atomic submarines, with their Polaris and newer Poseidon missiles, have been important foundations of United States strategic policy.

Of greatest importance, Polaris and Poseidon may have cancelled what was the Soviet Union's principal strategic advantage: the surprise attack. These missile systems are not vulnerable to sneak attacks, since the submarines are will-o'-the-wisps, forever on the move. Enemy missiles have great difficulty zeroing in on them, and, as the submarines carrying these weapons occupy the whole of the world oceans, it becomes impossible for any enemy to destroy a large proportion of our retaliatory strength with a sudden "first strike."

America's allies have exhibited less and less enthusiasm about allowing United States bases to be maintained on their soil, for fear of retaliation. The new, larger Trident submarine, with its long-range missiles, will eliminate the necessity for overseas land bases. It is larger than a heavy cruiser and costs more than a billion dollars.

The Soviet Union has developed similar awesome weapons. Furthermore, she has spent much energy developing systems to locate our Polaris/Poseidon subs, and to keep them under surveillance. Russia's submarines greatly outnumber ours; she has far more than Germany had prior to World War II. The sea will be the battleground in future wars, if they are fought. To prevent such wars, we must understand the sea.

Apart from the grim realities of defense, the necessity to be first in the race for knowledge about the sea stems in part from the contest for international prestige. We learned to our surprise and chagrin that the standing of the United States with the other countries of the world was greatly affected by the early successes of the Soviet Union in the development of space vehicles. Now we are faced with a similar contest in the science of oceanography. Starting far behind us, the Russians have advanced the level of their oceanographic research at an extremely rapid rate, so that they are now advancing much faster than we are. In ships and research facilities they are considerably ahead, and even though the total accomplishments of Soviet oceanographers cannot yet be compared with those of the Western world, they show every sign of being capable of catching up and surpassing us. The Soviet Union has four times as many ships capable of deep-sea research as we have; marine biology, oceanography, and fisheries are studied at nearly 100 Soviet institutions of research; by recent count they had 800 scientists engaged in the study of the ocean, compared to about 600 in this country.

This struggle for world prestige is not necessarily bad. A substitution of this kind of constructive intellectual struggle for the bloody carnage of a hot war as a method of international conflict would be welcomed by a fearful world. Science is a fruitful area for international cooperation. Like music, it is an international language; scientists in Canada build on discoveries made by their colleagues in Germany; Soviet advances help progress in the United States. A notable spirit of good will exists among members of the scientific community in different countries. If politicians would show the same spirit of friendliness, progress in international relations would progress far more rapidly. And oceanography has been one of the areas in which international cooperation has been greatest. The scientists of the sixty-six nations that took part in the International Geophysical Year showed a tolerance and cooperation unparalleled in history, and this same spirit is continuing. The current concerted attack on the mysteries of the Indian Ocean involves participation by European, Asian, African, and North American nations and Australia. This has included such opponents as Germany and France, Australia and Japan, and the Soviet Union and the United States.

17　The Future in Oceanography

Oceanography has lagged behind some other areas of scientific endeavor. There are a number of reasons—or excuses. It is difficult to study the seas; yet this is not a fully valid reason, since oceanography is no harder than space research. Still another reasonable explanation for slow progress is the shortage of the proper tools and apparatus. But to blame lack of progress on shortage of equipment begs the question, since if there were a burning desire to solve certain problems in oceanography the researcher would design the necessary machines. It is true that progress in science must finally await the advance of technology; developments in instrument design and research techniques produce spurts of progress in the theoretical and practical aspects of a science. But the interest must be there as a spur. Studies of the deep sea have lagged primarily because we are less interested in things we cannot see—another case of out of sight, out of mind.

In recent years the United States scientific community has taken a closer look at oceanography, has crystallized the problems to be solved, and has turned its imagination to the invention of new tools to help in the solution of these problems. The trend in modern science and technology, including the marine sciences, is away from the tradition of the past where a man usually worked alone, and toward cooperation by a group of specialists from many fields studying different aspects of a problem. Moreover, much of the research in oceanography requires that the scientists go to sea for extended periods on highly specialized ships, surrounded by masses of complex gear and equipment. Both of these characteristics make oceanography expensive, so that for most investigations it is only the federal government that can support them. It is becoming increasingly apparent that this country cannot avoid the responsibility for a vigorously pursued research program in oceanography. Some of these reasons have been discussed in the previous chapter—the need for food, minerals, energy, atomic-waste disposal; the urgency of knowledge of the sea for national defense; the necessity to learn about the ocean as a major part of our globe. We need oceanographic research in pursuit of knowledge both for its own sake and for the practical reason that it can lead to a better, more comfortable world.

CURRENTS AND WEATHER

Research on movements of the ocean must be a major part of this enlarged program in oceanography. "The quiver and beat of the sea" has fascinated poets and ordinary men alike ever since they were capable of thought. In recent years there has been a growing realization that the flow of ocean waters has a profound effect on man, on where and how he lives, on his future comfort and prosperity. The pattern and the character of ocean currents, in combination with the air masses, determine the climates of the various areas of the world, making some fruitful and pleasant places to live and rendering others deserts. There are deserts in the sea too, created by ocean currents that prevent the surface waters from being supplied with nutrient chemicals. Ocean currents can also be vehicles for radio-active chemicals; whether debris from shore-based atomic plants and from wrecks of atomic-powered ships accumulates in lethal concentrations or is dispersed in harmless dilutions depends on the behavior of the sea in motion.

We know much about the surface currents of the ocean, but we need to know a great deal more; and our knowledge of the currents of the deep sea is so fragmentary that we are really only beginning the long and complicated studies that will be required to fill in this part of the puzzle.

Climate and weather are subject to changes in the atmosphere, the envelope of air surrounding the earth. The temperature of the air, its moisture content, and its movements determine weather and climate; cycles and patterns of these variables are reasonably constant and predictable. Variations in the patterns occur, however, and these are sufficiently capricious to make weather forecasting a notoriously hazardous occupation. The lot of the weatherman would be easier if we understood the laws that govern the behavior of the atmosphere.

We will be very much closer to understanding the laws, and consequently to forecasting the weather better, when we learn the effects of the sea on the atmosphere. The atmosphere and the ocean form a vast heat machine with an enormously complicated system of wheels and gyres. Some of these move quickly and some are ponderous; some cover considerable portions of the globe and others are tiny, whirling in back bays or small mountain valleys; some move parallel to the surface of the globe and others rotate vertically. The great geared wheels of the ocean help to spin the currents of the air, and the air currents in turn urge the water masses into vast patterns of motion whose energy is transmitted to other water masses in an endless series of whorls.

Air currents and water currents are thus simultaneously parent and child of each other. The initial thrust is given to these currents by the radiant energy of the sun, nearly three quarters of which falls on the sea. Solar energy is prodigious: the energy shed on Bikini Atoll by the sun in 2 days is greater than that released by the two atomic bombs that brought World War II to an abrupt end in Japan. About a quarter of the energy falling on the face of the oceans is used in evaporating sea water. When this energy is in turn released as heat by condensation, density changes occur that are a principal factor in creating and maintaining the winds. The winds pushing steadily over the sea create the major features of the current systems of the oceans. The surface transport of heated waters carries ameliorating warmth to bathe the land masses in high latitudes, and this heat is again released into the air so that some fortunate areas of the world have genial climates that would otherwise be frozen wastes. By this means the Gulf Stream makes Norway a country of green grass and fat farms, while Labrador, in the same latitude but far from the influence of the tropic-warmed waters, is a notoriously icebound land.

When air currents shift their patterns so that areas of evaporation are changed, weather may be affected for long periods—up to months in duration. If the quantities of cold water sinking into the ocean depths around the Antarctic Continent and the temperature of this water vary from year to year, the whole oceanic circulation may be affected for perhaps as long as decades, perhaps altering the climates of the world. Understanding currents is a prerequisite to a better understanding of long-term climate fluctuations and consequently of how to predict them.

The famous Mark Twain quip that everybody talks about the weather but nobody does anything about it has lost its edge in the last few years. Rainmaking has allowed man to alter his weather to some extent; rainfall has been increased in some mountain areas of the western United States by 10 to 20 per cent. Control of weather on a more massive scale, to the extent of altering climate itself, has been discussed by scientists. This could be accomplished, at least in theory, by diverting ocean currents at strategic places or by altering the present patterns of distribution of warm and cold water and ice-caps. Polar regions might have their climates ameliorated by melting the great sheets of ice that keep them refrigerated. At least two ingenious methods have been suggested for this. One is to dust the white glaciers, which now reflect the heat, with black carbon, which would absorb the radiant energy of the sun and melt the ice. The other is to create a cloud of fog over the polar regions by exploding a "clean" nuclear bomb in the sea. This cloud would create a blanket over the land that would prevent heat from being radiated into space and thus warm the earth so that the ice would melt.

There is actually no reason why some such procedure could not be carried out, and the result would almost certainly make the polar regions warmer. Whether this would be to the general advantage of mankind is open to serious doubt. The climate in one region is influenced by that of other areas of the world. If the climate near the poles were wrenched out of its normal pattern it would push conditions out of balance in adjacent regions, and this in turn would change the climates of areas next to them. The climates of the whole world, to the equator and beyond, would be knocked out of kilter like a row of dominoes falling. Scientists even talk about a new ice age being triggered by the freeing of moisture from the polar caps into the atmosphere. Furthermore, there is enough ice locked up in the Antarctic Continent to drown New York, London, Tokyo, Buenos Aires, and many of the other major cities of the world if it were melted; the oceans would rise 200 to as much as 500 feet above their present levels.

Soviet scientists are said to have considered thawing the frigid northern coast of their country by exchanging the cold water of the Arctic with warmer water from the Pacific. Here again the local effects might be beneficial, but the impact on the rest of the world could be disastrous. The same may be said about a visionary plan to construct a dam across the Strait of Gibraltar. Its purpose would be to create hydroelectric power, but its effect on climate could be of great consequence. Henry Stommel of the Woods Hole Oceanographic Institution has described how such a dam might change the whole future of mankind. A deep current flows out of the Mediterranean carrying heavy saline waters that help make the Atlantic the saltiest ocean in the world. If the saline waters were prevented from entering the Atlantic, in as little a time as 30 years this ocean would come to be no more salty than the Pacific. The resulting loss of density of the Atlantic surface waters might halt their sinking into the Arctic, leaving only the waters chilled by the antarctic ice to supply the abyss. Over some centuries of time the deep circulation might be profoundly altered. If the North Atlantic water stopped sinking, much of the warm surface water of the Gulf Stream might not reach northern Europe, and the climate there would be greatly changed. Moreover, Stommel wrote, "As a result of the reduction of heat transport to the Arctic, the ice packs covering the sea would grow. According to a new theory suggested by the oceanographer Maurice Ewing, this would lead to a decline of the glaciers on land and to a general warming of the earth." Stommel calls this proposed sequence "an entertaining fantasy," and expresses grave doubt that it could really happen. In reality, he says, the exact opposite might occur. The fact that a pair of opposite consequences can be argued equally well from the same supposition illustrates once more the poverty of our knowledge of ocean currents and of their influence on our environment.

Prediction of the behavior of the currents would enable meteorologists to make far better weather predictions and for much longer periods in advance. Whole trends in climate would be subject to prognostication. Even the eventual control of the productivity of ocean waters is contemplated, through the creation of artificial currents to produce vast upswellings of nutrient-laden waters. Such prediction and eventual control depend on the establishment of ocean weather stations. Some of these stations would report the usual data of meteorology, such as wind force and direction, temperature, barometric pressure, and precipitation; but they would also record data that have not ordinarily been recognized as pertaining to weather: water temperature, current force and direction, salinity, and other characteristics of the ocean waters. Some stations would be submerged and would report only the hydrographic data.

As in the air, where conditions may vary greatly from level to level, the temperature, currents, pressure, and other physical and chemical conditions change with depth in the sea. It would be ideal

FIGURE 17–1. *FLIP ship (FLoating Instrument Platform),
designed to serve as a steady, quiet platform for recording
underwater sounds. It can be turned upright by flooding its
ballast tanks. The upper end* (lower right) *contains living
quarters, machinery, and laboratories.*

Lawrence Barber

if continuous and simultaneous recordings could be made of these conditions throughout the whole water column, from the surface to the profoundest depths of the ocean. Since that is impossible, the longest practical structure will be used as an observation station. It will probably be at least 300 feet in length and may stretch down to 1,000 feet or more.

One imaginative kind of recording station called "FLIP" operates on the Pacific coast. Its name comes both from its official designation as a "FLoating Instrument Platform" and from its curious design, which allows it to be towed to station horizontally and then to be flipped upright by the flooding of ballast tanks. It does not serve as a permanent oceanographic observation point but instead is put in position for as long as 45 days at a time, after which it is towed back to its shore base. Its function is to serve as a stable and noise-free manned platform for delicate underwater acoustical observations.

In operating position, FLIP is a spar buoy 335 feet long, varying in diameter from 20 feet to 12½ feet in its submerged portion. At the upper part of the device, which remains above the surface when it is upended, there is room for living quarters, laboratories, and machinery. Other U.S. spar buoys with missions like that of FLIP include SPAR (Seagoing Platform for Acoustical Research), *Totem,* and POP (Perpendicular Ocean Platform). The French operate another one, *Isle Mystereuse,* in the Mediterranean.

Many of the conclusions of science must be inferred from things and events the scientist never sees. This is nowhere truer than in oceanography, where the mantle of the waters effectively blankets from view nearly everything beneath it. The oceanographer's reliance on instruments is therefore great. These instruments are rapidly becoming better.

In the latter part of the nineteenth century only the scantiest data were available about the depths of the sea, let alone the characteristics of the bottom. Since it took scientists on the *Challenger* as much as 2½ hours to make a single sounding with the clumsy hemp lines then in use, it is not to be wondered that only areas close to shore, where safe passage of boats required that water depths be known, were even partially sounded. Wire sounding lines that did not kink like those used on the *Challenger* improved this situation, but what really permitted a surge of new knowledge in this area was the advent of the echo sounder. Now ships traveling at full speed can continuously trace the contour and depth of the ocean bottom instantaneously and accurately. Moreover, the bottom type can be inferred, for muddy bottoms show up as a broad and fuzzy line on the trace, rocky areas as sharp and distinct records. As the echo sounders were improved they began to detect schools of fish and then even to distinguish different kinds of fishes. As still better echo sounders probe deeper they will add to our knowledge of the bottom, the overlying water mass, and the creatures on and in them.

But it is not enough to know the shape of the ocean bottom. We need to learn more about the nature of both the rocky basal structure and the overlying sediments of the floor in order to solve questions about the origin of the earth and the rest of the universe. Techniques are being improved for measuring the refraction of sound impulses through the rocky crust and mantle of the earth into its molten heart; the strange squiggles that are traced on recording paper when the signals are received yield information on the structure of the various layers. Similar data come from magnetic and gravity measurements, which are becoming more precise. Of great assistance in this work is the use of submarines from which to make the gravity measurements. Suspended in quiet water in the ocean depths, they can serve as instrument platforms unaffected by the motion of a rolling, pitching surface ship.

RESEARCH SUBMARINES

Submarines have been used in recent years for many other scientific purposes besides making gravity measurements. In 1959 the Soviets converted the *Severyanka* into an oceanographic research vessel, and other countries—notably France and the United States—followed suit. One special requirement for underwater research is operating depth. While military submarines can probably operate only to about 1,000 to 1,300 feet, for some purposes oceanographers would like to go much deeper. While most research submarines operate much shallower, some are designed to work down to 15,000 feet since this depth includes some 60 percent of the ocean floor. One deep-diving submersible has achieved fame by descending far deeper than this. The bathyscaph *Trieste* performed the stupendous feat of descending to 35,800 feet in the Challenger Deep off Guam, the

FIGURE 17-2. *Bathyscaph* Trieste *suspended from a crane. The observation sphere, 6 feet 7 inches in diameter, holds two cramped passengers and a tangle of instruments. It is suspended from a 50-foot flotation tank filled with gasoline.*

Official photograph, U.S. Navy

FIGURE 17-3. *RUM: the initials stand for Remote Underwater Manipulator. By remote control it makes observations of the sea floor, collects rock samples and living specimens, and assembles and maintains deep-sea instruments. The long jointed arm at the rear does the work, while television cameras serve as eyes for the scientist on shore.*

Official photograph, U.S. Navy

profoundest depth of the abyss. It was invented by Auguste Piccard, who had already achieved world fame through his balloon ascents into the rarefied air of the upper atmosphere before turning in the opposite direction, to the darkness and pressures of the ocean depths. To combat these pressures he built a sturdy steel sphere with walls over 3½ inches thick. The sphere, 6 feet 7 inches in diameter, is barely large enough to hold two cramped passengers and a tangle of instruments. Such a heavy device would sink like a stone in the water, so it is suspended from a big flotation tank, 50 feet long and 11 feet 6 inches in diameter, filled with gasoline, which, being lighter than water, is buoyant. Ten tons of iron pellets provide the ballast to sink the bathyscaph. These are held in place by an electromagnet; when the time comes for the ship to rise, part of the ballast is dropped. This arrangement provides a safety device too: if the power fails and the magnet loses its pull, the iron pellets drop, and the ship rises to the surface.

The epic dive into the Challenger Deep took place January 23, 1960. The *Trieste* spent 5 hours falling gently through 7 miles of Pacific Ocean water with two men inside: Jacques Piccard, tall son of the inventor of the bathyscaph, and Lieutenant Don Walsh of the United States Navy. Sitting tensely on stools in the cramped cabin, they eventually felt their ship settle gently onto ivory-colored sediment at the bottom of the sea. At that point nearly 200,000 tons of water pressed savagely against the shell of the cabin protecting the explorers. A fish of the sole family and a shrimp swam past the windows of the gondola while it was on the sea floor. After 20 minutes at the bottom of the world the explorers dropped their ballast and the *Trieste* rose slowly to the surface of the sea. From a calm unaffected by winds or waves for millions of years they came back to a storm-swept surface and world acclaim for a remarkable feat of modern exploration. Oceanographers had reached the limit of the abyss.

In 1963 *Trieste* was involved in another drama of the sea when she located the torn and scattered remains of the ill-fated nuclear submarine *Thresher*. *Thresher* ran into trouble during a deep test dive off the coast of Maine and sank to the bottom in a mile and a half of water—far below where her hull could withstand the crush of the abyss. Rushed from San Diego, *Trieste* searched the area of the disaster for months, finally locating and photographing the scattered fragments of the *Thresher,* and transmitting television pictures to a ship at the surface.

This experience stimulated the United States to a vigorous campaign to develop deep-diving rescue vessels, and gave impetus to efforts already under way to build small subs for research.

Aluminaut, constructed by the Reynolds Aluminum Company in 1965, is one. of these. She is 51 feet long and is designed to carry 3 to 7 men to depths of over 15,000 feet. Powerful lights illuminate the inky depths of the abyss, and direct observations can be made of things never before viewed by man in their natural haunts, except in rare and isolated pictures. During a dive in *Aluminaut* to the bottom of the Gulf Stream off Miami, I saw large, handsome crabs (*Geryon*) and big squid (*Illex illecebrosus*) at 2,600 feet. On the way down I saw bottle-nosed dolphins (the same engaging animals that are the stars of many marine aquarium shows) swimming curiously around the aluminum hull of the ship at depths much greater than it had been supposed that they could go.

A smaller research submersible, *Alvin,* has allowed scientists to make underwater observations of phenomena never seen before, gradually changing and improving our understanding of the world of the abyss. *Alvin* is just 23 feet long, and carries only three men, but she can dive to 12,000 feet. Biologist Robert Edwards and geologist Kenneth Emery of Woods Hole saw abyssal ridges off Norfolk that are believed to be remnants of beaches and oyster reefs as much as 10,000 years old, suggesting that the coastline of Virginia was once much farther offshore. The variety and numbers of the animals seen in the depths provided surprises. Red hake were unafraid of the little sub, darting quickly to catch the animals disturbed when *Alvin* touched gently on the sand as it drifted along. "There is nothing silver about silver hake in their natural environment," say the scientists. "They looked very much like blotched tomcod—brownish, with 5 to 7 irregular, darker, vertical bars. All the fins, but especially the 2 dorsal fins, had a luminescent greenish border." A big crab, *Cancer irroratus,* was common. It apparently resented the presence of the sub, taking an aggressive stand and offering to fight, with claws raised. The scientists even saw a "sea monster," 25 feet long and 6 inches in diameter, undulating slowly in midwater. The sub's mechanical hand caught a piece of this mysterious creature, which turned out to be a chain of unusually large salps, 5 inches long, and arranged in pairs. The chain was strong enough to be dragged along by the sub for more than an hour.

Alvin and *Aluminaut* are closely linked in the story of early research subs, since they were partners in a dramatic recovery of a hydrogen bomb. In early 1966 United States prestige was at stake when it became absolutely vital to recover an H bomb lost in the Mediterranean off Spain. A plane carrying four bombs collided with its refueling tanker, and both planes were destroyed. Three bombs fell on fields near the town of Palomares; the fourth splashed into the sea. Fortunately a fisherman saw it fall, and this gave searchers at least a chance. A 3,000-man, 25-ship search resulted in the eventual, triumphant recovery of it from a steep, rugged section of ocean floor. *Aluminaut* and *Alvin* were the most important ships in this fleet, and without them the bomb probably would not have been found. It was *Alvin* which actually located it and brought it to the surface.

In 1968 *Alvin,* with its hatch open, sank in 5,000 feet of water 120 miles southeast of Cape Cod when a cable broke as it was being lifted into the water. No one was in the ship but its own life seemed ended. In September of the next year, however, the Navy submersible *Mizar* located *Alvin.* Then *Aluminaut,* using manipulator arms, crawled up the hull of the stricken ship and attached a toggle bar to the open hatch. Seven thousand feet of nylon towing cable was attached to the bar and *Alvin* was hoisted and towed to safety.

Alvin continues to be in the scientific and everyday news. Her latest major exploit was to team with the French submersibles *Cyana* and *Archimède* in a three-year study of the Mid-Atlantic Ridge. In project FAMOUS (French-American Mid-Ocean Undersea Study) the three subs made fifty-one dives into the heart of the ridge, adding substantially to the information that now convinces scientists of the validity of the theory of continental drift.

American submersibles include *Shelf Diver,* built by the Perry Company in Florida. The design of this submarine allows divers to emerge from it at depths up to 800 feet, returning to a built-in decompression chamber. Canadian fishery scientists have used the ship to watch the reaction of queen crabs to various kinds of baits and to trace the pattern of scallop concentrations. Both of these activities can lead to more efficient use of seafood resources. The measurement of the success of herring spawning has been greatly improved by the use of *Shelf Diver.* Unlike Pacific coast herring, which spawn in the intertidal zone, those on the Atlantic attach their eggs to rocks and seaweed in water 40 to 60 feet deep. With the new sub, Canadian scientists swim out and make their estimates of the number of eggs far more accurately than with the old method of guessing from grab samples. An interesting sidelight is that, having descended in an enormously expensive underwater ship to the spawning area, biologists find the best tool to take samples of the gelatinous layer of herring eggs is a 69-cent cookie cutter.

Other countries have progressed in the development and use of research submersibles. The French Centre National pour l'Exploration des Oceans have put *Argyronète* ("water spider") into action. Like *Shelf Diver,* the French ship combines the functions of a submarine, underwater house, and decompression chamber. Divers can emerge at considerable depths, perform their work, and return to the ship for decompression and rest.

The Soviets have been in the game early, converting the W-type navy submarine *Severyanka* for service as a fisheries research vessel. Aboard the *Severyanka* scientists have floated in the midst of enormous herring schools, observing schooling behavior and watching as herring drifted, sometimes upside down, apparently lifeless or asleep or at least oblivious to the enormous strange fish in their midst. The action of trawls as they swept through the schools helped to design better gear. *Severyanka*'s usefulness has encouraged the Russians to build *Sever II,* a bigger and better research submarine.

In Canada the little two-man sub *Pisces* has had a colorful career. In 1969 it was the eyes and hands of a salvage crew that raised a 95-ton, 51-foot tug sunk in 670 feet of water. The Canadian government was willing to pay handsomely to recover the tug intact in order to judge what had caused the disaster—but to pay nothing if the ship was not raised in its horizontal position. Working in water thick with mud, the men in *Pisces* succeeded in attaching toggle bars and cables to the sunken tug and guiding its slow and careful rise from the bottom.

Newer and larger versions of *Pisces* have been built, and one was tested in 3,500 feet of water in Loch Ness, home of the famous Scottish "sea serpent." *Pisces* played a part in the movie *The Private Lives of Sherlock Holmes,* but her future seems to lie in more practical things. The latest design is able to dive to 6,000 feet.

The *Ben Franklin,* a descendant of the famous *Trieste,* gained its own fame in the summer of 1969 when it drifted from Florida to New England. For 30 days the *Franklin* allowed herself to be carried by the Gulf Stream, sometimes at speeds approaching 4 knots. Jacques Piccard, leader of the expedition and designer of the vessel, has stated that this accomplishment "represents a big step into the sea equivalent to Neil Armstrong's giant leap to the moon."

This seems considerably to overstate its significance, and *Ben Franklin* has since been retired from service, as has *Aluminaut*. But submersibles have a role to play in oceanography, even though it may be a lesser one than pioneer enthusiasts claimed. After some of the early spectaculars, it was believed that subs would be widely used in research, and in the 1960s United States industry invested over $100 million to develop more than 50 submersibles. Most of these were retired or scrapped since the demand for their use failed to materilize. This was partly because of the high cost of operating these vessels and their support ships, and because many of their contributions can be made better by surface ships and other devices. But about 1972, submersibles began to make a comeback as new ones were designed and built to match specific missions. In particular, the offshore oil industry finds submersibles highly useful, and 14 were operating in the North Sea in mid-1975. At that time, 70 submersibles were operational or under construction world-wide; of these, 29 were owned by the U.S., 8 by England and 7 each by Canada, France and the U.S.S.R.

ROBOT SHIPS

Some of the new oceanographic vehicles will not require the presence of man. New robot devices are more efficient for many purposes than the present trawls and dredges used in deep water. Observations of the living animals and of the inanimate environment can be made with a combination of television, photographs, and microphones. Arms guided from the surface can manipulate spades and trowels, nets and hooks, or a dozen other tools to collect specimens. Heavy robot machines can handle gear and equipment-assembly jobs, manipulated by a man far removed on a ship at the surface. One device of this kind, with the nautical-sounding name "RUM" (for "Remote Underwater Manipulator"), has been used off the coast of California. Looking like a small tank with a bewildering clutter of gadgets, RUM is capable of performing many skilled operations for the oceanographer, who remains safely and comfortably aboard ship.

TAKING THE SEA'S PICTURE

There comes a time for any scientist when he must check his theories and see how accurately his reconstruction approaches actu-

FIGURE 17-4. *The face of the abyss. These two remarkable photographs were taken in August, 1956, 24,600 feet deep in the Romanche Trench off Africa's Ivory Coast. Of enormous interest are the clear images of living animals at these great depths. In the top picture are three white animals: the largest is at the left edge of the dark splotch in the lower center; a second is on a small oval-shaped area near the upper-left corner; and the third is near the edge of the dark area in the upper-right corner. Animals in the lower photograph include a clearly visible brittle star.*

Photos by Harold E. Edgerton, National Geographic Society

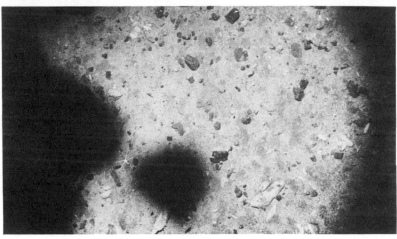

FIGURE 17–5. *Photograph of the deep-sea bottom on the Mid-Atlantic Ridge, at a depth of 11,200 feet. A compass is shown attached to a hose and wire from the ship.*

Woods Hole Oceanographic Institution

ality. The biologist becomes impatient with the limp and muddy-colored specimens in the museum bottle and yearns to see the flash of the bright colors and quicksilver movements of the live creature as it really is. The geologist wants to look at the bottom of the sea, to see how the sediments and rocks look in position and how animal tracks and burrows arrange and rearrange the ocean floor. This actual viewing of the undersea world is made possible by photography. There cannot be many environments that present more problems to the photographer than the sea: water, darkness, and pressure. But the challenge has been enthusiastically accepted. The first underwater color pictures were made on slow plates that required so much light that they had to be exposed for a full second at f/8 even in bright sunlight. For underwater work extraordinary amounts of artificial light had to be supplied: each exposure used a pound of magnesium flash powder. These pioneer underwater pictures were taken in 1926 by Charles Martin and Dr. W. H. Longley in the clear waters near the Dry Tortugas, west of Key West, Florida. Since that time enormous strides have been taken in underwater photography. Sophisticated equipment and techniques have permitted thousands of amateur and professional divers to take hundreds of thousands of high-quality pictures.

But the depth limit for most swimmers, even with an Aqua-Lung, is somewhere between 100 and 200 feet, with a handful of the most skilled being able to penetrate to about 300 feet for a brief period. When photography of the deep sea is attempted, the camera must be sent down without a man attached to it. This requires that the camera be made more complicated through the addition of remote-control devices. The problem of guarding the camera from water damage is greatly aggravated by the ever increasing pressures in deeper water. The complete darkness of the depths requires that large amounts of illumination be supplied under difficult circumstances.

[365]

One solution to the problem of a suitable light was provided by Dr. Harold E. Edgerton of the Massachusetts Institute of Technology. Dr. Edgerton developed a high-speed electronic flash lamp capable of producing brighter-than-sun light for exceedingly brief periods. This "stroboscopic" light had been used to stop the action of hummingbirds' wings, splashing water drops, and whirling machinery, and Edgerton adapted it to underwater cameras, taking pictures at three-thousandths-of-a-second exposure.

Photographs have been taken in very great depths. In the Atlantic Ocean Dr. D. M. Owen has made pictures on the ocean floor at 18,156 feet, clearly showing a 28-inch sea spider, smaller sea spiders, and numerous marks that were apparently the tracks of living animals. Another of Owen's pictures shows a 30-inch eel swimming across a section of the Gulf of Mexico whose floor is abundantly marked and pitted, making it quite clear that at 2,100 feet there are a considerable number of residents. Dr. Edgerton's deepest pictures were made at 24,600 feet in the Romanche Trench off the Ivory Coast of Africa. These historic photographs showed living organisms, the first to be depicted at such depths.

TELEVISION SPECTACULAR

Dr. H. Barnes of the Millport Marine Station in Scotland is given credit for the first use of television in marine science. In 1948 he demonstrated underwater television in the London Zoo, of all places, and carried on the first experiments in the sea in 1950. But it was difficult to arouse much interest in this new idea among people with money to help the development, and it took a tragic naval accident to demonstrate television's underwater potential. In June 1951 the British submarine *Affray* was sunk in the English Channel in 280 feet of water. Salvage operations were delayed because the precise position of the wreck could not be determined; sonar could not distinguish the sub from the many other derelicts cluttering the bottom. Someone proposed television. The British Admiralty went into action and in the amazingly short time of 3 weeks had solved the many technical problems: the name *Affray* was thrown dramatically on the television screen of a searching vessel at the surface, identifying the wrecked sub in the most positive manner. Underwater television was on its way.

The progress of fishery science has been aided by motion pictures, made by frogmen, of commercial fishing gear in action. A picture made in Scotland of a net called the Danish seine made history by revealing how fish react to the presence of the net; as a result it was possible to improve the design of the gear. The television camera can take the place of the frogman. Some of the most interesting of such pictures were made in 1958 by scientists of the United States Fish and Wildlife Service. The camera was fastened so that it spied on an "otter trawl" being dragged over the sea bottom off Massachusetts, and views of the strains and wobbles of the net under fishing conditions have been of great value to gear specialists in improving design. Of even greater interest is the behavior of fish pursued by the trawl. Movies show the haddock casually swimming in and out of the mouth of the net, apparently unaware of their danger. Later, when they are deeper in the throat of the trawl, they show the effect of efforts to keep up with the movement of the net, which they now cannot avoid. Eventually they struggle, exhausted, against the end or, if they are small enough, pop through the meshes to freedom, perhaps wiser for the experience.

BORING INTO THE PAST

One of the great research tools of the geologist is the core, a cylinder of sediment reamed up by a device like a giant apple corer. Its layers represent successively older strata deposited in ancient eras. Cores from the deep sea are especially valuable to the student of fossils. Over millions of years, as the sediments collected layer on layer, the plants and animals that lived in each era left their representatives buried in the rocks and ooze. The kinds of creatures appearing in each age tell the scientist a great deal about how life progressed and what conditions were like in each time, and will eventually lead to an understanding of how life began. In some places the sediments may have been left undisturbed since earliest geological times, providing the scientist with an unmatched opportunity to trace the history of the earth. Because of the amazingly slow deposition of sediments, a fraction of an inch of material from the sea floor can represent thousands of years of the earth's history, while a 10-foot core penetrates back through fantastically long periods of time. Probing through more and more of these layers of sediment is like flipping back through the pages of ancient history books, and geologists are seeking longer and longer cores.

Scientists have been puzzled about the depth of sediments on the ocean floor. They seem to be much too thin, judging from the known rate of deposition and the apparent age of the ocean. But now geologists believe that this is evidence for the movement of the great plates that comprise the earth's crust, and of the continents that ride on them. Up to now we have been able to obtain a fossil record on land only as far back as the Paleozoic Era, some 500 million years ago, when all the phyla, or great groups of animals and plants, had already appeared on the earth. In the sea and in "oceanic" islands the record is even less complete: so far we have recovered fossils only about 100 million years old—only a tenth of the 1,000-million-year period during which life is believed to have existed in the sea. That we have not found the older fossils is believed now to be because the sediments are carried atop the gigantic conveyer belts of the moving plates, to be swallowed up eventually in the disposal system of the abyssal oceanic trenches. In any case, we lack records of the immensely long period during which the simple animals and plants were branching off to develop into the more complex groups.

Fossils chipped from the rocks on land are derived mostly from areas that were once under shallow seas and have since risen. In shouldering above the surface of the sea, the rocks have exposed themselves to the eroding forces of the land, and many thousands or even millions of years of fossil records have been worn away. In the deep sea, on the other hand, where the bottom has never been above the surface of the ocean, it was once hoped that there would be a continuous and uninterrupted series of fossils in their proper order, and perhaps it remained only to probe deep enough to bring these to light. But now it seems likely that no matter how many cores we might obtain we will not be able to get an uninterrupted record back the full way to the beginning of life, since the older ones have long since been destroyed in the trenches.

As recently as the nineteen forties deep-sea cores were only 6 or 7 feet long. These were obtained by allowing a hollow tube to fall onto the sea floor, penetrating it by gravity. Swedish oceanographers improved on this considerably. The round-the-world *Albatross* Expedition of 1947, which had as its principal objective the securing of long cores, employed a piston corer developed by Hans Pettersson, the leader of the expedition, and Dr. Borje Kullenberg, who was later Director of the Oceanographic Institute in Göteborg, Sweden. The new corer consisted of a long, hollow steel tube about 2 inches in diameter, incorporating a piston to overcome internal sediment friction. With this device core samples of 50 to 60 feet have been taken in many of the oceans of the world. Longer cores are becoming common; the Soviet scientists Sysoyev, Kudinov, and Zenkevich recovered cores up to 112 feet long in the early 1960's. The piston corer has a disadvantage in that it may distort the sediment core and thus mislead the geologist who attempts to interpret history on the basis of the relative positions of the various sediment layers. In addition, the sediment penetration is limited to cores of only about 100 feet. To overcome these problems, new tools have been developed. Deep-sea geological projects have abandoned the hydrostatic corer, instead taking sediment samples by drilling. Enormously long cores can now be taken by this method. The deepest penetration made at the time of this writing was 4,631 feet in the autumn of 1975, by the *Glomar Challenger*. Such long cores are difficult and expensive to handle, and if they penetrate some strata which are relatively uninteresting geologically, that part is washed away and not preserved for study.

INSIDE STORY

Beneath the soft sediment layers, the primordial rocks have another story to tell. From many kinds of evidence men have built up concepts of how the great depressions that now contain the seas were formed, what kinds of rocks form the earth's crust and the underlying layers, how old these rocks are, how mountains and ocean trenches are built. But these remain largely speculations assumed from indirect evidence, and the theories can only be finally checked by direct observations. Earth scientists itch to get their hands on bits of rock from deep in the earth's interior, to weigh them and measure them in a dozen other ways. Only by knowing for sure the density of these rocks and their chemical composition, radioactivity, magnetism, and electrical and thermal conductivity will geologists be confident of the kind of earth we sit on.

The earth is layered like an onion, as shown in Figure 1-2. The outer layer, the crust, is by far the thinnest, averaging only about 10 miles thick. Under the ocean the layers are, successively, the soft sediments about 2,000 feet thick in the Atlantic and about 1,000 in the Pacific; a "second layer," which may be consolidated sediments or other kinds of rock of some 9,000 feet thick; a layer of deep crust;

and then the Mohorovicic Discontinuity (the Moho), where the speed of earthquake waves suddenly increases from about 24,000 feet per second to 28,000, indicating an abrupt increase in density. This is considered the limit of the crust. Next comes the mantle, a layer of dense rocks 1,800 miles thick. The core of the earth consists of an outer section of molten rock about 1,360 miles thick, and an inner section of immensely dense solid rock that is about 815 miles thick.

One of the fundamental unanswered questions of science is whether the rocks beneath the Moho are of different chemical composition from those that overlie it, or whether the difference is no more than a phase change resulting from different combinations of temperature and pressure. Despite the relative thinness of the crust compared to the diameter of the earth (about 10 miles compared to some 8,000), its rocks are the only ones that scientists have so far been able to examine.

There was great excitement in the scientific community in the early 1960's over a proposal to drill through the crust of the earth into the elemental rock of the Moho—to drill a Mohole! This bold idea was first proposed by Walter Munk of Scripps Institution of Oceanography, and it led to the formation of the whimsically named American Miscellaneous Society. Despite their apparent lighthearted approach, as exemplified by their motto: "The ocean's bottom is at least as important as the moon's behind!" the AMSOC members had a serious and soundly based scientific interest in obtaining information on the ancient history of the earth. For several years the National Science Foundation financed the activities of the Mohole Project, which sought to drill a hole perhaps as deep as seven miles into the bottom of the ocean, in water three miles deep. The choice of the ocean rather than the land as the place to penetrate the Moho was made because the discontinuity is eighteen to thirty miles deep beneath the land but only three to seven miles under the deep ocean, beneath dense basaltic rocks.

After a great deal of hard work, scientists and engineers running Project Mohole succeeded in devising a workable system for anchoring a ship over a single spot on the ocean, and to drill a number of test holes. The anchoring problem was a difficult one. The ship had to be kept very steady over the drill hole; the slightest shift under the influence of wind or current could wreck the drill pipes and other complex apparatus. Apart from the prevailing currents at the surface or those induced by storms, there are in many parts of the ocean

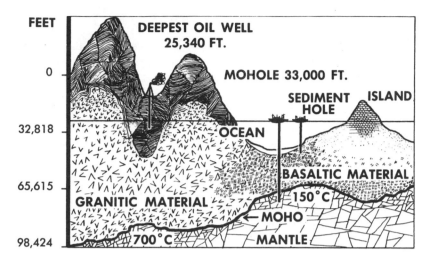

FEET

DEEPEST OIL WELL
25,340 FT.

0

MOHOLE 33,000 FT.

SEDIMENT ISLAND
HOLE

32,818

OCEAN

BASALTIC MATERIAL

65,615

GRANITIC MATERIAL 150°C

← MOHO

98,424

700°C

MANTLE

FIGURE 17–6. *The Mohorovicic Discontinuity, or Moho, divides the earth's crust from the mantle, which could reveal much about the origin of the earth. The Moho comes considerably closer to the bottom of the ocean (3 to 6 miles) than to the surface of the land (18 to 30 miles).*

deep currents, running in directions different from those at the top; in fact, there may be several different currents at various levels, all of different strength and going in different directions. They could bend a drill pipe into a mass of useless junk. It was estimated that the ship could move no more than 2 per cent of the depth of the water (about a 240-foot radius in 12,000 feet of water) without doing damage to the drilling equipment.

The first ship used in the Mohole Project was a sea-going barge called *CUSS 1*—the initials standing for the four oil companies that jointly owned it (Continental, Union, Superior, and Shell). Deep-moored radar and sonar buoys were set in a circle around the drilling site and any movement of the ship from the center of the ring was immediately revealed to the pilot on screens. He corrected for the movement by proper throttling of four very large outboard motors placed at the corners of the barge.

In 1961, *CUSS 1* drilled a 750-foot hole eighteen miles off San Diego, in water 3,000 feet deep. Later, off Guadaloupe Island near Baja California, holes 110, 234, 560, and 601 feet were drilled. Fossils ranging from 11,000 to a million years old were brought to the surface.

Later, Project Mohole fell into administrative and political bogs.

The cost of the project began to rise in a way that alarmed the National Science Foundation and Congress. Eventually officials were informed that the price tag would be three times the original estimates and the time of completion four years more than the first schedule. One congressman labeled the Mohole scheme "a rathole," down which he opposed pouring any more money. Congress killed the project in 1966. By that time it had also lost a good deal of scientific support.

But before the demise of Project Mohole its successor had been born, and the same important scientific objectives are now being pursued. The new research is partly an extension of Project LOCO (for Long Cores), conducted by geologists at the School of Marine and Atmospheric Science of the University of Miami. This program was designed to drill sediment samples several hundred feet long. In 1962 a 186-foot core was drilled in the ocean bed beneath 2,000 feet of water southwest of Jamaica, bringing up fossils several million years old.

Two years later LOCO was expanded into a bigger scheme called JOIDES: Joint Oceanographic Deep Earth Sampling. JOIDES research was planned and carried out by four oceanographic institutions: Lamont Laboratory of Columbia University, the School of Marine and Atmospheric Science of the University of Miami, Scripps Institution of Oceanography, and Woods Hole Oceanographic Institution. The National Science Foundation supplies the research funds.

The objective of JOIDES was not only to obtain very long core samples of the sediments, but also rocks of the deep ocean floor. From examination of these samples it was hoped to deduce the history of the ocean basins and the fundamental structure of the earth —including the characteristics of the rocks below the Moho. These are the same objectives which launched the Mohole Project, but the approach of JOIDES is different, and in the view of most scientists, more rational. It is to drill a series of deep holes, but not the super-deep holes of Mohole, in many places in the earth's crust beneath the sea.

The first drilling by JOIDES scientists was done in 1965 on the Blake Plateau off the coast of Florida. Long sections of pipe 3½ inches in diameter were fastened end to end and lowered to the sea floor. A diamond drill on the end of the pipe bit into the bottom, and cores, one as long as 1,050 feet, were obtained. Drilling was successful in water as deep as 3,500 feet. A much more sophisticated position-keeping apparatus was available now, a computer being used to command the adjustments necessary to keep the ship in position over the drill hole. A stable platform was maintained even at the edge of the Gulf Stream, where a current of three knots pushed against the ship.

The JOIDES project changed its name in 1968 to the Deep Sea Drilling Project (DSDP), and the *Glomar Challenger,* a ship owned by Glomar Marine Corporation, was adapted to the highly specialized job of penetrating the floor of the ocean. Her work started in the Gulf of Mexico and off the Atlantic coast of Florida. One of the most spectacular results was the discovery of oil in the sea bottom under very deep water, far offshore. Oil is usually associated with nearshore deposits and it had never been found under the deep sea. The discovery was made in the Gulf of Mexico, about halfway between Louisiana and the Yucatan Peninsula of Mexico. In that part of the ocean, which is nearly 12,000 feet deep, salt domes push up from the sea floor. These are called the Sigsbee Knolls (after the captain of Alexander Agassiz's research ship *Blake*). Traces of oil and gas were detected under nearly 500 feet of sediment. This find is probably of greater scientific than practical interest, since deposits of oil in the deep sea would be difficult to exploit.

But beyond the remarkable discovery of oil, the drilling of the *Glomar Challenger* has brought up cores of great scientific value. Samples—the oldest ever collected—obtained beneath 15,000 feet of water off Columbus' landing place, San Salvador Island in the Bahamas, lend credence to the theory of continental drift. Other samples taken in the Gulf of Mexico will help solve the question of how that body of water was produced—perhaps by the sinking of part of the North American continent, or as a gap in the earth's surface left when continents drifted apart. By mid-1975 the *Glomar Challenger* had completed forty-four legs of drilling patterns across most of the oceans of the world.

Cores taken in the Mediterranean Sea will be used to straighten out the geological record there and to determine whether the history of that inland sea is as geologists now speculate: that it was once a vast ocean called the Tethyan Sea, stretching all the way to the Himalayas. It seems to have been squeezed to its present limits by the slow movement of the continents, particularly Africa, whose enormous bulk is apparently pivoting slowly to the northwest, pushing the Mediterranean inexorably upon itself.

Phase IV of DSDP began in 1975. In this phase the plan is

to drill for the first time in the oceanic basement under the sediments on the ocean floor. This is what Mohole planned to do, so the scheme to penetrate the earth's mantle, abandoned when the Mohole project was stopped, is alive again. Mohole proposed to drill in water 14,000 feet deep, into the bottom to as much as 17,000 feet. *Glomar Challenger* has drilled in water over 20,000 feet, and has penetrated 4,631 feet into the bottom, so DSDP is well along the road to accomplishing the Mohole objectives. Unlike Mohole, the project continues to be enthusiastically supported, since it has achieved exciting scientific and potentially practical results.

Deep-sea research has a short history. It goes back only to 1872, when the voyage of the *Challenger* marked the first systematic assault on the problems of the ocean depths. For various reasons, including the expense of oceanic studies and scientific preoccupation with land areas and the easier-to-see skies, the science of oceanography has progressed more slowly than its earlier history promised, and much more slowly than is good for the welfare of mankind generally and the safety and comfort of our nation in particular. In the last very few years this neglect of the oceans has been brought forcibly to the attention of the public and of government leaders. No other branch of human endeavor offers more adventure, intellectual or physical, than oceanography. The Age of the Abyss is upon us, and it promises not only scientific advance and economic gains but even a new era of international cooperation. If nations must compete, let it be in the constructive and dignifying areas of science and not in the obliterating and degrading ways of war. Science can lead to a better world; war will lead to no world at all.

18 *Pollution of the Sea*

After World War II the total world catch of food from the sea increased year after year, at a rate that even kept ahead of the rapid increase in the human population. Then in 1969, for the first time since the war, the catch fell. Almost certainly, one reason is pollution. While we have been straining to harvest more of the ocean's food resources we have perversely been dumping more junk, more filth, and more poison into the very cradle of those resources. The sea is the ultimate dump: almost everything that man spreads on the surface of the earth, pours into rivers, or spews into the atmosphere ends up in the ocean.

There is disagreement as to how much damage has already been done and as to the future of the oceanic environment. Thor Heyerdahl, after his spectacular drifting journey across the Atlantic on the frail reed vessel *Ra II,* reported indignantly that he and his crew repeatedly encountered trash, oil slicks, and "black lumps" in the open ocean hundreds and thousands of miles from land. Jacques Cousteau, the colorful French popularizer of ocean exploration, declared in 1970 that "the oceans are in danger of dying. The pollution is general." In 1967 Lamont C. Cole, at a meeting of the American Association for the Advancement of Science, expressed concern that pesticides dumped into the ocean may be poisoning marine diatoms. By thus killing diatoms, he maintained, we might "bring disaster upon ourselves" by cutting off 70 per cent or more of the supply of oxygen produced through photosynthesis by plants. In 1968 Charles F. Wurster, Jr., of New York State University at Stony Brook, reported that DDT reduced the ability of four kinds of phytoplankton to carry on photosynthesis. Something approaching

[375]

panic followed. For example, Paul Ehrlich, the Stanford University population biologist, imagined a historian of the future recording that "the end of the ocean came late in the summer of 1979, and it came even more rapidly than the biologists had expected."

The DDT and oxygen cases have probably been overstated. Referring to laboratory evidence of the inhibitory effects of DDT on photosynthesis by one-celled marine plants, the panel for the Study of Critical Environmental Problems convened by the Massachusetts Institute of Technology in 1970 said: "It is doubtful that these results are ecologically meaningful. The concentrations necessary to induce significant inhibition exceed expected concentrations in the ocean by ten times its [DDT's] solubility (1 part per billion) in water." John H. Ryther of the Woods Hole Oceanographic Institution reports that although pesticides do affect the biological processes of marine plants, "several chlorinated hydrocarbons exhibit a selective effect on marine phytoplankton, strongly inhibiting photosynthesis and growth of some species while exerting no effect whatsoever on others. It thus seems more likely that pesticides and other toxic pollutants may influence the species composition of the phytoplankton than eliminate them entirely." Moreover, Dr. Ryther does not believe that man is dependent on the ocean for his oxygen supply. On the basis of an analysis of oxygen sources and the oxygen cycle, he says that "if all photosynthesis in the sea were to stop today, the total quantity of oxygen in the atmosphere could decrease by 10 per cent in no less than one million years."

But if the case has occasionally been overstated (sometimes deliberately, to jolt the public into paying attention to a critical problem) and if the real reason for the alarm has sometimes been stated wrongly, the gravity of the issue facing mankind is in no way diminished. There is nothing theoretical about the fact that DDT in the low concentration of 6 parts per million kills commercial shrimp, that this potent insecticide has been found in the bodies of penguins in the remote Antarctic, and that from 1½ to 6 million fish are killed in the United States by pesticides each year; nor is it overstatement to report that nearly a million cases of tuna were withdrawn from the market in late 1970, and that the American swordfish industry was nearly destroyed because mercury was detected in the flesh of these fish. Of the potential shellfish areas of the Atlantic coast of Canada, 25 per cent have been closed because of contamination. European scientists predict that sewage, pesticides, mercury, oil, and thermal pollution will soon

FIGURES 18 – 1 , 1 8 – 2 . *Waste water pollution from a paper mill in Oregon (above) and its results (below): dead fish, and a mat of sludge that has been lifted from the bottom by the gases of decomposition.*

C. P. Idyll

make the bottom of the Baltic Sea lifeless. And these spectacular disasters may be less serious than subtle, unseen damage. Pieter Korringa, the director of the marine laboratory at Ijmuiden on the Netherlands North Sea coast, says it is "the general trends, the stealthy deterioration of the environmental conditions in the sea that are of vital importance for its living resources."

The fact is that man is faced with a serious problem, and we must find better ways to dispose of waste than by simply dumping it into the sea. The runoff from the land carries waste to the rivers and thus to the ocean; waste discharged into the atmosphere falls or is rained into it; and man deliberately discharges sewage, garbage, industrial poisons, and a monstrous variety of other pollutants directly into it.

In the past people did this with serene confidence that the broad reaches and the abyssal depths of the ocean were vast enough to absorb whatever they were offered. And clearly, if the wastes— enormous as they are and increasing as they are—were evenly diluted throughout the 331 million cubic miles of the ocean, they would be undetectable and innocuous. But such dilution does not take place. Instead, by far the greatest quantities of mankind's noxious wastes are concentrated in the shallows next to land. Gradually, insidiously, the estuaries—the keys to much of oceanic production—are being rendered unaesthetic and unproductive, destroyed by pollution.

Many things foul the sea: sewage, industrial wastes, fertilizers, pesticides, silt, heat, trash, oil, and radioactive materials. These cause infections, rob the water and the living creatures of oxygen, cause frenzied outbursts of plant growth, cloud the waters with silt, poison animals and make them unfit as food; they render the habitat unlivable in scores of ways.

Pollution of the sea is increasing at a cancerous rate. The human population is doubling every thirty-five years, yet pollution of the sea is gaining much faster—some three times as rapidly. Despite recent efforts to control discharges, industrial pollution is rising at the rate of 4.5 per cent per year. The crowding of humans at the edge of the sea also contributes to the fouling of ocean waters. One third of the entire population of the United States lives in the coastal counties and one third of its industry is located there. Two thirds of the factories making agricultural pesticides are on coastal sites, as are two thirds of those that make organic chemical products, 58 per cent of those that make inorganic chemicals, 52 per cent of the petroleum refining plants, and two thirds of the nation's pulp mills. About the same pattern applies to other lands: seven of the world's largest cities are on estuaries and more than half of the world's population lives within 100 miles of a seacoast.

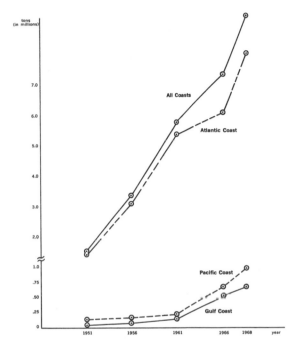

FIGURE 18–3. *Dumping of sewage, trash, and chemical wastes into the sea has increased enormously in recent years, and the future looks far worse. The highly industrialized East Coast feels the main impact of this pollution.*

Council on
Environmental Quality

PESTICIDE POLLUTION

In pesticides, as in so many other kinds of pollution, one man's pollutant is another man's useful—even necessary—product. The dilemma is particularly sharp in the case of that famous chemical, DDT. On the one hand, DDT is a highly effective destroyer of insects and other harmful pests. It has saved billions of dollars' worth of crops and livestock, helping to release millions of people from hunger. It has saved countless lives which would otherwise have been snuffed out by malaria, typhus, cholera, and other terrible diseases. And it has done these things more cheaply and efficiently than would have been possible with anything else that we know.

The trouble is that DDT does not stop at killing man's enemies. It has wiped out billions of other animals and has already pushed some species to the brink of extinction. It has poisoned our living space and threatened our environmental balance. Because it is almost indestructible, DDT piles up in soil and water and accumulates in the bodies of plants and animals. In the mere thirty-five years since it came into general use this chemical has spread over nearly the whole of our planet, even to areas once considered nearly immune to man's indiscretions: the Antarctic and the open expanses of the world ocean. And there is evidence that the long-term effects of DDT on the environment and on the reproduction, behavior, and genetics of the world's living creatures, including man, may be even more damaging than the deaths of fish and wildlife that have come to our attention.

DDT (dichlorodiphenyltrichloroethane) is the most important of the chlorinated hydrocarbons, and once was the most widely used class of pesticides. Its degradation products, DDD and DDE, and such related compounds as chlordane, dieldrin, aldrin, and endrin share many of its characteristics and have much the same effect on the environment. The chlorinated hydrocarbons kill by interfering with the enzyme acetylcholinesterase, whose activity is essential for the transmission of a nerve impulse from one nerve cell to another; in other words, they act somewhat as nerve gases do. The important thing about them is that they are "hard," or "nonbiodegradable" pesticides; they are slow to break down chemically and retain their killing capacity for years.

DDT was originally synthesized in 1874 but it was not until 1939 that the Swiss chemist Paul Mueller discovered its enormous potential as an insecticide—and received a Nobel Prize for his work. It was immediately put into use against the insects that carry such diseases as malaria and typhus—with electrifying success. It has wiped out malaria in many parts of the world, for example. As early as 1953 a health expert estimated that five million human lives had been saved and a hundred million illnesses prevented by DDT. And by killing pests that attack food and fiber crops it has increased farm yields the world over. Not only is DDT significantly more effective as a pest control than any other known chemical but it is also a great deal cheaper than the others.

If the benefits of DDT are tangible and enormous, so is the case against it. There is abundant evidence that it has killed hundreds of millions of animals besides those at which it was aimed—animals valuable to man economically or aesthetically or as vital links in the food chain.

The sea has been accumulating damaging amounts of pesticides. Their effects are worst in estuaries at the edge of the sea (because the greatest concentrations occur there from land run-off and because these are the richest parts of the sea), but there is disturbing evidence that minute but active quantities of pesticides occur even in the open ocean.

Animals carry DDT widely over the earth, and many of them accumulate it in high concentrations in their bodies. It is nearly insoluble in water, but highly soluble in fat—eighty million times more so in olive oil than in water. As a consequence it is stored in animals' fatty tissues—in fat deposits, in the adrenal glands, the ovaries, the testes, liver, kidneys, and thyroid. By accumulating DDT in their bodies, animals "magnify" its quantities and pass increased stores along in the food chain when they are consumed by predators. For example, an oyster continuously exposed to water with the extremely small concentration of 0.1 part per billion can concentrate the chemical in its tissues to 7 parts per million in about a month—a 70,000-fold magnification. If plankton containing 50 parts per billion is eaten by mullet, the DDT in the mullet can be as high as 1 to 15 parts per million, and if the fish are eaten by porpoises the level can go to 800 parts per million.

The consequence of such action has been that distressing numbers of the world's fauna now have accumulations of the poison. Of 169 oceanic fishes tested several years ago, 157 had DDT residues in their fat. Sea birds that feed only in the mid-Atlantic and in Antarctica had substantial residues; all ninety eggs tested from thirteen species of sea birds contained DDT; gray seals in Scotland had DDT in their fat; Weddell seals in the Antarctic showed traces of it. Humans have not escaped either, of course. As long ago as 1962 every meal analyzed by a group of researchers contained this chemical, and nearly every human showed some residues in his body. The levels were in the range of 10 to 12 parts per million—by U.S. Department of Agriculture standards, too high to be acceptable food for cannibals. Human mother's milk in this country has levels so high that it could not legally be sold for human use.

It is easy to see what the pesticides can do when millions of fish are destroyed and washed up on beaches or gathered in windrows in the water, as happened in the lower Mississippi River in the 1960's. Evidence is accumulating, however, that there have also been many unseen cases of destruction. Laboratory experiments show that minute quantities of DDT can kill aquatic organisms: as little as 0.6 to 6 parts *per billion* can kill or immobilize shrimp in two days. Many cases of damage to marine organisms go unnoticed because the dead or damaged animals are concealed by the water or because the effects are too slow to be detected. In laboratory tests, when oyster meats containing 4 ppm of DDT were fed to shrimp and fish, damage was slow but in some species half the population was dead after about four weeks. In nature such cumulative effects of DDT would be serious, but they would probably not have been noticed in a natural estuary.

FIGURE 18-4. *Gulf of Mexico oysters, like these in the Gulf Breeze Laboratory near Pensacola, Fla., stopped feeding and showed erratic shell movements when they were exposed to less than one part per million of DDT and other chlorinated hydrocarbons.*
Environmental Protection Agency

Public concern over the environmental effects of DDT was first expressed in the early 1940s, but it was not until Rachel Carson's *Silent Spring* was published in 1962 that real alarm was sounded. An expert committee appointed by the Government, and chaired by Dr. Emil Mrak of the University of California, concluded that the harm done by the hard pesticides is greater than the enormous benefits they offer. The Committee recommended that DDT and certain other persistent pesticides be eliminated over a period of two years, except for "uses essential to human health or welfare."

As a consequence, in 1972 the Environmental Protection Agency banned the use of DDT on crops. The ban did not include prohibition against export of DDT or its use in public health.

Immediately after the order was issued it was appealed—from

two directions. The industry, which had sold $1.7 billion worth of hard pesticides in 1970, sued to have the ban set aside, and the Environmental Defense Fund wanted the order to include prohibition against the few uses still allowed. A Federal Court ruled that there was "substantial evidence" to support EPA's action.

In mid-1975 EPA released a review of new evidence of DDT's environmental and health effects, and of the economic impact of the 1972 ban. According to the report ". . . none of the findings of the Administrator could be denied on the basis of new data." The economic effects had been substantial, but not crippling. At the time of its ban, 80 percent of the domestic use of DDT was for cotton. The cost of growing cotton rose $7.75 million per year in 1973 and 1974, about 0.5 percent of production costs. Each consumer of cotton in the U.S. paid 2.2 cents more per year.

As a consequence of the decreased use of DDT there has been a slow decline of the amounts of the pesticide in estuaries and the sea, and in animals, and this will continue in the United States. But DDT is still widely used in other parts of the world, so that it is still a global problem of considerable magnitude. In developing countries trying desperately to increase agricultural yields, DDT and other chlorinated hydrocarbons constitute over half the pesticides used. These chemicals are essential until effective, inexpensive substitutes are provided.

Apart from the economic considerations, and carrying far more weight with some concerned observers, a strong reason for caution against immediate total banning of DDT is that according to health authorities many countries simply have no substitute as a control for malaria and other diseases. When malaria seemed crushed in Sri Lanka (Ceylon), spraying was stopped. The result was two million cases of the disease in 1967–1968. Dr. M. C. Candau, Director of the World Health Organization, said that "the concept of eradication of malaria rests completely on constant use of DDT. . . . Limitations on its use would raise grave health problems in many developing countries."

There are a number of alternatives to the use of hard pesticides for the control of some of man's enemies. The most obvious is to substitute other chemicals that do not accumulate to poison the earth. Organophosphates and carbamates have proved to be effective pesticides, and they break down quickly in the environment. But they have strong disadvantages too: they are much more expensive than DDT or most of the other pesticides, and they are highly toxic to man. But the Mrak Committee held that it is "well within the capacity of the chemical industry, which annually screens some 100,000 new organic compounds as potential pesticides, to develop biodegradable pesticides of suitably narrow spectra of action."

No matter what advantages chemical insecticides may have, many experts believe this mode of control can only be a temporary solution and that other methods must be developed as quickly as possible. The alternatives to chemical controls are cultural methods and biological controls.

Cultural methods include such practices as crop rotation to reduce susceptibility to a given pest, or reduction of the area allotted to a single crop to encourage the presence of specific predators. Varieties of plants can also be developed that have high resistance to pests.

Biological controls depend on natural life processes. They are complicated and expensive to develop but they can be extremely effective. The biggest success story has been the control of the screw-worm. This pest is the larva of a fly that lays tiny eggs in the skin of cattle; when the maggots are hatched they eat the animal's flesh and can destroy a full-grown steer in ten days. After much experimentation, a method was devised of sterilizing large batches of male flies by irradiation. The flies are released, and when they mate with wild females the resulting eggs are infertile. Moreover, since the flies are monogamous, those females never produce young, and so the wild population is eventually decimated. In 1954 a test program on the Dutch West Indies island of Curaçao wiped out the screwworm there. In Florida the pests were eliminated in two years. Now a plant in Texas produces 200 million sterile male flies a day for release along the Texas-Mexican border.

Other biological methods include the importation of predators and parasites known to combat a specific pest; the use of sound, light, food, or even synthetic sex attractants to lure pests to their destruction; and the administration of synthesized hormones to interfere with the growth of some kinds of insects.

Chemicals have offered the easy and cheap solutions, and have turned us aside from the more difficult and slower search for biological controls. Now it is obvious that the cheap solutions are not the best ones; that they are, in fact, unacceptable to society. We have no choice but to give far more attention to research on biological controls.

[384]

In recent years a new chemical pollution threat has arisen, from the PCB's. These initials stand for the polychlorinated biphenyls, mixtures of synthetic hydrocarbons. The PCB's have many valuable industrial uses—in heat transfer systems, electrical devices, plastics and paints—but they also have some of the deadly characteristics of the hard pesticides as well.

Many of the PCB's are highly toxic to living things. Their presence in the unbelievably small concentrations of a few parts per trillion can be lethal to some species of shrimp and fishes. Moreover, even if they are not killed, animals exposed to minute concentrations of these chemicals may be rendered more sensitive to diseases and to changes in the environment.

The PCB's are like DDT and related pesticides in another unhappy respect, they are very resistent to degradation. They are therefore accumulating in an alarming manner in ocean waters, in sediments, and in the bodies of animals. PCB's concentrate in the fatty tissues, being magnified as much as 100,000 times the amounts present in the surrounding water.

Hence, much of the concern and research attention that was focussed on DDT is now turned to the PCB's. The nation has another series of hard decisions to make about the wisdom of the use of highly valuable chemicals in the face of grave threats to man and his environment.

OIL POLLUTION

Two catastrophes jolted the public into the realization that oil pollution poses a threat to the ocean environment. The first event was the sinking at sea of the enormous tanker *Torrey Canyon* in 1967, dumping oil on a great expanse of ocean and over beaches in England and France. The *Torrey Canyon* wreck was the costliest marine disaster in history. The *direct* cost of the cleanup to the British government was $8 million. The second was the blowout in 1969 of a well being drilled from an offshore platform in the Santa Barbara Channel. A runaway geyser of oil erupted 100 feet into the air, then streams of oil issued from breaks in the sea floor over an area of 50 acres, spreading a thick black scum on the beaches of the handsome California coastline.

There are many oil spills because a great deal of oil is moving around the world. In 1966, 3,218 tankers of over 600 tons plied the seas, carrying 700 million tons of oil; 20 to 30 million tons are at sea on any given day, with one vessel in every five engaged in this trade and one third to one half of all goods transported by sea consisting of petroleum products. The quantities are increasing rapidly every year. Estimates of the amount of oil spilled range from 3½ million tons to 10 million tons in a year.

But despite the drama and impact of the *Torrey Canyon* and Santa Barbara disasters, such accidents are relatively infrequent. Most oil pollution is not accidental at all: it is from oil pumped into the sea along with ballast water from empty tankers and washed out of the fuel tanks of many kinds of ships. A 50,000-ton tanker (not a large one, nowadays) can dump overboard as much as 1,200 barrels of oil at a time in this way.

FIGURE 18–5. *Collapsible 500-ton oil container inflated with air. This could be dropped beside a stricken tanker and the oil pumped from the ship into the container to prevent it being spilled into the sea.*

Uniroyal, Inc.

There is also some pollution from tankers sunk in World War II, from natural oil seeps in the ocean floor, and from spills near shore, both from ships unloading and from shore plants.

There is still uncertainty as to the damage caused by oil in the ocean. Petroleum is a mixture of a large number of components, principally hydrocarbons of different densities and chemical complexity. The most poisonous components are the "high" or light fractions, those with the greatest volatility. When these are present (either as components of crude oil or as purer fractions after refining), marine animals exposed to them are likely to be poisoned.

In some cases there may be massive kills of sea animals, as occurred off West Falmouth, Massachusetts, in September of 1969, when a tanker went aground and spilled about 170,000 gallons of fuel oil. Scientists from Woods Hole reported that 93 per cent of all marine life in the region was killed in three days. On the other hand, biologists believe there was no permanent damage to marine animal communities following the Santa Barbara incident.

If the oil loses its high-fraction components, so that only the thicker fractions remain, the toxicity to most marine animals is reduced, although many will still be killed by the direct physical effects of coating, asphyxiation or contact poisoning. In the *Torrey Canyon* spill, for example, the wind kept the oil at sea for nearly a week before it reached the intertidal zone and the beaches; many of the lighter components of the mixture had evaporated or dissolved in the sea water and the remaining components caused less harm than would otherwise have been the case.

The fact remains that the effects of oil on planktonic animals, including fish larvae, are hard to judge, and it may be that much more damage may result than is suspected.

There is certainly no doubt about the bad effects of oil on sea birds, which are destroyed when their feathers are covered with oil.

FIGURE 18–6. *Oil-soaked sand, and dead crabs and fish at Point Puntilla, Puerto Rico, following the wreck of the* Ocean Eagle.

M. J. Cerame-Vivas

The heaviest mortalities take place among diving birds such as guillemots, razorbills, puffins, scoters, murres, and penguins. The losses may run into millions of birds a year. Attempts to save oil-fouled birds have been nearly useless. Detergents and other agents used to clean the feathers interfere with the waterproofing properties of the plumage, so that the birds often die of exposure. The Royal Society for the Protection of Birds estimated that in the *Torrey Canyon* incident at least 10,000 birds were killed. About 6,000 had been treated, but after four months only 400 of these were still alive and some of them were dying.

For some time it was assumed that the best way to clean beaches was to flood the area with detergents; after the *Torrey Canyon* spill 3 to 5 million gallons were spread on the water and the fouled beaches. Long milky tongues washed out to sea and into adjacent coves in layers three feet thick. But now it is realized that detergents are more harmful than the petroleum itself, and that they intensify and prolong the damage done by oil. The detergents do not destroy the oil but merely disperse it, sometimes by sinking it into the sand or gravel, thus preventing it from being degraded by oxygen, and sometimes by killing bacteria that would otherwise break it down. One report states that "littoral life was relatively unaffected by the oil, even on grossly contaminated beaches. There was some damage to sea anemones but most molluscs were apparently unaffected and it was not uncommon to see limpets and winkles browsing and crawling on oil-contaminated surfaces. When the enormous quantities of detergents were dispersed along the shore, however, the scene changed dramatically. It was common to see vast areas of milky white sea and this, of course, affected areas of beach which had received little oil. Molluscs, crustacea, rock-pool fish, worms, sea anemones, seaweeds and other littoral fauna and flora were decimated."

The only satisfactory cure for oil spills is to prevent them in the first place. Once the material gets into the water and onto the beaches damage is inevitable. Stricter laws and enforcement are required. In 1970 the Canadian government proclaimed jurisdiction over vast areas of the Arctic Ocean off her northern shores in an effort to protect these vulnerable regions from possible oil-pollution damage. The development of the enormous Alaskan North Slope oil reserves, which seems to be on the verge of explosive activity, could result in the spilling of oil on frigid waters and shores where temperatures are not high enough for bacterial breakdown to occur. The specter of oil films remaining intact, gradually building up into deeper and more destructive layers, persuaded the Canadians to take a drastic unilateral action which may be fully justified for the sake of environmental protection even if it upsets international custom and law.

Many kinds of booms or barriers have been used around oil spills in attempts to contain them physically so as to prevent spreading. This system works only in calm weather; if the seas rise only a few feet or the current runs faster than about one knot, booms are useless. And unfortunately disasters usually take place during storms. Various materials have been employed to disperse or absorb floating oil, including sawdust (the French used 7,500 tons on the *Torrey Canyon* effort), volcanic ash, expanded polyester chips, chalk, cork chips, and foam rubber. The British Royal Navy and the

FIGURE 18 – 7. *An inflatable boom is used to prevent the spread of oil from the* Torrey Canyon *spill. Only in calm water like this are booms effective in containing oil.*

Frederick J. Warren

[389]

RAF tried very hard but with limited success to burn the *Torrey Canyon* oil. They dumped enormous quantities of incendiary materials on it: 160,000 pounds of high explosives, 10,000 gallons of aviation kerosene, 3,000 gallons of napalm, but failed to start any substantial fires. The lack of success was partly because of the length of time the oil had been in the water; less than 4 per cent of the light fractions that boil below 100° C. remained. Moreover, the detergent treatment had produced emulsions of oil in water that greatly reduced the flammability of the mixture.

Left to itself, oil gradually breaks down under the influence of oxygen and bacteria, and so if conditions permit, the best way to get rid of it is to let it decompose naturally. Spills far at sea should be left alone; even on beaches this is the best treatment if it is practical. After the *Tampico Maru* spill in March 1967 there was no clean-up program, but a year later the oil had disappeared.

Bacterial destruction of oil is up to ten times faster than that caused by atmospheric oxygen. Various kinds of bacteria and several other kinds of microorganisms attack the hydrocarbons. The rate of destruction depends on environmental conditions, especially temperature; the rate is slow below 5° C., so that there is little if any such decomposition in Arctic or Antarctic regions. Thin layers of crude oil are colonized by bacteria in one or two weeks and decomposed in

FIGURE 18-8. *Modernistic designs are formed on the surface of the water in Porthleven harbor, England, from black oil and white detergent oil emulsion, and an ugly black stain of oil covers the lower half of the harbor wall. This is part of the oil from the wreck of the* Torrey Canyon *in 1967.*

Geoffery W. Potts

[390]

two or three months; it is possible to speed up the colonization, but not the decomposition. Emulsification increases the surface exposed to oxidation, but if it is accomplished by detergents the active bacteria may be killed, and then the advantage is lost.

Control of marine pollution is an international problem. Great impetus for cooperation here was given by the UN Conference on the Human Environment, held in Stockholm in 1972. It resulted in the formation of a permanent organization, the UN Environmental Programme, to coordinate international activity. Then in 1973 the International Conference on Marine Pollution, held in London, led to the International Convention for the Prevention of Pollution by Ships. Marine pollution control has also been receiving close attention by the UN Conference on the Law of the Sea during its sessions in 1974, 1975 and 1976. Thus, while these problems have not yet been solved, they are receiving intensive study.

RADIOACTIVE WASTES

The nuclear era spawns radioactive waste, much of which will find its way to the ocean unless heroic preventive steps are taken. Radioactive substances wreak their damage differently from other poisons. Their effect comes from hurtling particles—alpha, beta, and gamma rays—released when the isotopes "decay" to other elements. These particles damage living tissues as they fly through them, killing cells or stimulating them into runaway cancerous activity or affecting the genetic material in germ cells so that their hereditary instructions are altered, usually with harmful effects that may be insignificant or lethal, immediate or long delayed.

Some radioactive pollutants have an exceedingly long life—years or decades, and there is no known way to neutralize these wastes. Radioactive garbage can only be handled in two ways: by either isolating it so that it does not come into contact with man or diluting it so that its effects are too small to be harmful.

The principal source of radioactive waste is the "burned" residue from fissile fuel elements in nuclear reactors. Smaller quantities are produced by neutron bombardment of otherwise neutral material inside reactors: reactor metals, coolant fluids, air, and other gases. The quantities of radioactive wastes are increasing rapidly, and will continue to do so as we become more and more dependent on nuclear energy.

[391]

Most radioactive wastes are in the form of a liquid or slurry. Material containing more than one curie per gallon is rated as "high level" waste; the intermediate levels contain between one curie and a microcurie (10^{-6} curies) per gallon; low-level wastes contain less than that. These wastes are stored in steel or concrete tanks for preliminary cooling. Nonliquid radioactive wastes include contaminated laboratory equipment, uniforms, rags, containers, reactor and machine parts. Since some high-level wastes generate as much as 200 watts of heat per gallon, they must be cooled a year or more before they can be put in permanent storage.

Many people consider the sea an attractive place to dump radioactive wastes. The danger to man of injudicious disposal in the sea stems primarily from the contamination of seafood eaten by humans. There is also a lesser but still present risk of direct contact through such accidents as the recovery of packaged wastes by fishermen or salvage crews, or contact with radioactive substances that find their way onto beach sand or onto shallow-water sediments. (In harbors the accidental sinking of nuclear-powered vessels constitutes another possible source of contact.) But all these direct sources are of less danger than the ingestion of isotopes from fish and other seafoods.

The poorest place to dispose of nuclear wastes would be in the shallows of the continental shelves and in the estuaries at the edge of the sea. These are the richest parts of the ocean, where the most abundant supply of food organisms occurs and where isolation and dispersion of the wastes is least likely. The upper layers of the open ocean would provide a better place for disposal; the depths of the open sea would be the safest, and may indeed be the best place on earth for this dangerous garbage. Because of the exceedingly long life of some radioactive substances, however, their effects may outlive even concrete or metal containers. In the deep sea, ocean currents are weakest, and there is least exchange with other parts of the ocean. Depending on the region, estimates of the probable time it would take a particle of water to return from the deep sea to near-shore areas range from years to decades. For most kinds of radioactive substances even the shorter periods would be long enough for the damaging effects to wear off. The difficulty is that we do not know enough to be certain about the rates of water exchange for many parts of the sea. Before dumping unrestricted quantities of radioactive materials in the abyss, man would therefore be wise to conduct research to determine accurately what chance there is for dangerous chemicals to be returned to his environment.

Furthermore, even after radioactive substances have been diluted to safe levels, they may be concentrated once more by the remarkable capacity of some marine organisms to accumulate them in their bodies. From very dilute concentrations in the surrounding water, some marine organisms can concentrate radioactive substances many thousands of times.

With the cessation of nuclear testing in the air, radionuclide pollution of the environment is diminishing. So far there is no conclusive evidence that living marine resources have been damaged by radioactive pollution. Present knowledge suggests that the levels of radiation now accumulating in the estuaries from wastes of the nuclear power industry are not adversely affecting that environment.

Despite these calming circumstances, it is possible that subtle long-range changes are taking place in the marine environment that are harmful to marine organisms. Since our knowledge of this complex problem is so limited, we have to make very conservative assumptions that may lead to much more restrictive, and therefore more expensive, regulations than would be necessary if more information were available. This is a strong economic argument for the expenditure of money on oceanographic research to provide the answers to the unknowns about biological, chemical and physical ocean processes related to the disposal of radioactive wastes.

THERMAL POLLUTION

One of the newer ways man has devised to foul the aquatic environment is to heat it. Some people have questioned whether "thermal pollution" is a fair term, since the hot water may be uncontaminated by bacteria, chemicals, or other traditional pollutants. Various euphemisms have been suggested, including "thermal loading," "thermal alteration" and even "thermal enrichment." In some special and limited cases the addition of heated water into a bay or estuary may indeed be beneficial, and with ingenuity the benefits of such heating may be increased. But in the great majority of cases a rise in the temperature of natural water is detrimental to its normal and desired use by man, and "pollution" is therefore an all-too-proper term. There is no longer any doubt as to whether heated effluents from industry and other sources are harmful to the aquatic environment, including the edge of the sea; the only question is how much harm is being done in a specific case.

Water is used for cooling by many industries, and the coolant water—now heated—is then discarded into the environment; among such industries are those manufacturing steel, chemicals, paper, and petroleum. It is the electric power industry, however, that gulps the most water for cooling and pours the most heat into fresh and ocean waters; it accounts for 70 per cent of the heat pollution. In the United States almost all power is manufactured by the burning of coal, oil, or natural gas, or by the heat produced by nuclear reactions. Thermal power plants operate by turning water into superheated steam that drives turbines. The steam must then be condensed by the cooling action of very large volumes of water so that the cycle can be repeated. After the cooling water flows over the condensing tubes it is usually dumped back into the river or estuary from whence it came— at a much higher temperature. The increase may run from 10° to 30° F. and is ordinarily about 10° to 12° F.

Thermal plants are not very efficient in their use of heat energy. Fossil fuel (coal, oil and gas) plants are only about 40 per cent efficient, which means that they waste about 6,000 BTU's of heat for every kilowatt of power they produce. Because they must operate at lower temperatures, nuclear plants are even less efficient, rating only about 32 per cent; they waste up to 10,000 BTU's per kilowatt.

To carry off this excess heat requires astonishing volumes of water. A large nuclear plant of 1,000 megawatt (a million kilowatts) capacity uses about 2,000 cubic feet of cooling water every *second,* requiring as much water as is used every day for domestic purposes by all the people in Texas! Large plants use the equivalent of the flow of a large river for their cooling, and this has forced power plants to move from the banks of rivers to the Great Lakes or to the edge of the sea to obtain sufficient volumes of water. In 1950 only 22 per

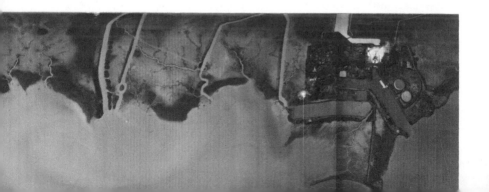

cent of power plants were on the banks of estuaries; by 1980 about a third of them will have to be located on the edge of the sea.

By that time the total number and power capacity of these plants will have increased enormously since the demand for electrical power in the United States doubles every 6 to 10 years. In 1958, 90 billion gallons of cooling water were used per day; by 1980 an estimated 200 billion gallons will be needed, and by the year 2000 it will be 600 billion gallons. The 200 billion projected for 1980 represents one sixth of the total daily average of fresh-water runoff of the United States! Moreover, it represents half the runoff in the nonflooding seasons of the year, which last about nine months. The

FIGURES 18–9, 18–10. *Light areas are the warmest in the aerial photograph at left of Biscayne Bay opposite the plant of the Florida Power and Light Company at Turkey Point. Water taken into the plant at 69°F. by the diagonal canal at right emerges into inland canals at 81°F. and is dumped back into Biscayne Bay by the canals at left at 75°F. All plants and animals in the small area opposite the two discharge canals have been wiped out; the amount of damage declines as the temperature returns to normal outward from this. Adjacent to the power plant, seen in background of photo above, are tanks of a University of Miami aquaculture research station. Studies here include the possible use of heated water from the plant to accelerate growth of shrimps and pompano.*

Bendix Corporation, C. P. Idyll

[395]

120 plants already planned for operation by 1976 will use the amount of water flowing in 150 Connecticut Rivers. The Florida Power and Light Company plant on Biscayne Bay near Miami will use 1.8 million gallons of cooling water a minute—all by itself, a Connecticut River.

The water discharged from power plants will affect the aquatic environment in several ways: by producing alterations in the speed and pattern of water currents, by changing the salinity, by dumping toxic chemicals and, most seriously, by elevating the temperature. Of all the environmental variables affecting the well-being of aquatic organisms, temperature is by far the most critical. A rise in water temperature can kill fish and other organisms outright or have other damaging effects. Spawning, hatching, and the development of young may be altered; behavior may be modified; swimming powers, and thus the ability to escape predators, may be impaired; the metabolic rate of the organisms may be increased, creating more demand for oxygen—just when there is a decrease in the oxygen supply because of the increase in water temperature. The food supply may be decreased in quantity and quality; an increase may take place in the sensitivity of organisms to disease or to toxic substances, and at the same time the toxicity of some poisons may be enhanced by the rise in temperature. And communities may also be adversely affected through alterations in the food-web relationships, community structure or genetic selection. Finally, as serious as the overt and obvious effects are, they may not be as significant as some subtle, long-term, sublethal impacts, whose results may not be seen until it is far too late to correct the damage.

There are a number of ways that better control could be gained over the temperature of power-plant effluents. The efficiency of existing plants could be improved so that less waste heat is thrown off and new power-producing systems, including "breeder" reactors, can be designed and built. These developments will come only slowly, however.

The temperature of water can be lowered before it is returned to a river, a lake, or the sea by evaporation and radiation in cooling ponds. Very large pond areas are required—more like lakes. A 1,000-megawatt plant (and some are planned four times that size) requires a 2,000-acre lake as deep as 50 feet.

Evaporating cooling towers are being used in some power plants and have been proposed for many more. In these structures the heated water is sprayed down inside the tower and is cooled by air moved upward by convection or by large fans. These towers have the disadvantage that they are enormous, ugly structures, dwarfing the rest of the power plant. Moreover, the spray from such towers can cause fogging in the adjacent area. For this reason they cannot be used where salt water is used for cooling because the salt spray would ruin farms downwind. Closed cooling towers have also been designed that eliminate both the water loss and the spray of the open towers. They operate by radiant heat exchange much as an automobile radiator does. Enormous electrically powered fans drive forced air through the towers to accomplish the cooling. This is very expensive, requiring about 3 per cent of the power produced by the plant.

Yet most ecologists and many other citizens are beginning to believe that the costs of cooling devices must be paid. And in proportion to the cost of generating electricity, these costs are not usually very large. One estimate has put the added cost of cooling towers at 3 per cent of capital costs and the additional cost to consumers at about $1 per year.

The discharge of enormous volumes of heated water in future years would represent a very large loss of energy. Perhaps instead of fighting it, we could put it to good use. There have been many suggestions. Beaches might be warmed so that swimming could be enjoyed beyond the usual season. Harbors that freeze in winter might be kept open longer by heated water from power plants. Buildings might be heated, industries supplied with energy, water desalted. Farm crops might benefit from warm water, which might also be used for sea farming, to create better growing conditions for culturing fish, shrimp, oysters, and other animals.

Unfortunately the use of waste heat for heating or industry does not appear to be promising. Agriculture, however, might be well served by heated water from power plants under some circumstances. In a demonstration project east of Springfield, Oregon, heated water pumped from an industrial plant at temperatures from 90° to 130° F. is carried two miles to the project site by underground pipes. From there branch lines lead to hydrants on seven ranches, where each farmer controls the supply of water to his crops. Orchard crops, including pears, peaches, apples, sour cherries, and filberts, are protected from early spring frost by overhead sprinklers; and beans, corn, potatoes, tomatoes, and other crops are under irrigation.

Waste heat also appears to be promising as an aid to the commercial culture of marine organisms. In Japan oysters, clams, shrimps, and a number of species of fishes are being raised with the help of warm water from power plants. In experiments in Britain, plaice and sole have been brought to market size in two years instead of the usual three with the help of warm water and some new power plants in Britain are being designed to incorporate culture facilities. One successful oyster-culture operation in the United States is making use of heated water.

It seems possible, then, for marine farmers to recover some of the energy that would otherwise be wasted, but this does not overcome the threat to the marine environment posed by the dumping of heated water. If heated effluents are to be of real value to the aquaculturist, there must be an assured and steady supply of heated water. If this can be restricted to the farm ponds, no harm will result to adjacent ecosystems; if it cannot, the thermal problem is still unsolved.

SEWAGE AND SOLID WASTE

Every person in the United States produces an average of about 120 gallons of liquid waste per day. This is usually treated in some manner, but more often than not the treatment is inadequate to render the wastes innocuous. Moreover, the load on sewage treatment plants is becoming so heavy that their efficiency is seriously lowered. And the load is expected to multiply at least four times in the next fifty years.

FIGURE 18 – 11. *Every man, woman, and child in the United States produces about 5⅓ pounds of trash a day; the country receives a shocking 190 million tons a year. This New Jersey estuary is being overwhelmed by garbage.*

Ralph A. Schmidt

In many areas domestic wastes are dumped into rivers from which much of the material eventually gets into the sea. Coastal cities, which contain a large proportion of the population, often dump waste directly into the ocean through sewers or by barging it offshore. New York City has been dumping sludge from sewage-treatment plants into the Atlantic Ocean for more than four decades. Every year 4.5 million cubic yards of sludge are dumped over 15 square miles of the ocean floor. In the U.S. 4.5 million tons of sewage sludge are dumped and millions more are piped into the sea annually.

Domestic wastes do their dirty work in the ocean in several ways. The simple physical blanketing of the sea bottom by sewage and sludge covers and kills marine organisms and their food, and destroys the habitat. Formerly productive shellfish beds off big Eastern cities such as New York, Boston, Norfolk, and others have long since been smothered.

Sewage in itself is usually not toxic or only slightly toxic to fish (although at some sewage outfalls in California fish have been observed to be dull-colored, listless and flabby, and to exhibit tumor-like sores). The greatest direct harm produced by sewage is reduction of the oxygen content of the water when oxygen is used up by bacteria as they destroy organic matter. Reduction of the oxygen concentration below 2.5 parts per million is lethal to many forms of marine life, and sewage may reduce the oxygen content far below this. In some areas off New York where treated sludge is dumped the oxygen concentration is less than 1 part per billion; sometimes it is zero. Sewage has robbed the estuary of the Thames of its oxygen so that fish migration is blocked. In 1966 the President's Science Advisory Council predicted that the oxygen demand of domestic waterborne sewage effluent in this country would soon be great enough to consume the entire oxygen content of water equivalent to the dry-weather flow of all twenty-two United States river basins!

Potential Value of U.S. Shellfish Catch, 1969
$320 million

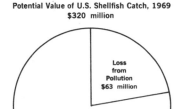

Loss from Pollution $63 million

Actual value $257 million

FIGURE 18-12. *Sewage dumped into the shallows at the edge of the sea render shellfish dangerous to eat because of bacterial or virus contamination. Close to a quarter of the potential value of the United States shellfish catch in 1969 was lost through pollution.*

Council on Environmental Quality

[399]

FIGURE 18–13. *Fishing in a eutrophic (overfertilized) lake. Fishermen once made larger catches than this from Lake Constance of Germany, Switzerland, and Austria, before sewage pollution reduced its productivity. There is evidence that some parts of the ocean are suffering the same fate, as excess phosphates and other nutrients are poured into it.*

C. P. Idyll

Sewage usually carries bacteria and viruses that can contaminate seafoods, especially shellfish. Infectious hepatitis, a dangerous, sometimes fatal liver disease, has been transmitted by oysters and clams exposed to sewage. In 1961 after an outbreak was traced to the consumption of raw shellfish from Raritan Bay near New York, harvesting was forbidden there and three thousand fishermen were put out of work; the loss of the industry cost $8.5 million annually. The gathering of shellfish is now prohibited in an estimated 1.2 million acres, or 8 per cent of the potential grounds along the United States coastline. Swimming beaches are sometimes closed because bacterial and virus counts are too high for safety—and this probably should happen oftener than it does.

Eutrophication or overfertilization is another undesirable result of sewage pollution. Excesses of nitrogen and phosphorus trigger the uncontrolled growth of algae. This upsets the ecological balance by changing the composition of species and abundance of the animals present, the noxious species usually taking over as desirable kinds are shouldered aside. Oxygen may be eliminated completely as the proliferating algae decay, and all but useless, low-oxygen-tolerant forms such as mud-dwelling worms survive.

While we work hard to develop real solutions to the sewage problem, even our existing technology permits great improvement over present practices. At the present time less than a third of our sewage is subjected only to primary treatment—physical removal of solid matter. Approximately 60 per cent is subjected to primary plus secondary treatment—biological oxidation, a controlled version of the degradation that occurs slowly in nature. Less than 10 per cent of the sewage receives tertiary treatment, from which water emerges with all noxious substances removed. This water is safe to drink. If all sewage were carried to this point an enormous step forward would be taken in the protection of the sea and the fresh waters.

Another kind of garbage is solid waste: oil drums, old mattresses, wire, bottles, rags, dead animals, paper, plastic articles, discarded automobiles, refrigerators, and a hundred thousand other kinds of scrap. Every man, woman, and child in the United States produces about 5⅓ pounds of junk *a day,* for a mountainous total of 190 million tons a year. By 1980 each person is expected to discard 8 pounds a day, for an annual total of 340 million tons.

New York City tried barging solid wastes to sea, but bits and pieces kept floating onto the New Jersey beaches, and in 1933 the U.S. Supreme Court put a stop to this practice. Now some of the solid is incinerated and the ashes dumped at sea. Investigators from Harvard and the University of Rhode Island have maintained that well incinerated residues, even in concentrations up to two or three times the normal expectation, have produced no ill effects. "It appears practical," their report says, "to utilize the vast assimilative capacity of ocean waters and the ocean atmosphere to solve a troublesome urban problem. Studies show that this can be done without polluting the environment, decreasing the recreational use of the waters or interfering with commercial and sport fishing." This conclusion is not accepted by all environmentalists. A study by biologists of the Sandy Hook New Jersey Laboratory of the Bureau of Sport Fisheries and Wildlife leads them to doubt that dumping is harmless. Their report states ". . . some of the urban problem being deposited in the New York Bight is coming right back to the beaches of the same urban areas." The sea is not a really satisfactory sink for solid wastes because it is practical to deposit this material only in the shallows near shore—where it will do harm—rather than far out at sea, where it would disintegrate in the abyss before it could affect living creatures.

CHEMICAL POLLUTION

Heavy metals and other chemical pollutants of the sea have been projected into the news in recent months as the public has been frightened by the threat of mercury poisoning from fish. The scare goes back to the 1950's, when Japanese fishermen and their families who lived around Minamata Bay on the island of Kyushu developed a severe and often fatal neurological illness. The waters of the bay were found to be highly contaminated by mercury from a plant on the shore that manufactured acetaldehyde. From 1953 to 1960 there were 111 documented cases of neurological disorders and 43 fatalities, and 19 brain-damaged children were born to mothers who had eaten fish from the bay. The flesh of the fish contained 5 to 20 parts per million of mercury. The U.S. Food and Drug Administration established an upper limit of .5 parts per million of mercury as the permissible concentration, and when some canned tuna in this country was found to exceed these limits, it was condemned. So many samples of swordfish showed high levels of mercury that the FDA recommended that this fish not be eaten.

Mercury occurs naturally in the sea, as does every chemical element. It occurs in innocuous concentrations, and in its original inorganic state it is unavailable to fish and other organisms, and does not enter the human body. Two things happen, however, that can eventually deposit an injurious form of mercury in human food. First, bacteria change the inorganic metal into organic methyl mercury, a form in which it is assimilated readily by marine organisms. Because it is retained, not excreted, methyl mercury may be concentrated 500 times above levels occurring in the water; as organisms are consumed by larger ones, the mercury is passed up the food chain, the concentration increasing at each predator level. Long-lived fish at the top of the food pyramid, like tuna and especially swordfish, exhibit the largest concentrations in their flesh and are therefore most likely to be dangerous. Short-lived seafoods like shrimps, or fish like herring and even salmon which feed at various levels below the top, are less likely to have dangerous accumulations of mercury or any of the other heavy metals.

It is still uncertain what level of mercury concentration is safe. The .5 ppm of the FDA may be high; Sweden fixes the limit at 1 part per million. However, given our present state of ignorance, it is just as possible that the FDA level may be too low; some scientists have suggested a maximum of .2 part per million. Of course, it is the total quantity of mercury consumed by an individual that really determines the effect on his health. Even at high mercury concentrations substantial amounts of fish must be eaten (as in the case of the Minamata Bay Japanese) for sickness to occur.

F I G U R E 1 8 – 1 4 . *Skipjack tuna, like the Hawaiian fish pictured here, is one of the two principal kinds canned by the United States industry. In 1970 a few cans of tuna were detected with higher than legal levels of mercury, and the public was made nervous about this product. The Food and Drug Administration later declared that canned tuna was safe to eat.*

National Marine Fisheries Service

Other heavy metals including lead, arsenic, cadmium, chromium, and nickel also pose threats to the marine environment. The amount of lead in the sea has increased measurably over the years, mostly from its wide use in antiknock gasolines, and perhaps also from the lead in defoliants and weed killers. The amount of lead in surface waters has increased twenty times since the first use of tetraethyl lead in gasoline.

DREDGE AND FILL

With the rapid disappearance of salable waterfront property, enthusiastic land developers often create more by dredging up the bottoms of bays and piling the fill over marshlands and estuarine grass beds. This is exceedingly damaging to the environment. It covers feeding and nursery grounds, alters the nature of the bottom and thus changes the kinds of animals and plants that live there, increases the turbidity, reducing the amount of photosynthesis that can take place, and changes the salinity. Since 90 per cent of the seafood caught by

[403]

United States fishermen is accounted for by species that inhabit the continental shelf, and two thirds of these depend on the estuaries in some way, this destruction of the shallows at the edge of the sea is deadly serious. Already 4 per cent of the coastline of the Gulf of Mexico, including 20 per cent of the west coast of Florida, has been adversely affected by shoreline alteration; 300 of the 700 square miles of San Francisco Bay have been sacrificed; 7 per cent of all the tidal marshes of the United States have been lost.

THE COST OF RESPONSIBILITY

Neither the open ocean nor the abyss is threatened with destruction by pollution in the near future, nor is the world's oxygen supply in imminent danger. If our wastes could be spread throughout the whole vast ocean—if the deep sea could easily be used as a dump— even the enormous volume of material that future generations threaten to cast off could be assimilated. But it is not possible to put even a tiny fraction of our wastes in the abyss; the increasing use of the sea as a sewer therefore means piling garbage in the estuaries and on the continental shelf. As a consequence, these ecologically important areas are rapidly being fouled and their biological processes warped and impoverished by swelling tides of waste. This practice threatens our expanded use of the entire ocean for the extraction of food, for recreation and for other desirable purposes.

After being nearly neglected through all the previous ages of mankind, pollution of the environment has become a major public issue of our time. Whether one considers pollution by sewage, heated water, radioactive materials, oil, pesticides, or something else, it must be clear that a staggering task confronts us—one that will be expensive and technologically difficult. It is not at all certain that society will accept the costs quickly enough to avoid irreversible damage to the environment. It has been estimated that it will cost the people of the United States an additional $750 billion in the next five years, on top of the present expenditure of $3.7 billion a year, to cope properly with solid waste disposal alone.

Unless society is willing to accept these costs and thereby to launch a campaign of the magnitude and intensity that took man to the moon, we may fulfill the horrible prophecy of Albert Schweitzer: "Man has lost the capacity to foresee and to forestall. He will end by destroying the earth."

FIGURES 18–15, 18–16. *Waterfront land has great value, which increases as the amount available becomes less, and fortunes can be made by creating more. Here, in central Boca Ciega Bay between Tampa and St. Petersburg, Florida, dredge and fill operations have changed the bay from a nearly unaltered estuary in 1949 (top) to the condition seen in the lower photograph taken in 1963. The biological productivity of the bay has been reduced to a fraction.*

National Marine Fisheries Service

Appendix

CLASSIFICATION OF ANIMALS

MENTIONED IN THE BOOK*

PHYLUM PROTOZOA [UNICELLULAR
ANIMALS]
Class Mastigophora
  Order Dinoflagellata
    Ceratium
    Gymnodinium [red tide
      organism]
    Noctiluca [night light]
    Pyrodinium [fire whirler]
Class Rhizopoda
  Order Foraminifera
    Globigerina
  Order Radiolaria
    Oroscena

PHYLUM PORIFERA [SPONGES]
Class Hexactinellida [glass
    sponges]
  Desmascedon
  Euplectella [Venus's
    flower-basket]
  Monorrhaphis [one-needle
    sponge]
  Pheronema
Class Demospongia [includes bath
    sponges]
  Geodia

PHYLUM COELENTERATA
[COELENTERATES]
Class Hydrozoa
  Order Siphonophora
    Physalia [Portuguese
      man-of-war]
    Velella [before-the-wind
      sailor]
Class Scyphozoa [jellyfishes]
  Order Cubomedusae
    Chironex [fire medusa]
    Chiropsalmus [fire medusa]
  Order Discomedusae
    Agliscra
    Atolla
    Aurelia [moon jelly]
    Crossota
    Cyanea
    Halicreas
    Lobenema
    Nausithoe
    Pelagia
    Periphylla
    Rhizostoma
    Rhopilema

* Common names are given where they exist.

Class Anthozoa [corals]
Order Pennatularia [sea pens]
Funiculina
Pennatula
Umbellula
Order Madreporaria
Calliactis [sea anemone]
Epizoanthus [sea anemone]

PHYLUM CTENOPHORA [COMB JELLIES]
Beroë
Cestus [Venus's girdle]
Mnemiopsis
Pleurobrachia

PHYLUM GORDIACEA
Gordius [horsehair worm]

PHYLUM CHAETOGNATHA [ARROW-WORMS]
Sagitta

PHYLUM POGONOPHORA [BEARD WORMS]
Lamellisabella
Zenkevitchiana

PHYLUM ANNELIDA [BRISTLE WORMS]
Class Chaetopoda
Acholoe
Eunice
Euphione
Odontosyllis [fireworm]
Tomopteris

PHYLUM MOLLUSCA [MOLLUSCS]
Class Monoplacophora
Neopilina
Class Gastropoda [snails]
Cassis [helmet shell]
Clio [pteropod]

Gastropoda, *cont.*
Clione [pteropod]
Diacra [pteropod]
Carinaria [heteropod]
Class Pelecypoda [clams]
Pholas [boring clam]
Class Cephalopoda [squids and octopuses]
Order Tetrabranchia
Nautilus
Piloceras
Order Dibranchia
Argonauta
Order Vampyromorpha
Vampyroteuthis
Suborder Decapoda [squids]
Spirula

Alloteuthis
Architeuthis [giant squid]
Bathothauma
Benthoteuthis
Calliteuthis
Chiropsis
Chiroteuthis
Cucioteuthis
Dosidicus [Humboldt Current squid]
Galiteuthis
Histioteuthis
Loligo
Lycoteuthis
Octopodoteuthis
Ommastrephes
Rossia
Sepia [cuttlefish]
Taonidium
Watasenia [firefly squid]
Suborder Octopoda [octopuses]
Cirrothauma
Grimpoteuthis
Octopus
Opisthoteuthis
Tremoctopus

PHYLUM ARTHROPODA [JOINT-
FOOTED ANIMALS]
Class Pycnogonida [sea spiders]
Anoplodactylus
Class Insecta [insects]
Phrixothrix [railroad worm]
Photinus [firefly]
Class Crustacea [crustaceans]
Subclass Ostracoda [ostracods]
Cypridina
Cypris
Gigantocypris
Subclass Cirripedia [barnacles]
Balanus [acorn barnacle]
Chelonobia
Coronula
Lepas [goose barnacle]
Pollicipes
Sacculina [parasitic
barnacle]
Subclass Copepoda [copepods]
Acartia
Bradycalanus
Calanus [brit]
Candacia
Centropages
Cyclops
Euchaeta
Eudeuchaeta
Haloptilus
Lucicutia
Megacalanus
Pseudocalanus
Subclass Malacostraca [lobsters,
crabs]
Order Isopoda [pill bugs]
Bathynomus [deep-sea
isopod]
Gnathia
Order Amphipoda [sand hoppers]
Cystosoma
Phronima
Phthisica
Thermisto

Order Euphausiacea [euphausids]
Euphausia
Nematoscelis
Order Cumacea
Diastylis
Order Decapoda [lobsters,
shrimps, and crabs]
Acanthyphra
Alpheus [snapping shrimp]
Aristaeopsis
Gnathophausia
Hippolyte [grass shrimp]
Hoplophorus
Hymenopenaeus [royal red
shrimp]
Notostomus
Pandalus [northern com-
mercial shrimp]
Penaeus [southern com-
mercial shrimp]
Phronima
Sergestes
Systellaspis
Xiphocaridina [fresh-water
shrimp]

Galathea [squat lobster]
Homarus [northern lobster]
Munidopsis
Nephrops [Norway lobster]
Nephropsis
Willemoesia [blind lobster]

Anamathia
Dromia [sponge crab]
Ethusa
Lithodes
Ocypode [ghost crab]
Portunus [velvet fiddler
crab]
Uca [fiddler crab]
Melia [hermit crab]
Pagurus [hermit crab]
Parapagurus [deep-sea
hermit crab]
Polydectes [hermit crab]

PHYLUM ECHINODERMATA
[ECHINODERMS]
Class Crinoidea [sea lilies]
  Rhizocrinus
Class Asteroidea [starfishes]
  Asterias
  Henricia
  Nymphaster
  Pycnopodia
  Solaster
  Styrcaster
Class Ophiuroidea [brittle stars]
  Gorgonocephalus [basket
  star]
  Ophiacantha
  Ophiolepis
Class Echinoidea [sea urchins]
  Diadema
  Hygrosoma
  Salenia
Class Holothuroidea [sea cucum-
  bers]
  Elpidia
  Enypniastes
  Galatheathuria
  Myriotrochus
  Pelagothuria
  Periamma
  Scotoplanes

PHYLUM CHORDATA
Class Thaliacea [salps and tunicates]
  Oikopleura [tunicate]
  Pyrosoma [fire body]
Class Chondrichthyes [Cartilagi-
  nous fishes]
Order Squaliformes [sharks]
Family Chlamydoselachidae
  Chlamydoselachus [frilled
  shark]
Family Lamnidae
  Cetorhinus [basking shark]

Family Squalidae
  Centroscyllium [black dog-
  fish]
  Scyliorhinus
  Somniosus [Greenland
  shark]
Order Rajiformes [rays and
  skates]
Family Rajidae
  Raja [ray]
Order Chimaeriformes
  [chimaeras]
  Chimaera
  Harriotta
Class Osteichthyes [bony fishes]
Order Coelacanthiformes
  Latimeria [coelacanth]
Order Clupeiformes
Family Salmonidae
  Oncorhynchus [Pacific
  salmon]
  Salmo [Atlantic salmon]
Family Gonostomidae
  Cyclothone [bristlemouth
  or roundmouth]
Family Sternoptychidae
  [hatchet fishes]
  Argyropelecus
  Polyipnus
  Sternoptyx
Family Stomiatidae [deep-sea
  scaly dragon fishes]
  Grammatostomias
  Stomias
  Ultimostomias
Family Chauliodontidae
  Chauliodus [viper fish]
Family Malacosteidae
  Aristostomias
  Malacosteus [loosejaw]
Family Melanostomiatidae
  Bathophilus
  Chirostomias
  Melanostomias

Family Idiacanthidae
Idiacanthus [stalk-eye fish]
Family Argentinidae [argen-
tines]
Winteria
Family Opisthoproctidae
Opisthoproctis [barrel-eye]
Order Myctophiformes
Family Myctophidae [lantern
fishes]
Aethoprora
Dasyscopelus
Diaphus
Electrona
Gonichthys
Lampanyctus
Myctophum
Neoscopelus
Rhinoscopelus
Tarletonbeania
Order Scopeliformes
Family Sudidae [ray fins]
Bathymicrops
Bathypterois [tripod fish]
Bathysaurus
Benthosaurus
Ipnops
Family Cetomimidae [whale-
fishes]
Ditropichthys
Order Giganturiformes
Family Giganturidae
Gigantura [giant-tail]
Order Lyomeri
Family Saccopharyngidae
[swallowers and pelican
eels]
Saccopharynx [pelican eel]
Family Eurypharyngidae
[gulpers]
Eurypharynx

Order Anguilliformes
Family Anguillidae [fresh-
water eels]
Anguilla [common eel]
Coloconger
Histiobranchus
Order Notacanthiformes
Family Notacanthidae
Notacanthus [spiny eel]
Order Gadiformes
Family Gadidae [cods and
hakes]
Gadus [cod]
Family Moridae
Harargyreus
Haloporphyrus
Mora
Family Macrouridae [rat-tails
and grenadiers]
Bathygadus
Coelorhynchus
Gadomus
Hemimacrus
Lionurus
Macrouroides
Macrourus
Malacocephalus
Nezumia
Shalinura
Trachyrhinchus
Order Lampridiformes
Family Regalecidae
Regalecus [oarfish]
Family Trachipteridae
Trachipterus [dealfish]
Order Beryciformes
Family Anomalopidae
Anomalops
Photoblepharon
Order Perciformes
Family Sciaenidae [drums]
Cynoscion [weakfish or sea
trout]

Family Chiasmodontidae
[great swallowers]
Chiasmodus [black swallower]
Family Triglidae [sea robins]
Peristedion
Family Brotulidae [brotulids]
Acanthonus
Bassogigas
Grimaldichthys
Leucicorus
Parabrotula
Tauredophidium
Typhlonus
Family Carapidae
Carapus [sea-cucumber fish]
Family Nomeidae
Nomeus [man-of-war fish]
Family Mugilidae
Mugil [mullet]
Order Pleuronectiformes [flatfishes]
Family Bothidae [left-eye flounders]
Bothus [flounder]
Paralichthys [flounder]
Rhombus [brill, turbot]
Family Pleuronectidae [right-eye flounders]
Azygopus [deep-sea flounder]
Chascanopsetta [deep-sea turbot]
Microstomus [lemon dab]
Pleuronectes [flounder]
Family Soleidae [soles]
Aphoristia [deep-sea sole]
Order Batrachoidiformes
Family Batrachoididae [toadfishes]
Porichthys [singing midshipman]
Order Lophiiformes [angler fishes]

Family Lophiidae
Lophius [goosefish, frogfish]
Family Chaunacidae
Chaunax
Family Photocorynidae
Photocorynus
Family Caulophrynidae
Caulophryne
Family Linophrynidae
Acentrophryne
Borophryne
Haplophryne
Linophryne
Family Oneirodidae
Dolopichthys
Lasiognathus
Family Melanocetidae
Melanocetus
Family Ceratiidae
Ceratias
Cryptopsaras
Family Galatheathaumidae
Galatheathauma
Class Reptilia [reptiles]
Brontosaurus [dinosaur]
Class Aves [birds]
Corvus [hooded crow]
Erithocus [robin]
Morus [gannet]
Oceanites [Wilson's petrel]
Phalacrocorax [guanaye bird]
Puffinus [shearwater]
Sturnus [starling]
Sylvia [warblers, blackcaps, whitethroats]
Class Mammalia [mammals]
Order Cetacea [whales]
Basilosaurus [fossil whale]
Balaenoptera [blue whale, fin whale]
Delphinapterus [beluga]
Physeter [sperm whale]

MORE BOOKS ABOUT THE SEA

—ESPECIALLY THE DEEP SEA

This list includes some of the many books about the sea, but no attempt has been made to include all those of special merit. Neither has an attempt been made to record the source of the facts given.

BEEBE, WILLIAM. Half Mile Down. New York: Duell, Sloan and Pearce, 1951.
A pioneer account of dives into the deep sea off Bermuda in the 1930's. Written in Beebe's vivid, lively style for the general reader.

BROWN, M. E. (ed.) The Physiology of Fishes. New York: Academic Press, Inc., 1957.
Two volumes, containing authoritative information on the senses of fish and other aspects of their physiology. Technical.

BRUUN, ANTON F., and others. The Galathea Deep Sea Expedition, 1950–1952. Translated from the Danish by Reginald Spink. New York: The Macmillan Company, 1956.
A fine account of a modern deep-sea oceanographic expedition. There are chapters on people and places visited as well as on the animals caught and the method used to catch them. For the general as well as the professional reader.

CARSON, RACHEL L. The Sea Around Us. New York: Oxford University Press, 1951.
A poetically written, enormously successful best-seller. For the general reader.

CHAPIN, HENRY, and F. G. WALTON SMITH. The Ocean River. New York: Charles Scribner's Sons, 1952.

An interesting account of the Gulf Stream, including its oceanography and history. General reading.

COKER, ROBERT I. This Great and Wide Sea. Chapel Hill, N. C.: University of North Carolina Press, 1947.

A good popular account of the ocean and its inhabitants.

COUSTEAU, J. Y., and F. DUMAS. The Silent World. Edited by James Dugan. New York: Harper & Row, Publishers, 1953.

Exciting reading about scuba diving and other adventures under the sea. General reading.

GRAHAM, MICHAEL. The Fish Gate. London: Faber and Faber Ltd., 1943.

A distinguished British fishery scientist writes popular accounts of commercial fishing and research on oceanic fisheries.

HARDY, ALISTER C. The Open Sea: Its Natural History. Part 1. The World of Plankton. New York: Houghton Mifflin Company, 1956.
—— The Open Sea: Its Natural History. Part 2. Fish and Fisheries. New York: Houghton Mifflin Company, 1959.

Two magnificent books by Sir Alister Hardy which manage with great success to combine useful material for the professional marine scientist with interesting reading for the layman. Excellently illustrated.

HARVEY, E. N. Bioluminescence. New York: Academic Press, Inc., 1952.

An interesting summary of Dr. Harvey's lifelong work on living light. Useful for the scientist and interesting to the layman.

HEYERDAHL, THOR. Kon-Tiki: Across the Pacific by Raft. Translated by F. H. Lyon. Chicago: Rand McNally and Company, 1950.

A best-selling and fascinating account of the courageous voyage on a raft across the Pacific.

IDYLL, C. P. (ed.) Exploring the Ocean World: A History of Oceanography. New York: Thomas Y. Crowell Company, 1969.

In addition to being a comprehensive history of oceanography, this book describes the present knowledge of the various branches of the science, written by specialists in the field.

IDYLL, C. P. The Sea Against Hunger: Harvesting the Oceans to Feed a Hungry World. New York: Thomas Y. Crowell Company, 1970.

An appraisal of the chances of overcoming world hunger by the use of food from the sea: the use of plankton and seaweed; the potential of fish "farming"; modern improvements in fishing methods.

LANE, FRANK W. Kingdom of the Octopus. London: Jarrolds Publishers Ltd., 1957.

A comprehensive and carefully documented account of the folklore, biology, and other aspects of the squids and octopuses. General reading.

MARSHALL, N. B. Aspects of Deep Sea Biology. London: Hutchinson and Company, Ltd., 1954.

A remarkably fine technical treatise on deep-sea animals. Especially useful to the biologist, but of interest to the layman as well. Excellent illustrations.

MURRAY, JOHN, and JOHAN HJORT. The Depth of the Ocean. London: Macmillan and Company, Ltd., 1912.

A comprehensive account of the pioneer work of the Norwegian research ship *Michael Sars*, told with the excitement of trail-breakers. A classic to which all accounts of deep-sea oceanography make frequent reference.

NORMAN, J. R. A History of Fishes. New York: Frederick A. Stokes Company, 1931.

An older but still highly valuable general account of the biology of fishes. For the general reader as well as the specialist.

RUSSELL, F. S., and C. M. YONGE. The Seas. New York: Frederick Warne and Company, Inc., 1928.

Two distinguished British marine biologists combined to produce this book. It is used at the Institute of Marine Science as an undergraduate textbook in marine biology, and is excellent for the general reader.

SCHULTZ, L. P., and E. M. STERN. The Ways of Fishes. Princeton, N. J.: D. Van Nostrand Company, Inc., 1948.

Another good general account of the biology of fishes.

SVERDRUP, H. U., and others. The Oceans. Englewood Cliffs, N. J.: Prentice-Hall, Inc., 1942.

This is the classic textbook on oceanography.

WILSON, D. P. Life of the Shore and Shallow Sea. 2nd rev. ed. London: Nicholson and Watson, Ltd., 1952.

A well-written and superbly illustrated book by the English marine biologist, who is also a highly talented photographer.

Index

abyss, 1, 17
  environment of, 49, 254
  photography in, 363-365
  population of, 90-95, 123-124, 253
Acholoe, 305, 376
Aethoprora, 269, 379
Agassiz, Louis, 235, 238, 239
age:
  earth, 3
  ocean water, 67-69
  sediment layers, 36
Agliscra ignea, 131, 375
air bladders, 173, 174, 255, 270-272
  and sound production, 264-265
Alaskan North Slope oil reserves, 388
Albatross Expedition, 24, 39, 112, 233, 368
Albert I, Prince of Monaco, 162, 177, 233
Alepocephalidae, family, 184-185
Aleutian Trench, 23, 31
algae, 71-72, 76
Allan, Joyce, 155
Alloteuthis subulata, 157, 376
Alpheus (snapping shrimp), 264-265, 377
Aluminaut, submarine, 360-361, 363
Alvin, submarine, 360-361
ambergris, 162
American Miscellaneous Society (AMSOC), 370, 372
ammonites, 236, 249

amphipods, 86, 138, 139, 260, 261, 377
anchovetas, 59-60
anemones, *see* sea anemones
angler fish, 196-203, 256, 258, 274, 297, 380
  light organs of, 201, 290, 298, 305
  reproduction in, 203, 277-280
Anguilla, see eels
Angulhas Plateau, 344
annelid worms, 81, 101-102, 115, 243, 300, 305, 376
Anomalops, 292, 379
Antarctica, 11, 58-59, 63-66, 68, 352
Antarctic Bottom Water, 63-65
Antarctic Intermediate Water, 63-65, 68
Antilles Current, 51
Aphoristia wood-masoni, 190, 191, 380
Appalachia, 13
Architeuthis, 149, 152, 153, 159, 162, 222, 376
Arctic Ocean, 26
Argonauta, 155-156, 250, 376
Argyropelecus, 218, 262, 378
Aristotle, 92, 95, 99, 108, 310
Armstrong, Neil, 362
arrowworms, 79, 81, 86, 319, 376
Arthropoda (*see also* crustaceans), 244, 377
Asia, continental shelf of, 16-17
Asteroidea, *see* starfish